Land Degradation

LAND DEGRADATION
Creation and Destruction

SECOND EDITION

Douglas L. Johnson and Laurence A. Lewis

ROWMAN & LITTLEFIELD PUBLISHERS, INC.
Lanham • Boulder • New York • Toronto • Oxford

ROWMAN & LITTLEFIELD PUBLISHERS, INC.

Published in the United States of America
by Rowman & Littlefield Publishers, Inc.
A wholly owned subsidiary of The Rowman & Littlefield Publishing Group, Inc.
4501 Forbes Boulevard, Suite 200, Lanham, Maryland 20706
www.rowmanlittlefield.com

P.O. Box 317, Oxford OX2 9RU, UK

British Library Cataloguing in Publication Information Available

Library of Congress Cataloging-in-Publication Data

Johnson, Douglas L.
  Land degradation : creation and destruction / Douglas L. Johnson and Laurence A. Lewis.—
2nd ed.
     p. cm.
  Includes bibliographical references and index.
  ISBN-13: 978-0-7425-1947-3 (alk. paper)
  ISBN-10: 0-7425-1947-3 (alk. paper)
  ISBN-13: 978-0-7425-1948-0 (pbk. : alk. paper)
  ISBN-10: 0-7425-1948-1 (pbk. : alk. paper)
  1. Land degradation—Environmental aspects. 2. Nature—Effect of human beings on.
I. Lewis, Laurence A. II. Title.
GE140.J64 2007
333.73'137—dc22   2006005525

Printed in the United States of America

♾™ The paper used in this publication meets the minimum requirements of American National
Standard for Information Sciences—Permanence of Paper for Printed Library Materials,
ANSI/NISO Z39.48-1992.

Contents

Illustrations

Figures

Tables

Boxes

Preface

LAND DEGRADATION is an ancient problem. It likely began hundreds of thousands of years ago when humans first controlled the use of fire to hunt game, create warmth, and prepare food. Through the millennia, control of fire gave humankind access to an increasingly wide range of habitats, initiated major environmental modifications, and encouraged more intensive use of local resources. The greater control over and pressure upon local habitats that these changes implied was accelerated by the agricultural revolution that began approximately 10,000 years ago. Success in agriculture often required significant changes in the vegetation properties of the utilized area. In particular, the substitution of annual crops for perennial vegetation acted as a catalyst for a multitude of environmental modifications that frequently resulted in degraded land resources. As more energy flowed from the environment into those segments of the ecosystem controlled by humankind and its domesticated plants and animals, both the intensity and spatial scale of human land use, and degradation, increased. The industrial revolution of the past three centuries, with its increasingly powerful technology has, in many cases, simply amplified trends long present in the human use of the Earth. In other instances, technology has initiated a new array of environmental problems often far removed spatially from the initiators of the degradational processes. Much has been written about this process of increasingly unwise exploitation of the Earth's resources, from Plato through George Perkins Marsh to Paul Ehrlich and an ever increasing host of contemporary commentators.

Given the plethora of publications in recent years addressing land degradation problems and advocating sustainable development, we risk redundancy in adding another volume to the discussion. However, our personal experiences of living in a postindustrial society, as well as many of our professional research encounters while engaged in fieldwork in Africa, Latin America, and the Middle East, convinced us that there are still fresh perspectives on the land degradation issue that merit consideration. Not least of these insights is the realization that there are numerous instances in which people purposely, occasionally with great success, degrade a portion of the environment to produce a positive advantage by which to meet specific societal needs. We define as *creative destruction* those intentional alterations of the existing habitat that impoverish one facet of the environment while erecting new, stable systems that are of greater benefit to humankind than the natural ecosystems that they replace. While land degradation is usually not an intended outcome of human activities, there are enough exceptions to this generalization to make exploration of their characteristics worthwhile. Of equal interest is the appearance of such "positive degradation" in a wide array of socioeconomic formations across both time and space. Indeed, the variety of these positive cases of land degradation is as impressive as the more common and more frequently recounted instances of unintentional negative land degradation.

A second inspiration for this volume arises from the conviction that land degradation is rooted in an interaction between physical and human systems. Much of the recent scholarship on land degradation stresses the political economy and structural features of environmental decay. This critique provides a welcome and important contribution to the environmental debate by emphasizing long neglected structural aspects of the problem. In this corrective development, the role of environmental processes in interaction with human systems has sometimes been lost. Many of the processes that contribute to land degradation are as much physical as they are human in nature. This is particularly true of accelerated soil erosion, where the conservation strategies and technological interventions for curtailing and reversing soil loss already exist but often are socially and politically difficult to implement. In other instances, use of many lands proceeds without managerial recognition that physical processes link regions together, and that changes implemented in one area may have an adverse impact on distant sites. Exploration of the many critical spatial and process interactions among physical and human systems that contribute to land degradation in both a historical and a contemporary context seems useful. By so doing, this book attempts to provide insight into both the causes of land degradation and the reasons why it has been so difficult to control.

In this second edition, content has been reorganized to illuminate the physical and human causes of land degradation. In addition, land degradation is examined in the context of local, regional, and global scales. Finally, new studies have been added as well as existing materials updated to incorporate some of the positive and negative environmental changes occurring since the publication of the first edition.

Douglas L. Johnson and Laurence A. Lewis

Land Degradation
Human and Physical Interactions

EVER SINCE the formation of the Earth, change has been the rule rather than the exception on the planet's land areas. Land areas are continually being altered due to both tectonic forces (energy derived from the internal heat of the Earth) and exogenic processes (those that result from the action of wind, water, ice, and human activities). Mountains increase in size due to tectonic processes, such as in the young and growing mountains of the Andes, the Alps, and the Zagros. Lands also are uplifted from beneath oceanic areas as occurred in recent geologic time in the southeastern United States, a process that resulted in the emergence of Florida. Similarly, volcanic eruptions can create new lands, such as the island of Surtsey (Iceland) or destroy areas. Examples include the Tamboro (Indonesia) eruption in 1815 when the mountain lost approximately 1,300 m (4,265 ft) of its upper portion or the extremely violent eruption of Krakatoa in 1883 that transformed six cubic miles of rock into pumice, ash, and dust and blasted the pulverized material into the atmosphere before the bulk of the volcano collapsed in upon itself and disappeared (Winchester 2003, 239). This eruption triggered devastating tsunamis. Conversely, terrestrial zones are also always exposed to exogenic forces that erode and ultimately lower surface areas. The lowering of lands takes place once tectonic processes that are uplifting land areas decrease in magnitude compared to the exogenic processes that are eroding these lands. The Appalachian and Piedmont areas in the eastern United States, the Urals in Russia, and the shield area of northeastern Brazil and Guyana represent tectonically stable areas that are gradually being lowered due to erosion processes. It is the net balance between the tectonic and exogenic processes that determines whether specific areas are growing or being destroyed.

Examination of the Earth as a natural system (devoid of the role of human activities on geologic processes) clearly documents that the dynamic interaction between tectonic and exogenic forces culminates in a complex set of results. At one end of the spectrum, land slowly evolves, allowing a complex array of earth science relations to change in a largely stable manner. The evolution of a lake to a wetland over hundreds or thousands of years or the reduction of a highland slowly into a lowland through fluvial erosion and slow mass movements represent examples of the natural geomorphic evolution of landscapes. In these cases change largely becomes evident only through a historical examination of the area's attributes, such as its lithology, vegetation, and soil properties. In opposition to these slow, chronic changes are rapid, acute alterations that occur when critical thresholds are exceeded. Under these conditions changes are rapid and the immediate destruction of existing natural systems makes way for processes that create new systems and landscapes. Major earthquakes, 1,000-year floods, immense volcanic eruptions, and massive landslides can completely eliminate the previous conditions in brief periods of time. History and folklore document the nonbenevolence of Mother Nature in a multitude of natural disasters that range from the Noachian Flood of ancient times to the contemporary widespread flooding in eastern Germany and the Czech Republic in the late summer of 2002; to the eruption of Mt. Vesuvius and Pelée that destroyed surrounding settlements; to the contemporary volcanic eruptions of Mt. St. Helens and Mt. Pinatubo.

Land Degradation: Its Definition

The term "land degradation" is a relatively new addition to our scientific vocabulary. Not until August 1994 was land degradation recognized as a separate category within the U.S. Library of Congress reference system. As a result, there is a lack of universal agreement on its definition. Is it but a synonym for environmental and resource deterioration?

Examination of its usage in literature and in the *Journal of Land Degradation and Rehabilitation* (first published in 1989) implies that there is general agreement concerning two critical aspects of land degradation. First, there must be a substantial decrease in the biological productivity of a land system; and, second, this decrease is the result of processes resulting from human activities rather than natural events. Thus the results of exogenic forces such as geologic erosion and climatic change, as well as natural catastrophic events such as earthquakes, volcanic eruptions, and flooding—unless exacerbated by human activities—lie outside the realm of land degradation, even though areas can become less productive biologically due to these natural changes.

On the basis of how the term is currently being used, we define land degradation as "the substantial decrease in either or both of an area's biological productivity or usefulness to humans due to human activities." Because biological productivity is determined not only by the attributes of the land resource but also its water properties, land degradation incorporates those aspects of the hydrologic domain that are significant in a given area. Usefulness is also a crucial attribute of land degradation. The biomass productivity in an area could remain constant yet land degradation might still occur. For example, overgrazing could result in a decrease in an area's soil fertility and a diminished ability of the environment to support the growth of palatable forage plants. When the vegetation cover reestablishes itself after grazing pressure is reduced, even though the grass cover (biomass) is similar in density to the initial cover, the species that replace the original vegetation often are not nearly as palatable. As a result, a smaller number of livestock can now survive on these lands. Thus, after this change, the area is no longer as valuable to its inhabitants. This decrease in an area's biological productivity and usefulness (land degradation) may be either reversible in the short-term or nonreversible in the long-term.

With our definition, not all resource and environmental deterioration falls within the rubric of land degradation. For example, if desertification—the impoverishment of arid, semiarid, and subhumid ecosystems (Dregne 1976)—is a result of increasing aridity due to climatic change, it is not a manifestation of land degradation. However, if the increased desertlike conditions are the result of overgrazing, water well drilling, or other human activities, then the desertification of the area is an example of land degradation (Reynolds and Stafford Smith 2002).

Any area, regardless of its climatic, topographic, or other environmental characteristics, can undergo degradation. Cape Cod, Massachusetts, represents a humid coastal setting where lands are degrading, sometimes due to natural changes and at other times due to human activities. The Cape's origin is due to Pleistocene glaciation. During this epoch large quantities of nonconsolidated glacial materials (outwash and till) were deposited. This deposition occurred in a nonmaritime environment as sea level was approximately 100 meters (328 ft) lower than today (figure 1.1). Since the end of the Ice Age, large areas of the Cape have been modified or lost to the sea due both to a rise in sea level and coastal erosion primarily resulting from storm waves. Wave action continues to erode the Cape, especially along the eastern margins of the outer Cape that face the open sea. It is inevitable that the lands located in zones of wave energy concentration will continue to be lost to the sea, regardless of human attempts to stop this natural trend (figure 1.2).

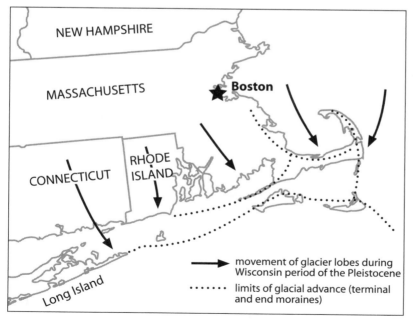

Figure 1.1. Southeastern New England's present shoreline and Pleistocene Coast. (Cartography by Anne Gibson, Clark Labs)

Figure 1.2. Coastal erosion due to wave action on Cape Cod. All of these houses in Chatham, Massachusetts, regardless of efforts to protect them with a riprap barrier, have succumbed to wave action. (Photograph by L. Lewis, autumn 1994)

In contrast to this natural erosion of the Cape, along some coastal stretches, especially in private beach areas, some human activities have accelerated erosion processes that are irreversible under current environmental conditions. Because private ownership of some beaches exists, individuals owning beach frontage often attempt to stop the natural migration of "their" beach sand. Previously, when legal, the most popular method was to build groins. While groins can stabilize beach location for the short-term, they actually increase sand loss to deep water and thereby decrease the sand supply available for beaches. Thus, the long-term net effect of groins is to accelerate beach erosion. The first example of coastal erosion typifies the dynamic nature of the Earth's lands due to natural processes, while the latter example of accelerated beach erosion represents an example of land degradation.

Long-term (Nonreversible) and Short-term (Reversible) Land Degradation

Stability and Resilience

Stability is the ability of a system to return rapidly to its previous equilibrium state after a temporary disturbance (Holling 1973). A critical aspect of stability is the speed with which a system returns to its previous state after a disturbance. Like a Cupie doll, a carnival game figurine with a low center of gravity that is difficult to knock over, the more rapidly the system returns to its previous state, the more stable is the system. A tropical forest that regenerates quickly after being cleared by shifting cultivators is more stable than a semiarid grassland that takes a long time to regenerate after its grass cover has been cleared for cropland. The difference in the behavior of the two systems is linked to the amount, frequency, spatial distribution, and variability of available moisture, the basic energy component in the system.

Resilience, on the other hand, is the capacity of a system to absorb change without significantly altering the relationship between the relative importance and numbers of individuals and species that compose the community. This characteristic of a system is analogous to a boxer whose opponent delivers numerous blows to body and head without apparent effect. The more blows the boxer can absorb without his ability to continue being impaired, the more resilient he is. However, continuing to box may mask sustained damage to his internal organs and central nervous system. Sudden and unexpected collapse can occur from even a light blow when the resilience limit is exceeded. Camels have a legendary capability to recognize when their master can no longer continue to add to the burden they bear. Of course, if the camel's owner does not recognize this limit of resilience, the proverbial piece of straw will break the camel's back! An environmental example of the same phenomenon is the use of rivers, lakes, seas, and oceans as dumping grounds for agricultural and urban wastes. The limit beyond which the water body can no longer absorb the pollutants is often suddenly reached, with consequent catastrophic collapses in fish and other aquatic populations depending on decent water quality and adequate oxygen levels for life. Resilience limits are not easy to identify, but once they are exceeded, the system rapidly seeks equilibrium at a different level of production.

When an area is disturbed resulting in a degraded state, two conditions must be met if it is to be able to return to a nondegraded state, that is, to the stability and resilience characteristics that the area formerly possessed. First, the disturbance must be only temporary. Second, it also must be of a sufficiently limited magnitude and duration that the changes resulting from the disturbance have not created fundamentally different environmental conditions for the area. If these two conditions are not met, the resulting land degradation type is considered to be nonreversible and long-term.

If the catalyst causing the land degradation is temporary and of low enough intensity that critical thresholds have not been reached, once the stimulus is removed, the area will evolve back to its previous nondegraded state. Under these conditions, the land degradation would be classified as both short-term and reversible. If the catalyst results in system changes that are too extreme for the area to absorb, a new balance will emerge at a totally different level of productivity. When productivity is lower, the system has degraded. For any land system, there are tradeoffs in the use of the system. Actions that ignore the stability and resilience limits of the area, and that discount the differing impacts that variable technologies can have on the system, are likely to initiate serious and long-term degradation. Also, they are likely to create cultural-ecological systems that are precariously poised on the brink of disaster.

Time

The time dimension, that is to say the period required to reverse any damage to the productivity of the land, is a crucial factor in land degradation. The feasibility of a particular rehabilitation ecology, applied to any land facet, is largely determined by both the economic costs and the time period required for improving the environmental situation. In most cases, no mater how severe a degradation process is, given enough time, natural processes will repair the damage without human assistance. An exception to this observation is the case of synthetic materials. Some synthetics are so stable that, in many cases, the time factor is ever increasing as technology creates more and more new materials that are less and less susceptible to decomposition. But in general the time scale implied in nature's rehabilitation ecology, when damage is excessive or when it occurs in environmental situations in which many natural processes are slow (e.g., cold and arid areas), can be geologic in duration. Likewise, to reverse degradation may be so costly economically—such as importing soil into Haiti to replace the eroded soil from many of its hills and plateaus (figure 1.3)—that the restoration processes are meaningless from the perspective of human needs.

Within the context of this book, because of the duration of the human life cycle and the demands of livelihood systems, any degradation process that is set in motion by human

Figure 1.3. Severe soil erosion in Haiti. Farming on steep slopes at les Platon in l'Acul river basin in southwestern Haiti has initiated severe soil loss and led to abandonment of much of the area. (Photograph by L. Lewis, winter 1985)

actions and cannot be reversed in less than fifty years is considered as long-term degradation. For practical purposes, this type of severe land degradation is a permanent change from a human point of view, even though it is a relatively inconsequential period from a geologic perspective.

Short-term degradation occurs in any cultural-ecological system. For example, the cultivation of most crops results in at least a short-term decrease in soil fertility and accelerated soil erosion rates. If these trends remain unchecked, yields will diminish. Under extreme circumstances farming will no longer be viable and land will have to be abandoned. In this situation, short-term degradation becomes long-term. Short-term degradation can be relatively easily countered in most agricultural systems from a technical perspective (i.e., by conservation practices). It would appear to be in the land manager's self-interest to do so, because usually it is less costly to invest in short-term, status quo maintenance rehabilitation than it is to abandon land altogether and begin anew elsewhere.

Short-term rehabilitation ecology is achieved by practices such as manuring and the spreading of chemical fertilizers, contour plowing, mulching, terracing, crop rotation, and a host of other strategies (OTA 1988). Yet for many reasons, ranging from accounting procedures to limited capital, these remedial technological interventions often are not implemented. For example, in corporate farming, the farm manager and stockholders often are primarily concerned with maximizing annual profit, a result that often determines the manager's annual bonus, promotion prospects, and job security. Technological interventions require investments. Most accounting analyses ignore environmental concerns. In particular, they rarely incorporate a declining soil resource as a decreasing asset. Thus, under current accounting practices, it is often in the immediate interest of corporate farm managers to minimize investments associated with environmental quality unless it shows up in the bottom line.

Individual farmers may wish to invest in conservation, but if they are short of capital, their economic constraints may be so serious that such an investment is impossible. This is the general case in many parts of the "developing world" and is also widespread for many small farmers in the United States. Developing a land-oriented, biodiversity-sustaining ethic that can be used to manage farmland responsibly and still support farmers economically is difficult, although efforts based on the "land ethic" advocated by Aldo Leopold appear promising (Jackson and Jackson 2002; Pretty 1998). When those land rehabilitation mechanisms either are not implemented or are not sufficient, more expensive interventions are required to rehabilitate a degraded area. In extreme cases, such as many rural areas in northwestern Haiti and Madagascar where the abuse of the land has been great and the damage severe, rehabilitation of the land is not a likely, viable option from an economic cost/productivity improvement perspective.

In arid areas, Dregne (1977b) has suggested that rehabilitation interventions should focus on high-productivity ecotypes and intensive agricultural systems (e.g., irrigation), rather than on low-productivity and extensive systems (e.g., livestock grazing). Acceptance of this philosophy by the World Bank as a guide to its investment policy for much of the 1970–1990 period did not prevent most extensive and low-productivity areas in regions prone to desertification from continuing to be exploited and degraded (Nelson 1990). Rather, adherence to a favored sector approach has resulted in a form of strategic amnesia that has starved peripheral livelihoods and regions of the resources they needed to cope with adverse change in their physical and social environment. Not only does this triage approach have serious ethical implications, but it also ignores the areal geomorphic and hydrologic linkages that exist in natural systems. A cost-effectiveness approach to rehabilitation ecology makes many zones that are capable of rehabilitation ineligible for governmental concern. It suggests that degradation may become functionally irreversible long before it is technically beyond restoration.

In theory it is easy to define the properties of long- and short-term degradation. In practice it is often difficult to classify which type of deterioration is occurring within a specific area. Sudan's Kordofan area (figure 1.4) illustrates some aspects of this classification problem. Beginning around 1955 and continuing until the middle 1970s, the population of the Kordofan area increased at an annual rate of about 2.7 percent (Stern 1985). Up to the early 1970s, the cultivated area simultaneously increased at approximately the same rate as the population. Throughout this period precipitation was generally greater than normal. Beginning in 1968, and continuing through the early 1970s, drier conditions returned to the Sahel. With the arrival of drier conditions in Kordofan, areas under cultivation decreased (Olsson 1985). One explanation for this decrease in cultivated area is that many lands were degraded and therefore abandoned. According to Horowitz (1981), in many parts of this area, social factors led to the desertification of lands due to overcultivation, grazing, and firewood demands. These pressures were often phrased in terms of population growth requiring more land to be developed and brought into production (Ibrahim 1978). Other social changes, such as an increase in urban population that needed more fuel for heating and cooking, were also seen as contributing to increased pressure on rural landscapes to generate more fuel wood for urban consumption.

All of these pressures were met during the 1950s and 1960s when rainfall amounts were generally higher than the long-term average. With the environment producing more, farmers and herders were able to cultivate more land and raise more sheep with greater success than ever before. But these increases in productivity could not be sustained when environmental fluctuation produced a return to drier conditions. Historically, these environmental fluctuations could be accommodated because the pressure of human resource use was less and could be sustained by existing cultural-ecological adaptations. Now, this perspective contended, good environmental conditions masked basic changes in nature-society relations. Because of the intense social pressures, serious land degradation occurred during this period of precipitation shortfalls (Horowitz 1981). The implication was that a long-term decrease in the land's biological productivity occurred throughout the semiarid area.

This is the general impression of what happened throughout the Sahel during the drought of the late 1960s and early 1970s. Lamprey (1975) concluded that the desert had expanded by between 90 and 100 kilometers in Kordofan during the previous seventeen years. Eckholm and Brown (1977) and other publications of the Worldwatch Institute likewise described continuing desertification throughout the Sahel. Other reports, both popular and professional, agreed with these findings (Dregne 1977a; Grainger 1990). They contended that in response to human activities and climatic stress (deficient rainfall), formerly productive farmland was evolving into nonproductive farm and grazing lands (i.e, long-term degradation: Ibrahim 1984). These observers viewed the trend as one that was rapidly approaching a level of long-term degradation from which the prospects of recovery were bleak.

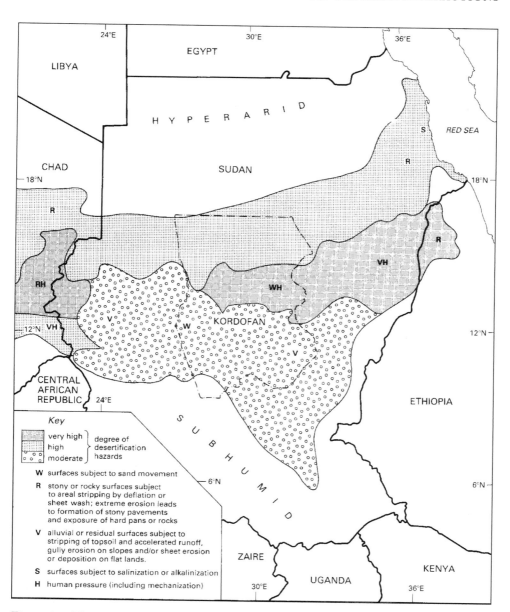

Figure 1.4. The desertification risk in Sudan's Kordofan Province and adjacent semiarid areas. (Reproduced from D. Henning and H. Flohn, "World map of desertification," in *UN Conference on Desertification*, Publication A/CONF, 74/31, 1977.)

Contradicting this perception of widespread land degradation throughout the Sudanese Sahelian areas, Olsson (1985, 147), through the analysis of Landsat data, simulation, and GIS techniques, concluded "reports on desertification have been very exaggerated." Most of the pessimistic studies based their conclusions on the comparison of data sets collected *prior* to the Sahelian drought and *during* or shortly after the drought. He deduced "not surprisingly, the environmental conditions degraded from the first (wet) to the second (dry) situation" (Olsson 1985, 147).

Olsson found that desertification exists primarily during the extremely dry periods. When rainfall returns to a wetter climate phase, both natural vegetation and agricultural crops return to more typical production levels. Other observers reporting on conditions elsewhere in the area also contended that a long-term perspective was essential, and was more impressed by the resilient ability of farmers to cope with adversity over time than with the collapse of their adaptations in the face of adversity (Hiernaux and Turner 2002; Mortimore 1998; Mortimer and Adams 1999; Robbins et al. 2002). These findings are based on an array of data gathered for conditions before, during, and after drought conditions. These include analysis of the area's albedo values; examination of changes in the soil textural properties of the study area; and statistical analysis of crop production and climatic factors. Thus the decrease in biomass production noted during the droughts is largely in response to natural conditions, and hence is not illustrative of even short-term land degradation. But the impacts of drought on humans and the landscapes in which they live do reflect weak Sahel institutional and policy frameworks to support people during droughts and assist their recovery when better conditions return (Olsson 1999).

This is not to imply that desertification is a myth or to underestimate the likelihood of desertification occurring in some parts of the Sahel. Worse yet, there is alarming evidence to suggest a substantial drop in precipitation in much of the region; if this reflects a change in climate rather than an episodic variation in rainfall, the implications for future land-use strategies are significant (Olsson 1998). However, this example illustrates the difficulty in determining whether land degradation is occurring or whether decreases in biomass production just represent normal adjustments to the variable but normal rhythms of meteorological phenomena. To substantiate desertification within the Sahel, data must be gathered over a time period of sufficient duration to include the range of climatological events that occur within this zone. Unfortunately, it is the norm for this semiarid area to have highly variable climatic conditions, both temporally and spatially (Nicholson, Kim, and Hoopingarner 1988). Drawing conclusions about land degradation based on time-limited data can result in serious misinterpretations of both status and trend.

In the most historically accurate context, the American Dust Bowl is restricted to portions of the states of Colorado, Kansas, New Mexico, Oklahoma, and Texas (figure 1.5). However, through years of general usage, the Dust Bowl has been extended to include all of the Great Plains. In either context, it represents the prototype example of ecological failure resulting from drought in areas where rain-fed agriculture has been extended beyond the climatological limits of humid areas (Worster 1979; Braeman 1986) into semiarid areas.

While precipitation variability in all climatic types is normal, as climates become increasingly deficient in moisture, variability increases. In semiarid areas, except in selected favor-

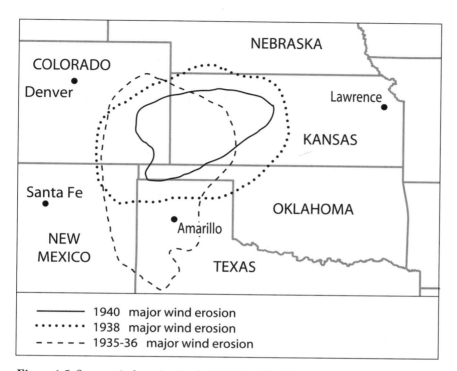

Figure 1.5. Severe wind erosion in the U.S. Dust Bowl, 1935–1940. (Cartography by Anne Gibson, Clark Labs)

able environmental settings such as floodplains, most crops are exceedingly sensitive to these moisture fluctuations. During more humid periods, the moisture deficiency is either extremely small or a small surplus may exist during a growing season in semiarid settings. During these wet episodes, agricultural yields are good and erosion problems, especially those related to wind, are minimal. This is particularly true in the American Great Plains where a significant proportion of the moisture supply falls as snow. Under typical conditions in the Great Plains, soil moisture content is high at the advent of the spring growing season, since it is the late winter and early spring snowmelt that recharges soil moisture. Under these conditions, adequate soil moisture and sprouting vegetation minimize wind erosion.

All semiarid and subhumid areas, including the Great Plains, experience dry cycles during which precipitation is notably less than average. Under pristine natural conditions, the continuous sod cover of these areas protects the underlying soil from extraordinary erosion during periods of moisture stress. The environmental situation maintains an equilibrium even though drought conditions exist and the grasses do not attain their normal growth. The sod cover and permanent vegetation protect the soil from the ravages of wind and water. With the introduction of agriculture, the natural conditions of semiarid and subhumid lands are disturbed. The ability of these areas to absorb environmental stresses resulting from drought conditions is reduced. In the Great Plains during the 1930s dry period, sod busting, required for agriculture, exposed the silt and clay components of the soil to deflation processes. Hence wind erosion, not significant in prior and similar dry periods when a complete sod cover existed, became effective once desiccation of the exposed cultivated soil occurred. During the dry years of the 1930s, windstorms were effective in removing clay and silt-size soil particles from the bare ground and transporting huge quantities of them in suspension as dust. Under similar climatic conditions in the 1750s, 1820s, 1860s, and 1890s, dust storms did not occur because the soil remained protected by the area's complete sod cover (Stockton and Meko 1983). The inference is that it was the exposure of the bare soils by agricultural activities, and not climatic change, that created the American Dust Bowl conditions of the 1930s.

The first great dust storm began on May 11, 1934. As dust from this storm traveled all the way to Washington, D.C., and beyond into the Atlantic, the nation was clearly alerted to the need to protect agricultural lands. During the great windstorms of March 1935, when clouds of suspended dust reached heights in excess of 3,000 m (9,843 ft) and estimates of the eolian load in central Kansas were in excess of 35×10^6 kg/km^2 (Skinner and Porter 1987), clearly massive human-induced wind erosion took place. During the 1930s, wind erosion removed over 1 meter of soil in some areas. This erosion rate is in comparison to long-term average regional estimates of only a few centimeters per thousand years under natural groundcover conditions (Skinner and Porter 1987).

In the early 1990s soil erosion and land degradation remained a problem within the Great Plains. Despite modern conservation practices, wind damage to land in the Great Plains has increased sharply in recent years. This is mainly because of lingering drought in North Dakota and the exposure of soil resulting from agricultural activities. The U.S. Soil Conservation Service has reported that 2×10^6 hectares (4.93×10^6 acres) were damaged between November 1989 and February 1990, while about 1.9×10^6 hectares (4.7×10^6 acres) suffered damage by wind in the same four months of 1988–1989. However, statements that contend that erosion is worse under contemporary conditions than during the Dust Bowl may be questioned (SWCS 1984). Plains climatic and crop data indicate that the scattered dry periods in the 1970s had fewer environmental impacts than those of the 1930s and 1950s (Riebsame 1986). This was due in large part to a series of agricultural adjustments after the Dust Bowl that resulted in many farmers leaving their farms for employment elsewhere, an increase in farm size, the withdrawal of marginal land from production, and the use of better adapted crops and crop varieties. However, in the late 1990s, the post–Dust Bowl reduction in erosion rates have been compromised by a drastic reduction in soil banking (and other programmatic changes), coupled with increased demands for American agricultural products overseas. This has encouraged farmers to once again plant land with a high erosivity potential, with a resulting increase in erosion rates throughout the northern Great Plains.

Unlike contemporary "dust bowls" in many other semiarid areas, such as the Sahel, the Thar, and the former USSR's virgin lands, where major capital constraints, educational limitations, nonprivate land ownership, poor governmental infrastructures, high population pressures, and a multitude of other economic, governmental, and social factors are cited as both the cause of and rationale for the lack of progress in arresting the problem, this is not the case for the American Great Plains. For example, not only have governmental policies been developed and implemented, but also a host of high-technology interventions have been widely available to the area's farmers for decades (Harris, Habiger, and Carpenter 1989).

Despite all of this intervention, evidence exists that the resource base is still declining in the Great Plains. Most national farm bills still do not make natural resource conservation and environmental protection central features of American agriculture and agricultural policy (Cook 1989).

Since the early 1970s, policy and conservation practices emphasizing the prevention of future major dust bowls and extreme soil erosion resulting from channelized runoff have been favored by most U.S. federal farm legislation. This policy deals with those aspects of the soil resource base where degradation is clearly manifested. This perspective officially became dominant with the implementation of the Soil and Water Resources Conservation Act of 1977. The goal of this act was to minimize soil loss from cropland. It ignored most other environmental aspects that have long been considered critical in the ecological sciences (Richards 1984). One result has been that the low-intensity, slowly accruing, chronic forms of land degradation have not been addressed for the Great Plains and other areas. The relative productive potential of any soil is determined not solely by erosion rates, but by a soil's ability to provide a good environment for root growth and development. These conditions are determined by a multitude of soil properties such as an adequate pH, available water capacity, nutrient characteristics, organic matter, and soil tilth (Lawson et al. 1981).

To date, the slowly accumulating environmental impacts of modern farming have been offset by huge energy inputs (chemical fertilizers), new hybrid crops, and the use of irrigation. But environmental danger signals associated with agricultural intensification—the use of large inputs of chemicals and pesticides to achieve high yields—are becoming evident. As but one example, soils in the rich French agricultural areas of Beauce, Brie, and Somme exhibit significant decline in their organic content. This has resulted in those soils becoming "a neutral medium to which everything must be added: fertilizers, mineral elements and, because the crops grown there are so fragile, huge amounts of pesticides" (James 1993, 17). The question remains as to whether continuous technological advancements will remain economically viable as the soil resource slowly degrades and associated pollution problems increase in intensity.

Human Causes of Degradation

Although environmental change and degradation have their origins in natural processes, the main contemporary factor in environmental deterioration is the impact of human activity. Identifying the level of human responsibility in any given situation is often a complex task. The difficulty arises because humankind and nature are linked in an interactive system in which cause and effect, and process and response, often blur. Unexpected natural events such as a prolonged drought, a flood, or a shift in the location of an ocean current (e.g., the Peruvian Current—El Niño) may overwhelm the adaptive capabilities of a particular resource use system. What appears to fishermen and government agencies to be a safe level of fish catch may, in the face of an environmental fluctuation of an unexpected magnitude, initiate a sudden collapse of the fishery. In this context, people often are less the purposeful initiator of degradation and change than they are accidentally in the way of environmental fluctuations that are (given our present state of knowledge) essentially random, and for which they are unprepared.

An example of this situation is found in many traditional pastoral nomadic communities, whose characteristic initial response to drought is to move their families and herds to the best known, most reliable local water source (Johnson 1973). The concentration of animals in this restricted zone initiates overgrazing and local degradation. As long as the dry spell falls within normal expectations, long-term degradation is not a threat either to the biomass production of the local habitat or to the survival of the herder and herd. However, if the drought is an extremely low-frequency event, occurring once or twice in a century, it may be of sufficient magnitude that the local "reliable" water supplies dry up. Under these extreme conditions, which, despite their severity, are within the range of normal climatic variability, the herding community faces serious problems.

Unless an early decision is made to move to better pasture and water conditions elsewhere, significant herd losses are likely to occur. Long before animals die in sufficient numbers to balance local forage and water availability, severe, but highly localized, overgrazing has likely taken place (figure 1.6). This places a heavy responsibility on community leaders

Figure 1.6. Overgrazing near a traditional well at Mazrub, Kordofan, Sudan. Use of hand-lift technology limits the number of animals that can be watered at the well and restricts degradation to the well's immediate vicinity. (Photograph by D. L. Johnson, September 1982)

to reduce grazing pressure at an early point in the drought/degradation cycle in order to avoid long-term deterioration of the resource base. Yet group experience, based on traditional ethnoscience knowledge, demonstrates that prolonged droughts are rare (100-year frequency) and that local resources usually can meet the demands placed upon them. For this reason, herders are reluctant to move too early, since they can never be entirely certain of grazing and security conditions elsewhere. The knowledge of local limits and how to prevent overuse is an integral part of resource use in many dryland pastoral systems (Draz 1974; Hobbs 1989; Galaty and Johnson 1990; Johnson 1993a). Of course, there are some critical problems: How is the magnitude of a dry period determined as it occurs? What are the appropriate measures to take once an area enters a dry period? Unlike the biblical story in which it was foreseen that seven good years would be followed by seven bad ones, it remains difficult, if not impossible, to assess the duration of a drought until after it has ended. Unable to react with sufficient rapidity as moisture conditions deteriorate, human-initiated degradation appears as more the unforeseen consequence of how a normally well-adapted livelihood is unable to adjust to unexpected environmental fluctuations rather than a ruthless and willful misuse of resources.

The Ecological Transition

Nature no longer is an external force, separated from human culture. It is increasingly incorporated into the matrix of human experience and action (Bennett 1976; McGibben 1990). This trend reflects the degree to which human societies now influence and alter natural systems, and is dubbed the "ecological transition" by John Bennett. Ever since humankind constructed its first simple tools, environmental alteration was possible, although at first these changes occurred on a microscale. Domestication of fire permitted the first major areal modification of nature. Coevolved ecosystems were the result, in which grasses and thick-barked trees were favored over more fire-sensitive species. In this process of altering the ecological balance between people and the environment, two types of societies emerge—those that, for longer or shorter periods, are able to create sustainable agroecologies, and those that are not.

The ecological transition is a product of both humankind's increasing technological capabilities and also its emergent anthropocentric value systems. These changes are increasingly viewed as producing an estrangement from the natural world that is reflected in an aimless art, a bankrupt philosophy, and a boring literature (Trussell 1989). A more powerful technology makes it possible to overcome the capability constraints to the exploitation of the local habitat (Carlstein 1982). Resources, to satisfy a wide range of needs, come from an ever-increasing distance to meet local site requirements for manufacture and consumption. Mines in Chile produce copper that covers church roofs or is drawn into wire in industrial communities in the United States. Coal mining and thermal-powered electricity generation in northwestern New Mexico sends clean electrical energy to southern California. Tall local smokestacks in Ohio and Pennsylvania lift pollutants high enough into the atmosphere to be carried out of the local habitat; the acid rain resulting from these airborne pollutants, which corrodes buildings and poisons lakes further to the east in Upstate New York, New England, and eastern Canada, is an impact that was unforeseen originally. High chimneys were built to minimize local pollution: the intent was never to transfer this problem eastward—the assumption was that the unwanted elements would disperse in the vast reservoir of the atmosphere without unwanted impacts locally or elsewhere. The problem continues to afflict the northeastern United States, although data show a slight decrease in acid rain in the region due to decreases in atmospheric deposition of SO_4 (Clow and Mast 1999).

The environmental change produced in mining districts is remote from the conscience and knowledge of the manufacturing populace. Likewise, the atmospheric pollution transported great distances from plants is not of immediate concern to plant managers and local residents. Yet the ability to cause environmental change, destroy vegetation, alter wild animal habitats, acidify lakes, remove unwanted overburden to reach mineral ores, and affect groundwater distribution and quality is great. Distanced from the activity site, with the costs of environmental change and degradation exported onto other peoples and places, the initiating population ignores the impact. Basking in the warm glow of self-satisfied security generated by elevated material well-being and by apparent mastery over the natural world, it is easy to view environmental degradation as someone else's problem. This rampant consumerism and exaltation of economic growth and materialistic gain has been attained at the expense of the spiritual values that sustain concern for the well-being of the biosphere (Light 2000; Raisch 2000).

Values are an integral part of the causal mechanism of degradation. The animistic attitudes toward nature characteristic of "primitive" peoples promote a less aggressive exploitation of environmental resources. Respect for the spirits inhabiting bush and beast does not prevent the hunter, gatherer, or cultivator from killing, collecting, or cutting. However, it does promote an ethic that recognizes limits to useful extraction, and it does undergird a sharing value system that is hostile to excessive accumulation and hoarding (Lee and DeVore 1968). This value system has been in retreat ever since the invention of agriculture. With the manipulation of domesticated plant and animal species, both surplus accumulation and storage become possible. The surplus sustains greater local populations on a sedentary basis and concentrates environmental impacts that a mobile hunting and gathering livelihood system spreads more evenly over a greater territory. Control over plants and animals helps to assure humankind of its mastery over the natural world, and generates the imagery found in Genesis whereby nature is placed at the service of *Homo sapiens*. The shift in value orientation places humankind at the center of the universe, created in the image of, and an earthly surrogate for, the divine. A type of species loyalty emerges that envisages the human species as the ecological dominant. This encourages the direction of energy toward our species to the detriment of other denizens of the biosphere. While this tendency probably existed in the first hominid, tool-using cultures, it was mitigated by both the animistic ethic characteristic of early hunting and gathering communities and by the relatively limited power of early technology. Contemporary efforts to create a more geocentric ethic that allows livelihood rights to and habitat space for other species exist (Lynn 1998) and reconnects people with nature (Pretty 2002), but the proponents of a broader ethic face an uphill battle to alter dominant values and affect policy change.

The emergence of first the Agricultural Revolution and, more recently, the Industrial Revolution destroyed or pushed into more remote and marginal habitats the more reverential animistic ethic. In its place was substituted the more anthropocentric value systems characteristic of modern civilization. These value structures favor an I/it orientation toward nature rather than an I/thou attitude. The significance of this shift was noted by numerous observers, many of whom follow Lynn White (1967) in attributing the change to the influence of the Judeo-Christian heritage. This perspective is manifested in the conflict between ". . . the idea of stewardship of land [being] pitted against the belief in soil exploitation for personal gain and that soil is merely an economic commodity in the marketplace" (Jenny 1984, 161).

Yet causation is more complex than this neat association between religious and cultural heritage appears to suggest. As White himself notes, reverential attitudes can be found within the Judeo-Christian tradition. Moreover, reverential images of nature in the literature of non-Western cultures often mask the reality of ruthless exploitation and degradation. Tuan (1970) points out that China has a very checkered record of deforestation and soil erosion, despite its literary and artistic motifs of reverence for and harmony with the natural world. Traditional use of upland forest environments for their plant and animal products constitutes a critical resource for local communities, and often places villagers and park managers at odds with serious long-term consequences for conflict and successful conservation (Xu and Jim 2003). It was the environmental degradation of the Greeks and Romans, which appeared long before Christianity, that inspired Marsh (1965 [1864]) to examine environmental change in the Mediterranean, and in so doing to serve as the founding father of the modern environmental movement. Jacobsen and Adams (1958) have shown that salinization in the fields of ancient Sumer in 2000 BC was responsible for significant decreases in yields and in a shift from salt-sensitive wheat to more resistant barley. When presented with modern technology, few hunting societies are immune to the value shifts that encourage a more wasteful and exploitative use of existing resources. Kemp (1971) demonstrates this with Baffin Island Eskimos who, after gaining access to the rifle, the outboard motor, and the snowmobile, clearly became more wasteful users of the wildlife in their area.

Ideological blindness can also promote an absence of ecological concern, as the recently acknowledged ecological impacts of the smokestack industries of the former socialist/communist societies of Eastern Europe graphically illustrate. By 1990, parts of Poland, eastern Germany, and the Czech Republic had perhaps some of the most environmentally devastated lands in Europe, a situation that developed under their former state socialist command economy regimes. So serious were conditions in this region, replete with mining operations and smokestack industries, that it was dubbed the "Black Triangle." This situation has begun to change dramatically (EST 2001). Over the last decade and a half, strenuous efforts have been made to remediate the pollution of the region's soil and air and bring it closer to the standards expected in the European Union. These efforts have resulted in a decrease in emissions of sulfur dioxide, nitrogen oxide, and particulates by more than 80 percent. The economic, utilitarian, and anthropocentric values that underpin degradation transcend political, cultural, and religious ideologies, and occur at all points in the time and space record of our species.

Intentional and Unintentional Change

The human dimensions of degradation are further complicated by two additional factors. First, environmental change is simultaneously both intentional and unintentional in its causation. Frequently, people set out to initiate change knowing full well that some of the consequences will constitute degradation. For instance, strip mining inevitably produces blighted landscapes and degraded land and water systems. Only recently has a feedback loop been created in the United States that requires the restoration of such degraded areas. These remedial measures invariably are limited to the reestablishment of surface vegetation and the soil conditions that promote plant growth, by no means an easy task. Subsurface conditions in which altered groundwater tables and deranged subsurface drainage patterns are typical are never addressed. In arid areas, mining operators, citing the high cost of reestablishing plants in water-deficit habitats, find even the restoration of surface conditions to be an intolerable economic burden. In most countries, even in wetter environments, the remedial measure of restoring surface vegetation to its premining condition is sacrificed to immediately perceived economic gains. On a smaller scale, in some semiarid areas (see chapter 2 for a detailed description) runoff farmers degraded the hill-slope zones in small drainage basins in order to increase surface water flow toward their valley-bottom fields. In so doing they sacrificed the productivity of the more marginal facets of their habitat, the hillslopes, in order to enhance the useful productivity of the higher potential elements in their land-use system, the valley bottoms.

African pastoralists often use fire as an environment-modifying tool, both to prevent the spread of trees and bushes, thereby increasing the amount of grass available for their animals, and to remove habitat that harbors the tsetse fly, the intermediate vector responsible for spreading trypanosomiasis (sleeping sickness) to cattle and humans. These strategies have one thing in common: they reflect a conscious commitment to environmental change even when those changes represent a biological degradation in part or all of their habitat.

Despite these instances of change that accept the possibility of degradation with some equanimity, most of the degradation that occurs is of the unintentional variety. In most cases when humans promote environmental change, they envisage the change as being one for the better. They expect that the land-use system they create will be more productive for their purposes than the unaltered natural system that is replaced. The increase in productivity that occurs is usually true in the short-term, but often is less true in the long-term. A classic example of this is the use of irrigation in order to improve the productivity of drier landscapes that are risky or impossible for rain-fed agriculture. No one intends to promote soil degradation through the spread of salinization and water logging. The intention is to make the desert bloom, to create a paradise in the midst of the wilderness, to intensify existing agricultural systems.

The Aswan Dam in Egypt is a controversial example of this problem, for its success in promoting multiple cropping, desert reclamation, and hydroelectric power is also associated with initiating serious land degradation problems (Kishk 1986; Ibrahim and Ibrahim 2003). Today on a global scale, irrigation agriculture is locked into a treadmill effect whereby every new land unit brought into production is virtually offset by a unit lost somewhere else (and often not very far away) to salinization and water logging. Farmers who open up a new field in hilly terrain by plowing a field with straight furrows perpendicular to the contour follow cultural tradition and plan to make their farms productive. They do not intend to promote soil erosion. Planners who insert deep bore wells into pastoral zones have no intention of encouraging overgrazing and wind-aided soil erosion. The frequent negative consequences of these changes come as a surprise to the perpetrators and are often expensive, difficult, or impossible to reverse. None of these degradation effects are necessarily the product of malicious intent.

There is a second factor that contributes to degradation. "Drivers"(in the jargon of land use/land cover change specialists) push landscape use and modification in harmful directions and cause degradation. These causal variables may be both internal and external to the production system and human actors most immediately affected by the change. For example, starting by at least 3500 BP, farming communities in the Andean Highlands in the Lake Titicaca basin of contemporary Peru and Bolivia began slowly but massively to modify their environment (Erickson 2000). In the process of this transformation, human activities concentrated on specific land facets with high agricultural potential: the lake shore wetlands, flat plains near the lake, and steeper slopes on hills and peninsulas adjacent to the lake. All of these habitats were well within the natural range of native crops, but proximity to the lake provided a more moderate climate and longer growing season than is typical of the cooler and drier conditions prevalent in the larger region (Denevan 2001). The technological innovations and landscape modifications developed by the region's Quechua and Aymara agriculturalists drastically modified the local habitat, replacing wetland, grassland, and hillslope plant communities with raised fields, canals, drainage ditches, and terraces. While a number of plants native to the region became central components of the substitute agro-ecosystem, the species less desired by humans for food and fiber were eliminated. In the former shoreline marshes of Lake Titicaca, for example, where raised fields became numerous, cultivation concentrated on the totora reed (*Schoenoplectus tatora*) as a major crop. This reed was an important resource for thatch, mats, boats, construction material, and animal fodder, while its starchy roots served as an important source of human food; equally important to the local population were the fish and aquatic birds that dwelt in and adjacent to the modified wetland habitat (Erickson 2000; Kolata and Ortloff 1996). With potatoes, oca (*Oxalis tuberosa*, valuable primarily for its edible tuber, but secondarily for its leaves and shoots), ulluco (the multicolored edible tuber of *Ullucus tuberosus*), and grains (most notably quinoa and canihua) serving as dietary staples on the raised fields, the result was the creation of a new artificial humanized ecosystem. These raised fields were remarkably productive and stable for long periods, perhaps because they acted to modify local microclimate by reducing frost hazard (Denevan 2001, 270) and could be fertilized by sediments, rich in plant and animal remains, that were dredged from the waterways that separated fields. The artificially raised fields created a deeper soil medium for plant growth while at the same time protecting crops from seasonal flooding; the canals separating raised fields provided a source of irrigation water as needed and, by storing solar energy in the water and reradiating that energy at night, the canal network provided crops with protection against frosts (Erickson 1994, 143). Even when a post–Spanish Conquest political ecology deemphasized crop production on these raised fields, they continued to be stable loci for a more extensive, lower intensity system that stressed the rearing of sheep and cattle. And these same fields, with the expenditure of rehabilitation funds and labor, can once again become productive agricultural sites (Kolata et al. 1996).

Cultivated terraces near the lake were also important contributors to the food supply of the dense indigenous populations typical of the Lake Titicaca basin. These terrace systems were created in part by intentional encouragement of soil erosion (Erickson 2000,

330–33), which was then trapped behind stone walls lower down the slope. That the small, flat fields created in this way, like the raised fields characteristic of the coastal lake habitat, were productive is the product of three useful principles. First, systems that evolve gradually in close association with and sensitivity to the local ecological setting are most likely to achieve long-term success. Intimate linkage to the spirit of a place (*genius loci*) insures that the agroecology created will be attuned to the rhythms and forces operating in a particular area. Second, *sacrifice zones* (the thin soils on the upper hillslopes) were created at the microscale. One portion of the environment (the sacrifice zone) was deliberately degraded or destroyed in order to increase the productivity of the other. Because both the benefits and the costs of this action were localized, it was possible to erect a sustainable agroecology. This long-term sustainable system illustrates the third principle, that of *creative destruction*. Creating production systems useful to humankind often requires the destruction—or at least massive alteration—of nature, with consequent changes in species diversity and habitat quality. The critical variable is the erection of a system that can withstand the test of time.

While creative destruction can establish long-term sustainable agroecologies, continuous maintenance of the system is required. When upkeep ceases, the system declines. An example of this process is provided in a very wet environment by the pond field agricultural terraces of central Luzon in the Philippines (Conklin 1980). These fields are masterpieces of ethnoscientific engineering that transform incredibly steep slopes into productive fields by destroying tropical rainforest. By creating a totally artificial habitat with great stability characteristics, an extremely productive agroecosystem is established. The wet rice terrace farms created in central Luzon rely entirely on local labor, management, and initiative. Yet the terrace systems are vulnerable to forces operating outside their immediate locale. Maintenance of the terrace fields is intimately linked to having an adequate labor supply to repair terrace walls, to level fields, to transplant rice, to channel water, and to perform a host of other tasks in the agricultural system. Withdraw male labor by migration to coastal cities and the elaborate terrace systems, the product of half a millennium of effort, could quickly collapse.

The Limitations of Economic Evaluation

If land degradation could be solved by technological interventions alone, by today it should be declining in areal extent. Even with the multitude of international agencies, national government agricultural departments, and private nongovernmental agencies helping to facilitate and implement conservation practices worldwide, the processes of land degradation continue to spread their havoc throughout the world (Blaikie and Brookfield 1987). In 1960 Gilbert White published an important overview of the state of scientific knowledge about arid lands and the technologies and management practices that could be employed to cope with dryland degradation and rehabilitate damaged lands. Forty-five years later, an international group of scholars convened in Berlin to discuss desertification, the complex of land degradation processes afflicting arid lands; armed with significant technological improvements in monitoring, managing, and ameliorating dryland degradation, scholars continue to lament the same litany of problems, blighted prospects, and missed opportunities (Reynolds and Stafford Smith 2002). Along with other human factors, before land degradation will be solved, it will be necessary to develop widely accepted economic analyses that realistically evaluate the various costs of environmental degradation, including both aesthetic as well as production factors.

One reason for widespread "mining" of renewable natural resources is that accounting procedures rarely evaluate the deleterious environmental effects of these actions. In the short-term, at the individual or institutional level, it is often economically rational to pollute and destroy resources. Per-Olov Johansson (1990) believes that there are three critical reasons why adequate environmental costs must be incorporated into any activity that impacts resources, such as land quality. First, all of humanity can benefit from the direct effects of many environmental resources. For example, breathing clean instead of dust-laden air has both health and economic ramifications for the population at large. Thus the direct costs of soil erosion go beyond the immediate user of the resource—the farmer—and the consumer of the crops. Second, citizens may value the opportunity of using a resource in the future more than the income they would gain from using it in the present. For example, many people are in favor of protecting the gorilla, rhinoceros, or American buffalo and their natural environments in order to retain the future possibility of seeing them. This is true even if there is a high likelihood that the individuals favoring preservation will never actually see them in the animal's natural habitat; that is, there is a perceived value for their existence. This is also true for preserving unique landscapes for their aesthetic values without regard to wildlife. Jules Pretty argues that it is possible to calculate a value of approximately UK£140 (US$243) for the mixed agricultural and natural landscapes of the Norfolk Broads (2002, 66). Actual numbers of visits to the region is another measure of the economic value placed on the resource by a society, since visitors must spend substantial sums to reach, subsist in, and explore the region. In another case, the perceived value of the Canyonlands in their natural state has been a critical reason why major dam construction in the American Southwest ceased in 1963 with the completion of the Glen Canyon Dam on the Colorado River (Martin 1989; Russell 1999). Opposition to the construction of large dams, in the United States and other countries, has produced a comprehensive assault on the presumed benefits of large dams and a demand for both wiser use of existing water supplies by planning on a watershed scale and a return to unimpeded, natural flowing (McCully 1996). Despite the region's strong traditional economic development forces, the campaigns to prevent further major dam construction on the Colorado were successful, giving credence to Johansson's second reason. His third and final point, and this is especially a crucial component in the Green movement, is the desire to preserve the Earth's environmental heritage for future generations. Society grants us the right to exist, regardless of whether we are useful. If for no other reason, the same privilege should be extended to both plant and animal species.

Today techniques to assign costs to environmental impacts are incorporated in an ever-increasing proportion of cost-benefit analyses. However, the counter to this positive trend is that there is no widespread acceptance of any single technique as the most appropriate way to evaluate environmental costs. With the multitude of techniques available, in most cases it is possible to justify diametrically opposite outcomes while using identical data. Thus it is still difficult to incorporate—let alone resolve—environmental conflicts using numerical accounting procedures.

In contemporary domestic and international markets, competition is a deterrent to linking environmental impacts with resource utilization. Some arguments against the creation of a Canadian-United States-Mexican free market are based on the different environmental laws in these three countries. Weaker laws are perceived to give an economic advantage for attracting industry from one country to another. Without incorporating environmental value into every product's or service's cost, the presence of externalities indicates that, in attempting to maximize their utility or profits, households or businesses will not generally make socially optimal decisions (Fuchs 1986). Most consumers are not willing to pay for additional production expenses associated with environmental preservation when similar, lower-cost products are available, and these items are less expensive because they are produced with minimal concern for environmental degradation.

If the State of Iowa passes legislation that requires its farmers to minimize soil loss while Indiana farmers are not governed by similar regulations, the cost of Indiana corn (maize) might be less due to lower conservation costs than those experienced by Iowa farmers. The result could be that land degradation is arrested in Iowa and long-term benefits for Iowa are clearly greater; but Iowa farmers could be out of business, or at least might earn less money than their Indiana counterparts. Grocery chains and food manufacturing companies would likely purchase the less expensive Indiana corn at the expense of the Iowa producers.

Most politicians are not willing to resolve environmental conflicts and environmental regulations under the present limitations inherent in current environmental and costing evaluations. The debate over acid rain in both Europe and North America during the 1980s and 1990s reflects this point (Torrens 1984; Bunyard 1986). While almost all parties agree that acid rain causes environmental damage, to date it has been difficult to obtain a consensus on the environmental or economic ramifications of this problem. One reason why political accountability is minimal is the shortcomings in environmental and costing techniques. In the United States, the coal-producing areas, such as West Virginia, are minimally affected by the acid rain problem; but it is the burning of this fuel that is the major cause of acid rain. Under current practices, if controls are placed on air pollution emissions, which in terms of the acid rain problem largely originates in the Midwest, in order to protect New England's land, water, and forest resources, the Appalachian and Midwest areas will be burdened with higher production costs and yet receive minimal environmental benefits. This is clearly a high-risk situation for any Appalachian or Midwestern politician.

In the United States, as in most other countries, reversing land degradation (as well as other environmental problems) is further complicated by contradictory policies among governmental agencies, including the federal (national), state (regional), and local levels. For example, the federal government both subsidizes the exploitation and the conservation of land and water resources. In the Food Security Act of 1985, the soil and water conservation behavior of farmers leveraged against crop price supports, credits, and the availability of crop insurance, clearly is an attempt to encourage conservation. Yet at the same time, federal tax laws—in reality subsidies—have prompted the overpumping of groundwater reserves in the Southwest. In this dry region, this tax policy has encouraged excessive irrigation and in most cases wasteful use of water. A consequence of this policy has been the mining of the region's groundwater reserves. A renewable resource is becoming nonrenewable. Widespread deterioration in the quality and quantity of groundwater is one result of these subsidies. The resulting shortage and the poorer quality of the groundwater are two factors contributing to the increasing salinity of soils in these areas. Many of the region's existing irrigated lands likely will go out of production in the future due to the resulting soil degradation encouraged, at least indirectly, by government policy.

Summary

Land degradation is not a new problem; nor is it limited to any particular political or cultural ideology. Its "uniqueness" is that it exists virtually everywhere to one degree or another. One estimate is that 26 billion tons of soil are being eroded from crop and grazing land annually worldwide (Brown and Wolf 1984). Soil erosion is an expensive business, with annual estimates for both onsite and offsite damage in the United States ranging between $30 billion and $44 billion (Uri and Lewis 1999). Besides soil loss, an equal if not more important aspect of land degradation is the deterioration of soil fertility, including chemical, biological, and structural properties. Estimates of the magnitude of this component of soil degradation are not available.

The causes of land degradation are complex. It is a response to a multitude of complex interacting physical processes along with human values and constraints. From a purely technical perspective, the technology exists today to prevent, arrest, and rehabilitate the majority of degraded lands. Yet all indications are that the problem is getting worse as land resources remain continuously under attack. In widely diverse settings, ranging from the forests of Western Europe to the Amazon Basin, from the irrigated lands of the American Imperial Valley to Rajastan in India, and from the chemical poisoning of the land in Poland to the Valley of Mexico, land degradation continues to occur, and in many places it is accelerating.

In this introductory chapter we have discussed the basic dimensions of the land degradation problem. In the remainder of this book, we examine the degradation phenomenon with the intent of developing the understanding necessary to create a sustainable land resource use system.

Land Use and Degradation in Historical Perspective

Human Causation in Land Degradation

It is a widely accepted truism that the greater the density of the human population, the greater are its impacts on the surrounding environment. This principle also holds true generally in nature outside human-managed systems if concentrations of animals are encouraged by the distribution of scarce resources. A water hole is an example of this phenomenon. Large numbers of animals coming to drink in a specific locale will inevitably trample the vegetation around a water site. Limits on degradation are set by nature, because too extensive overgrazing near the water source will result in an absence of fodder for the animals. Lack of fodder makes the water resource unusable, and sets in motion processes of stress and movement that operate to conserve the basic land resource. Nonetheless, the vicinity of any water hole is always more degraded in nature than is the adjacent territory; one does not have to invoke human intervention as a causal agent to explain the deterioration.

However, human impact is the most frequent explanation for negative environmental change. Evidence for this assertion is easy to find in historical records. Even in small numbers and at low population densities, humankind initiated significant alterations in the natural environment. The most primitive tools, such as digging sticks, crude stone crushing and chopping axes, and butchering blades, gave small human societies increasing control over their local environment. The higher levels of caloric energy directed toward *Homo habilis* and away from other species gave early hominids and their descendants a comparative advantage. However primitive these tool technologies appear from a twentieth-century perspective, they were of tremendous ecological significance. The better levels of nutrition and higher rates of survival made possible by tool technology, combined with superior cognitive abilities, cooperative patterns of behavior, and great adaptability, enabled the descendants of *H. habilis* to spread gradually throughout the Earth's major ecological habitats. These attributes also enabled increased human populations to be sustained in local concentrations.

While these early technologies gave humankind a comparative advantage over other organisms in the quest for food and shelter, they were not sufficient to transform the Earth beyond a purely local scale. The domestication of fire altered this relationship (Sauer 1956; Pyne 1993). Full control of fire and its conscious use as a tool to modify the environment is at least five hundred thousand years old, although efforts to identify unequivocal human use of fire significantly further back in time are doubtful (James 1989). First used to provide domestic warmth, protection from predators, and a device—via cooking—by which to kill parasites and harmful bacteria, fire increasingly was used in hunting. By driving game into settings in which hunters could more easily dispatch their prey (mired in bogs, driven over cliffs, or directed into constrained "shooting galleries"), fire was an important part of the hunter's tool kit (Pyne 2001). In the process of setting fire drives, habitat was modified. The frequent use of fire discouraged the survival of most woody perennials. Fire promoted the growth of annuals and selected woody individuals that were genetically adapted to resisting or evading fire. In much of the dry tropics and subtropics, this meant that grasses were favored over trees and more open landscapes were created. Measured in terms of standing biomass, the result is a degraded ecosystem from an ecological standpoint, even though it was a more desired system from the standpoint of the human population inhabiting the space. This is the case because the spread of grassland at the expense of forest also meant that herbivores were at a comparative advantage. Since these animals were often the same species preferred by hunters, the result was the development of a coevolved ecological system of hunter—and later herder—grazing animals, and grassland. The land degradation that occurred as a result of this local-scale, controlled burning was desirable from the perspective of the hunter and later herder because it expanded the grazing habitat (Pyne 2001, 76-82). In Africa, the reduction of woody biomass additionally reduced habitat favorable to tsetse infestation. Thus through repeated use of fire, land which in its natural state had limited utility for humankind, evolved into habitat able to support more people. From the viewpoint of forest economies, whether of hunting or farming, this habitat change was distinctly undesirable. In environmental terms, modifications introduced by the widespread use of fire were negative insofar as they reduced the diversity of plant and animal species supported in a given area. But as long as the burning was not so extensive that it exclusively favored annuals over perennials, and created patchy, mosaic landscapes with increased edge habitats, little long-term loss in basic soil productivity took place through exposure of extensive surfaces to the erosive effects of wind and water.

If tools and fire gave prehistoric communities greater control over the natural world, their impacts were minuscule compared to the changes wrought by the Agricultural Revolution. The beginnings of plant and animal domestication some ten thousand years ago fundamentally altered the way in which humans were able to affect their environment. Increased mastery over the location and reproduction of favored food-producing species of plants and animals improved the quantity, quality, and availability of food. Small bands of hunter-gatherers were no longer required to move seasonally in search of sustenance. Instead, increasingly permanent settlement became possible and occurred.

More people in denser concentrations resulted in greater impacts on the local environment. In wooded tropical and mid-latitude environments, slash and burn agricultural systems developed that were extensive in nature and used long forest-fallow periods to regenerate soil fertility. Frequent movement was the rule in many such communities, although deficits in animal protein were as likely to occasion shifts to new locations as were soil fertility deficiencies (Denevan 2001, 81–83) and highly variable local adaptations, sometimes supporting surprisingly high population densities, did occur. Whenever shifting agricultural communities had to reduce the fallow cycle, whether due to internal population growth or the constraints placed on mobility by neighboring communities, and local land facets that could sustain more intensive use were absent, degradation in the form of grassland expansion at the expense of forest resulted (Conklin 1954). Where slopes were steep but soils were suitable, elaborate terrace systems often gradually emerged to support wet rice cultivation in northern Luzon (Conklin 1980). Increasingly intensive use of all facets of cultivated and uncultivated portions of the landscape mean that the entire rural space is transformed into a continually changing humanized landscape in which little forest survives. These normally very stable terrace fields, and their more risky extensions into increasingly high risk environments using bulldozers and other power construction machinery, carries with it the threat of increased slope failure, accelerated erosion, and declining soil fertility (Lewis 1992). To the outside observer, these changes often appear to be uniformly negative in result. But as David Kummer, Roger Concepcion, and Bernardo Cañizares caution, such observations can be misleading (2003). In Cebu, at least, a bias that assumes upland dwellers mismanage and degrade their habitat can blind outsiders to the reality of a high degree of environmental stability. Nonetheless, what modern technology can produce in relatively short periods, traditional technologies can achieve over longer time frames with greater chance of establishing a new, stable ecosystem. The agricultural impacts of native societies on adjacent forestland were the cause of the substantial grasslands found in the midst of the hardwood forests of eastern North America when European explorers and settlers first encountered the New World (Rostlund 1957). A similar process is occurring today in the Amazon Basin, where pioneer peasant farmers often are unable to sustain a viable agriculture once the forest is cleared. The outcome is more extensive rearing of cattle in the grasslands that develop following forest clearance and farm abandonment caused by declining soil fertility. In extreme cases, land degradation is of sufficient magnitude that even cattle rearing is impossible on the cleared lands (Denevan 1981). In the last two decades, this picture of pasture-dominated forest removal has been complicated by deforestation dominated by logging, mining, and industrial interests as transportation improvements and timber demand combine to threaten commercially valuable forest stands (Parayil and Tong 1998).

Through history, with each major change in technology, the number of people who can be supported increases. Concomitantly, the ability of those populations to have an adverse impact on their environment grows. Many other factors interact to favor changes in technology, food production strategy, and settlement pattern. But invariably innovation and altered population dynamics reinforce each other and a period of accelerated population growth results (Deevey 1960; Diamond 1999, 104–13). This "spurt" is very rapid for a relatively brief period, but the growth phase is masked in the longer term by the much slower rate of population growth that characterizes most of human history, particularly the periods of relative stability between the epochs of innovation and increase. These major population growth phases are characterized as the Tool-making, Agricultural, and Industrial

Revolutions. Eventually the rate of growth slows as a rough balance between nature and society is achieved, although at the present time contemporary populations at the global scale continue to grow as a consequence of the Industrial Revolution. But this pattern is only true for the overall population at a global scale. At more regional levels of analysis, local population-resource crises can produce dramatic declines in population (Whitmore et al. 1990, 26). In some places the combination of altered social and environmental conditions can result in degradation so severe that permanent production losses are produced. Under these circumstances not only are fewer individuals supported but also, on a human time scale, rehabilitation becomes impossible.

Up to the beginning of the Industrial Revolution, the pace of change fostered by new technologies was relatively slow. The less powerful the technology, the greater the amount of time needed for the practitioners of that technological tradition to have a negative impact on their habitat. Also, less powerful technologies were generally not able to affect distant areas. The combination of a slow pace of change and a local scale of operation meant that negative impacts could be seen and counteracted successfully. As a consequence, small-scale, slow-paced, low-technology human interventions tend to be self-regulating. They coevolve with the increasingly humanized landscape they create and the ecotechnology they employ. Such societies are not stagnant and static; rather, they exhibit strong stability and resilience characteristics because their agroecologies slowly develop in ways that progressively enhance productive capacity and their ability to support more people.

The gradual development of a sustainable ecotechnology is rooted in the fundamental principle of *creative destruction*. The production activities of human livelihood systems have a dual dimension: they contain elements of both creation and destruction. Often to create, it is first necessary to destroy, to remove what is present in order to clear the way for new activities. *Creative destruction* describes the process of altering the natural world, often in profound ways, and replacing what was present with a new system state that is both productive in human terms and sustainable for the foreseeable future. When this transforming process goes awry the creative elements are lost, overwhelmed by destructive dimensions and spiraling the system into a much less productive state. This debilitating process of destructive creation is called "land degradation."

The transformation of an agroecology is an ongoing process that inevitably involves changes that destroy and degrade some parts of the habitat. As long as these changes do not diminish the overall productivity of the environment, the net result is beneficial. But often even successful ecotechnologies are unable to maintain themselves indefinitely. Changes in society and economy outside the control of local communities may undermine the viability of an apparently successful ecotechnology. Thus, in assessing the durability of a given agroecology, it is essential to pay careful attention to the stability and resilience parameters of the system.

The dynamic tension between destroying and creating was early noted by Duke Frederick, William Howenstine, and June Sochen (1972), whose collection of primary readings about the American environmental experience, *Destroy to Create*, encapsulates this dilemma. A somewhat earlier, less literary, more ecology-oriented collection of readings complied by Robert Disch (1970) also calls attention to the creation-destruction dialectic in the writings of major mid-1900s figures in ecology. In the Frederick, Howenstine and

Sochen selection of documents, an unregulated, largely destructive, short-term, utilitarian perspective dominates America's use of its resources, the rise of the conservation movement at the end of the nineteenth century notwithstanding. Many of the destructive manifestations of the profligate American approach to resource use include soil erosion, deforestation, groundwater depletion, and farm abandonment. The use of the creative destruction concept as a way to explain how contemporary economies function has gained considerable currency recently (Denison 2001). The last decade has seen the appearance of three books with creative destruction in their titles, all of them written by economists and business managers (Foster and Kaplan 2001; McKnight, Vaaler, and Katz 2001; Nolan and Croson 1995). These authors draw their inspiration from the work of Joseph Schumpeter (1939), an economist whose theories about creative destruction in business cycles stress how obsolete products and production methods are swept away by new ones. Schumpeter's ideas about creative destruction have even been recognized in historic preservation by Max Page (1999), where cycles of renewal in the material fabric of Manhattan are recognized and accepted without abandoning commitment to preserving truly historic structures and the context that gives meaning to place.

Equally important in assessing land degradation is a time trajectory that enables the observer to place a given ecotechnology into its historical setting and to predict where trends in that adaptive system are heading. Fortunately, there are several historical examples that illustrate the complexity of creative destruction. From this array of potential studies, we examine two well-documented historical cases, the agropastoralists of the Negev and the irrigation farmers of Mesopotamia, each practicing a contrasting ecotechnology. Other historical and contemporary cases are explored in later chapters.

Run-on Farming in the Negev

The Negev is an arid region occupying the southern half of Israel (figure 2.1). It is a small segment of a much larger dry zone that stretches from the Sahara to Central Asia. Of the region's 12,500 km² (4,826 sq miles), only 350,000 hectares (864,859 acres) in the northern Negev ever supported ancient settlement (Evenari et al. 1961). Within this area's foothill and upland zones, only a small fraction of the total surface contains sufficient soil and water resources to provide foci for settlement. Lack of reliable rainfall is the primary reason for this paucity of resource use opportunities.

The Negev lies at the southern end of the Mediterranean climate zone, and therefore shares in the rhythm of winter-spring rainfall and summer-fall drought typical of that region. Across the Mediterranean region, precipitation totals decrease southward and eastward from the sea. Furthermore, rainfall variability increases toward the south and east as well. In the northern Negev just south of Beersheva, the annual average rainfall is between 200 and 350 mm (7.9–13.8 in). South of this transition zone, rainfall diminishes rapidly, vegetation becomes ever more sparse, and few agricultural and pastoral resources exist to attract human settlement. Even in the northern Negev, rainfall is an irregular resource, for as rainfall totals diminish, variability in interannual precipitation increases. Years with practically no rain are juxtaposed with years that exceed the average by considerable amounts. This variability in precipitation is the norm for almost all arid areas. In 1962–1963, at Shivta,

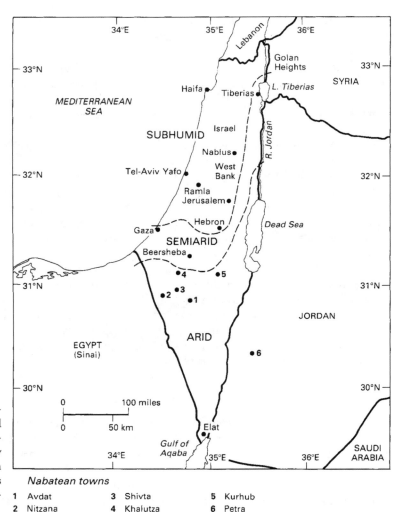

Figure 2.1. Nabataean settlement in the Negev. Small towns and farms were concentrated in the northern Negev where rainfall, collected from a larger catchment area, was adequate to support permanent settlement.

Nabatean towns

1	Avdat	**3**	Shivta	**5**	Kurhub
2	Nitzana	**4**	Khalutza	**6**	Petra

Michael Evenari and his colleagues (1971, 146) recorded a total precipitation of 28 mm (1.1 in). The next two years produced 153 mm (6 in) and 165 mm (6.5 in) respectively at a station that, over seven years, averaged 93 mm (3.7 in) of annual rainfall. The consequence of this small and unreliable water supply and highly variable environment with widely dispersed resources is a very hazardous habitat that most communities, especially agricultural ones, would shun.

These conditions of water resource poverty place severe constraints on those human groups required to live in the area. In order to survive in the region, inhabitants must first have a compelling reason to locate in what by any objective measure is an inhospitable and hazardous place. Second, they must develop sustainable patterns of resource use that are resilient and stable for long periods (Evenari 1981). The most successful livelihood systems are those that are able to adjust their activities to the natural ecological rhythms of the district (the *genius loci* principle) and to convert apparent negative features of the environment into positive contributors to human sustenance.

Both pastoral nomads and sedentary agricultural communities have managed to accomplish this feat. Nomadic pastoralists survive in arid environments by frequently moving from poor conditions to locales with better resources. This movement is dictated by an ecological rhythm that governs the general availability of grass and water. By using domesticated animals as an intermediate converter of pastoral resources, and by shifting these animals to seasonally favored sites, the Negev's nomads are able to capture spatially dispersed water and grass resources. Because these nomadic groups are characteristic of many dry regions, they are of only passing interest to us at the moment.

Of far greater interest are the settled agricultural communities that also occupied the northern Negev for substantial periods in the past. On several occasions, farmers and urbanites lived in the desert for periods of several centuries if not millennia (Evenari, Shanan, and Tadmor 1971; Taylor 2002). Because the earliest epochs are not well represented archaeologically, if only because so much of their settlement remains have been reworked by subsequent settlers, our attention is focused upon the period from 300 BC to AD 650 when settled agriculture flourished most extensively.

The settlement history of this epoch is closely associated with a group of people called the Nabataeans. They were an Aramaic-speaking Semitic people who occupied a broad swath of predesert zones from the northern Negev near Beersheva through Jordan to the Jebel Druze just south of Damascus. The Nabataean economy originally was based on two complementary activities: animal husbandry and caravan trading. The two activities supported each other, since animal products met many subsistence needs and animals provided beasts of burden for the mercantile operations. At the same time, consumption goods not provided locally could be purchased in more favored areas and carried into the Nabataean heartland by caravans passing through the area. A strategic location controlling the overland caravan trade provided the Nabataeans with considerable wealth, and over time much of this wealth was invested in a settled infrastructure. Heavily influenced by the urban Greek culture of the coastal districts, the Nabataeans created a spectacular capital at Petra in present-day Jordan and controlled a number of other urban centers, such as Avdat (Israel), of varying size throughout the region.

The growth of urbanization apparently was the product of a conscious policy on the part of the Nabataean monarchy throughout the first century AD (Bowersock 1983, 64 ff). A deliberate effort to promote sedentarization and the expansion of settled farming was part of this policy. All of the northern Negev urban centers exhibited considerable growth in this period. Bowersock (1983, 72ff) believes that a motivating factor in the process of agricultural expansion was the gradual decline in the viability of Nabataean overland caravan trading in the face of competition from more northerly routes through Palmyra and more southerly ocean traffic along the Red Sea. A succession of farsighted kings invested heavily in the development of a special form of agriculture that was ideally adapted to the climatic, topographic, and pedologic conditions of the local habitat.

This development process involved both special circumstances and an application of the principle of creative destruction. Most societies living in drylands with low agricultural potential do not possess significant resources to invest in agricultural development. This was not the case with the Nabataeans. Their mercantile activities generated the monetary resources needed to invest in agriculture. Roads, farmhouses, terraces, and urban centers all were made possible by the profits generated by the caravan trade. This same cash flow made it possible to obtain the labor resources needed for heavy construction. While some of the labor, especially in skilled categories, undoubtedly was paid, much of the manpower was forced to work. This uncompensated labor not only included slaves, who were obtained through the caravan trade and war, but also comprised the labor power of the military. After AD 106, the Nabataean kingdom was incorporated into the Roman Empire as the province of Arabia. From this point on, Roman legions not only provided security in the predesert frontier zones, but they also contributed much of the labor that constructed roads, forts, and settlement centers throughout the area. The traces of these infrastructural investments are widely visible today (Kennedy and Riley 1990). The increased security provided by imperial protection also encouraged farmers to settle in the frontier districts of the northern Negev. Indeed, it was a conscious policy of the Roman Empire and its Byzantine successor to promote the growth of a defense in depth zone by settling retired soldiers in potentially insecure districts where they functioned as a frontier militia. Thus, during the entire period in which agriculture flourished in the northern Negev, government investments in infrastructure, labor subsidies, and security were substantial and essential to the success of the agricultural regime.

The agricultural system that the Nabataeans created was based on the principle of creative destruction. Rich experiential knowledge of the desert environment enabled the Nabataeans to develop a simple system rooted in a sophisticated understanding of their environment. The secrets of this resource management system have been unraveled only in the past thirty years (Evenari, Shanan, and Tadmor 1971; Negev 1986; Rubin 1991). The system was based on the observation that, under natural conditions, the rainfall occurring in the region was too sporadic spatially and temporally, as well as of insufficient magnitude, to be of much use for agriculture. Although contemporary meteorological observation shows that most of the region's rains fall in small amounts of less than 10 mm (0.4 in) (Evenari, Shanan, and Tadmore 1982) with only limited runoff, larger storms do occur. The surface flow that accompanies these storms can produce destructive flash flooding that causes serious erosion. The essential problem was to figure out a way to concentrate the water that did fall into places where it could be useful, while minimizing the destructive potential of flash flooding. From this standpoint, most of the foothills and uplands of the northern Negev were quite unsuitable for agriculture, since their slopes were too steep, too stony, too thin soiled, and too lacking in water retention capacity to be of any use in crop production (figure 2.2).

The system that the Nabataeans evolved concentrated agricultural production in the valley bottoms of the Negev's sporadically flowing streams. These valley bottoms were the *critical zones* upon which the viability of the entire Nabataean agricultural system depended. The valley bottoms were critical zones in the Nabataean system for two reasons. First, they were the places where soil deposition took place naturally. As a consequence, these lowland sites were the locations where the soils with the greatest agricultural potential existed. Second, under natural conditions, the valley bottoms had the greatest water storage capability in the area. These were the places where runoff collected and, therefore, where perennial vegetation was most densely concentrated. Focusing agriculture in these critical zones

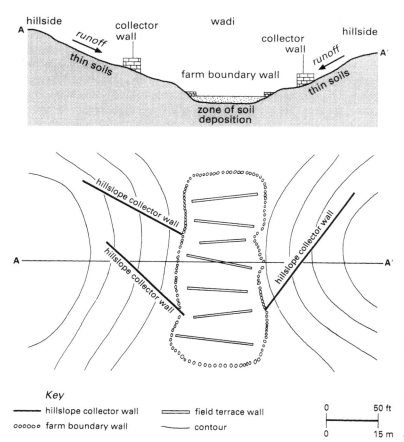

Figure 2.2. Nabataean field system. Water and soil movement from upland and steep slope areas were encouraged so resources could be concentrated on smaller but more productive land facets in the valleys.

reflected a sophisticated application of the genius loci principle that was based on a rich understanding of local environmental processes, rhythms, and potential possibilities. Instead of fighting nature and trying to conquer it, the Nabataeans devised a resource use system that operated within the constraints of the environment and was aligned with natural forces. By linking their agricultural system to natural processes, they were able to get nature to work for and with them. What the Nabataeans did was to direct, channel, and enhance the natural processes in ways that worked to their benefit. They were able to do this because they had lived in the area for a long time, and had gradually evolved a technology based on extensive experiential knowledge of the region. Individual elements of technology might be quite simple, such as the roughly coursed check dams used to impede water and sediment movement along the valley bottoms, but the technology was rooted in a sophisticated grasp of hydrology and other operational principles of the world in which they lived.

Much of their cultivation was located in the smaller tributary wadis ("wadi" is an Arabic word meaning an ephemeral, periodically flowing stream channel; the identical features are called "arroyos" in the American Southwest) of the region. In these areas, the available quantities of water and soil were manageable. Likewise, these small watercourses were also the places where concentrated runoff would first occur in the modest rainfall events that characterized the Negev. They were the sites that would respond quickest to management with the most limited labor inputs. In these wadi bottoms the Nabataeans built a succession of low stone walls that acted as leaky dams, and extended across the valley at right angles to the stream flow. When runoff occurred, the floodwater ponded up behind the crude stone barrier. This had three important consequences: (1) it reduced the velocity of the water flow, thus decreasing the risk of soil erosion along the valley bottoms where crop planting took place; (2) soil particles eroded and carried by the runoff from the surrounding hills settled to the bottom of the ponded water and were deposited on the valley floor. Over time a terrace of richer, agriculturally useful soil developed behind the check dam; and (3) water trapped behind the wall increased the time available for the ponded water to infiltrate into the soil behind the dam. Because of this increase in available infiltration time, soil moisture storage in the terrace soils often reached its capacity. This enhanced soil moisture store was usually sufficient to sustain a crop even if only minimal rainfall occurred during the remainder of the growing season.

The wadi terrace check dams seldom extended more than 30 cm (11.8 in) above the terrace surface. This height proved sufficient to retain enough water to provide the terrace with its soil moisture needs. Holding back more runoff water on a particular terrace with a higher wall served little purpose, since the oversupply of water would not increase the potential for crop utilization. The 30 cm dam height provided sufficient soil moisture recharge needed by available crops. Allowing surplus water to flow to a lower terrace once the base soil moisture needs of upstream terraces had been met distributed available water over the maximum amount of potentially productive land. Once sedimentation raised a terrace surface so high that inadequate water was being retained to replenish the soil moisture store, another layer of stones was added to raise the wall. In this way, terraces grew upward without changing the basic relationship of the agricultural land surface to the available water.

The runoff water produced by rainfall events to nourish the terraces came from the surrounding stream catchment. The hilltops and slopes that make up the land within the upstream areas were unsuited for agriculture due to their thin soils. However they played a critical role in the Nabataean farming system because it was water derived from these areas, once transferred to the adjacent lowlands, that maximized the benefits from the region's limited rainfall. Extensive surveys of the ancient farmsteads in the northern Negev by Evenari and his colleagues (1961; 1971, 104, 109) reveal that there was a consistent average relationship of 20:1 between the uncultivated water-producing catchment area and the agricultural area. The importance of this relationship is revealed by considering the water made available for the cultivated fields. If only one-fifth of the precipitation produced by a 10 mm (0.4 in) rain shower appeared as runoff, a 20 hectare catchment would generate 40 mm (15 in) of runoff. Combined with the 10mm (0.4 in) of rain that the shower directly produced on the 1 hectare (2.47 acres) of field to which the catchment was attached, the water available on the valley-bottom field was the equivalent of having received 50 mm (2 in) of direct rainfall. Since the areas around Avdat receive approximately 100 mm (4 in) of precipitation annually, the accumulated and concentrated runoff from the surrounding catchment could potentially produce available moisture equivalent to a rainfall regime that was five times as great. As long as the crops cultivated could withstand the rigors of a hot, desiccating climate, the field environment created by the Nabataeans reflected the soil moisture conditions of a much less arid habitat. For agricultural purposes, therefore, it was essential to view the area's two land-use components as part of an integrated system, and the Nabataeans did so. Moreover, wherever possible, the Nabataeans were interested in increasing the flow of runoff to their fields.

They did this by treating the hilltop and slope component as a *sacrifice zone*, the productivity of which was best exploited by the transfer of water and soil to higher potential land facets downslope. On the hillsides adjacent to their terraced fields they invested considerable labor to collect into large mounds the stones found on the surface. At one time it was thought that these mounds were intended to serve as dew collectors. Vines planted at the base of the stone mound and trained over the rocky surface were thought to have benefited from the increased soil moisture content of the soil beneath the mound. However, careful analysis of the amount of moisture concentrated by dew condensation on and beneath the mounds indicated that vineyard cultivation based on stone mound dew collection was not possible due to the insufficient quantities obtained (Evenari, Shanan, and Tadmor 1971, 134). Since documentary evidence indicates that vines were cultivated, the region's vineyards must have received their supplies from more conventional runoff sources. The real purpose of the stone mounds was to reduce surface roughness for the slope areas as a whole, diminish the amount of infiltration that occurred on the slope, and encourage greater runoff to the valley bottoms where the water could be utilized in farming.

In this effort the Nabataeans were aided by the characteristics of the local soil type. Many of the northern Negev's soils are fine-grained loess soils with limited pore space and a tendency for individual grains to swell when wet. As a result, the soil reacts to wetting by permitting relatively limited infiltration before it forms a sealed surface layer. Once this clogged surface develops, downward water infiltration into the soil proceeds slowly, and most of the subsequent rainfall flows away as runoff. By removing the stones from the soil surface, large pore spaces between the stones are eliminated, surface roughness that encourages retention of water for longer periods on the slope is reduced, and larger amounts of bare soil are exposed to rapid swelling. Contemporary experiments in the northern Negev demonstrate clearly that the closer the surface approaches bedrock and an increasingly impervious stony surface the greater the yield in terms of available runoff that can be directed to lower slopes and collected on valley bottom fields (Yair 2001).

What the Nabataeans did was to encourage water export and soil erosion from the slopes for the benefit of the valley bottoms. Soil erosion was not the prime purpose of the slope denudation; it was an ancillary benefit of the main objective of increasing water capture. Once soil removed from the slopes by the increased runoff reached the nearby wadis, it was trapped behind the valley check dams. Captured in relatively tranquil pools, the floodwater was able to infiltrate slowly into the terrace soil. The suspended soil particles carried in the water settled on the soil surface as a soil improvement addition. When all the captured water had infiltrated into the soil or evaporated from the surface, the resulting soil crust reduced the removal of water from the soil moisture store by evaporation and capillary action, thus locking up the runoff as available soil moisture for plant utilization.

Other water collection devices essential to the Nabataean agroecology also were developed. An important element in the water management system was the slope water channel. These water transportation conduits were built on hillsides at an angle to the slope gradient. Their purpose was to collect runoff from a particular portion of the slope or watershed and conduct that water to a designated terrace field in the valley bottom (Evenari, Shanan, and Tadmor 1971, 182ff). The main wadis, with larger and deeper channels than the tributaries, were more difficult environments to manage. Extracting water from them required the use of diversion dams in the stream channel in order to raise the water level to the height of the adjacent fields. The water diverted from the flood was directed into channels that led the water to the terraces. In this way fields had three water sources: floodwater extracted from the main wadi channel, the spillage of surplus water from upstream terraces, and the specific slope runoff from the adjacent watershed dedicated to a designated field. Within the field system, a network of canals and weirs insured the distribution of water to the individual terrace fields, and made it possible to shift excess water from field units saturated with moisture to deficit units. This cascading quality in the water distribution system guaranteed that upstream fields would receive the moisture needed to grow crops before any surplus was passed on to fields further down the farm and the wadi system. But upstream terraces also retained only the water needed for the crops grown on each field, thereby insuring that downstream terraces received a fair share of the available supply. Only the size of the individual flood event determined the number of terraces in the system that would receive water at any one time.

The more elaborate slope collection channels and terrace field distribution systems that were typical of main wadi agriculture required a much larger investment of capital and labor than did the simple check dam systems of the smaller tributary wadis. The more complicated field systems developed on the main wadi floodplains and terraces where bigger fields were possible, therefore producing larger yields to justify the added capital investment. These main wadi fields also needed much larger catchment areas, often measuring dozens of square kilometers, in order to generate sufficient water to supply the terraces. Because of their greater scale and larger capital requirements, the main wadi agricultural systems developed later than did the smaller-scale systems of the tributary wadis. These more elaborate agricultural enterprises represent a form of intensification within the Nabataean dryland farming system. It is not surprising that the major examples of these more elaborate farms are concentrated near the region's main towns. The more intricate infrastructure involved, and the larger amounts of floodwater with which they had to cope, made the main wadi agricultural systems more vulnerable to the destructive effects of episodic extreme floods. These floods required both more elaborate preventive protection devices and more costly reconstruction efforts than was needed in the subsidiary wadis. Eventually, the main wadi systems could not be maintained and they were abandoned. In contrast, many of the smaller-scale systems continued to be exploited by pastoral nomads after the settled agricultural system collapsed. In fact, many are still used by Bedouin who farm these relics of the Nabataean period.

The denizens of the northern Negev were equally adept at developing all available water sources for drinking water. These included wells, both shallow and deep, to tap aquifers and the gravel beds of the region's ephemeral streams (wadis) as well as an elaborate array of cisterns. There is no evidence in the northern Negev for the existence of the *qanat* systems (underground water collection galleries) that are found in many parts of the Old World drylands, although one "chain of wells" has been located at Yotvata in the Arava rift valley (Evenari, Shanan, and Tadmor 1971, 173ff). Cistern systems were invariably linked to centers of human habitation, since no household could hope to survive without adequate drinking water. The flat roofs of houses, cleared areas near farmhouses, the streets in villages and towns, the roofs and courtyards of public buildings—all were connected to cisterns that stored the precious water collected after each rainfall. Slope collector channels were also used to conduct water to nearby cisterns from small catchment areas that were not linked to agricultural terraces. Often equipped with sediment settling tanks, and protected from evaporation by their underground location, even small cistern collection systems could produce and store substantial amounts of high-quality water. Over a seven-year period, for example, a small 1.2 hectare (3 acre) catchment area near Shivta produced an annual average of 150 m^3 (5,297 ft^3) of water for its associated cisterns (Evenari, Shanan, and Tadmor 1971, 146; 166). A half-dozen pastoral nomadic families and their animals could survive for a year on this amount of water.

Relatively little is known about the specific crops grown by the Nabataeans and their Roman and Byzantine successors. The few local written records from the period are fragmentary, but they suggest that the crops used were the wheat, barley, vine, and olive complex commonly employed throughout the Mediterranean (Rubin 1991, 200–201). None of these crops, as well as minor associated species, could survive in the northern Negev on natural precipitation. All required the supplemental irrigation developed by the Nabataeans. Decay in the farming system rather than land degradation or climatic change is the reason why these crops are so infrequently encountered today. In sum, the crops cultivated in the northern Negev reflect an adaptation of the basic Mediterranean system to the more marginal and extreme conditions of the desert fringe.

The basis for this system's success was its diversity. Cereal crops, tree crops, and animals were a stable trilogy upon which the settled farming community rested. Each of the three elements was capable of withstanding or escaping drought to a considerable degree. Wheat and barley are relatively hardy grasses, and many of the traditional varieties in the region possess waxy surfaces that reduce evapotranspiration loss. Olives, vines, and other dryland-adapted trees, such as almonds, are deeply enough rooted so that they can reach groundwater and survive even in a year when rainfall and run-on water supplies are inadequate to grow cereals. By using microcatchments (Evenari 1977, 87–88; 1981, 11–13) that concentrate runoff from a small zone around each individual tree, modern experiments have demonstrated that sufficient water can be generated to nourish fruit trees in very desolate sites (figure 2.3). In fact, this very strategy is used in Israel today to sustain tree patches in isolated roadside sites along Negev Desert highways.

Animals, particularly goats, are the survival insurance par excellence, because they can be moved from areas deficient in water and fodder to districts unaffected by precipitation shortfalls. Many of the animals—especially sheep, who have a high water-consumption demand—must have been based around the cisterns located at a distance from the main zones of wadi farming. Here they could have survived on more than adequate water and just adequate rough forage, supplemented with postharvest residues from the cereal fields. The camels raised by the Nabataeans (Negev 1986, 40) were an essential part of their economy, for not

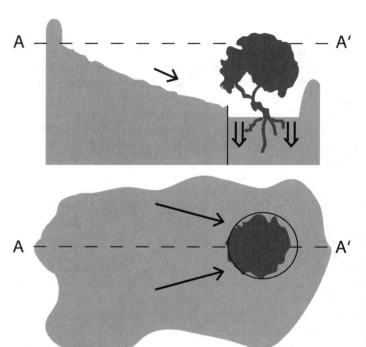

Infiltration from large area to irrigate single tree

→ run off   ⟹ infiltration

Figure 2.3. Microcatchment plot. This schematic representation suggests how employing the same principle as the Nabataeans at a different scale can produce dramatic results. (Diagram by Anne Gibson, Clark Labs)

only could they contribute hair and milk to domestic subsistence but also the camel was the mainstay of their overland caravans. When the importance of caravan trading declined in the Nabataean economy, the wealth of stables and votive horse figurines found in the Negev indicates that horse breeding became a principle source of local prosperity for the next six centuries (Negev 1986, 106). Whether or not cultivation of fodder grasses was common in the period is unknown, although it is clear that well-adapted perennial wild species do exist, including indigenous wild oats that give high yields and successfully resist severe drought. Thus an ecological trilogy of cereals, tree crops, and animal products could be produced in the desert fringe because a suitable set of water generation and conservation techniques existed.

A spectacular efflorescence of settled life based on the Nabataean system occurred in the fifth and sixth centuries AD in the Negev. This high point in the development of the region featured both rural agricultural and urban development (Negev 1986; Rubin 1991; Taylor 2002). What emerged was a *limes* (frontier) zone characterized by a defense in depth concept rather than a rigid fortified line of walls and forts. The idea behind the intensified settlement system was to provide protection for the high productivity coastal districts from raids conducted by more nomadic groups from the desert interior. Settlers during this period, in turn, depended on logistical support from the Byzantine Empire, for the imperial system of the Byzantine center promoted the steady expansion of the northern Negev periphery as a buffer. The ability of the frontier settler militia to act as a sponge, soaking up the aggressive attacks of tribal groups from outside the imperial system, was directly linked to the degree of support and security that the imperial military and logistical system could supply. When this imperial center came under increasing pressure from the forces of an expansionist Islam in the middle of the seventh century, the conducive conditions to settled life in the Negev deteriorated.

The termination of settled life did not occur as a cataclysmic collapse, but rather as a gradual decline in the viability of sedentary existence (Rubin 1991, 204). People slowly decreased their investment in field and tree crops, and increased their emphasis on animal husbandry. Towns and farmhouses were abandoned as sites of permanent residence, and a more mobile nomadic existence became the norm for the next thirteen centuries. In this shift away from settled exploitation of the region in favor of a less intensive livelihood mode, the post–Byzantine era mirrored the historic rhythm of alternating settlement and mobility. Nonetheless, even in the more nomadic episodes, the most manageable of the agricultural terraces were cultivated and the functional cisterns were still utilized. The same fixed points in the settlement system attracted the region's inhabitants. Only the manner and intensity with which these resources were utilized changed.

The Nabataean settlement system worked successfully for over nine centuries because it was rooted in the genius loci principle. By fitting into the rhythm of the local ecosystem and by building on specific features of the landscape, the Nabataeans promoted developments that constituted creative destruction. Erosion encouraged on the slopes (a sacrifice zone) increased runoff, which then ran on to adjacent lowland areas. The collection of water and soil in these lowland sites impoverished the steeper slopes, encouraging soil erosion and vegetation degradation in the least productive parts of the habitat. The Nabataean system possessed both *stability* and *resilience*, and as such it was able to exist for centuries. As long as a central government in a higher potential zone—whether Roman or Byzantine—was willing to subsidize activities in the Negev, recovery from any perturbation was possible and usually was accomplished quickly. A well-conceived, diverse agricultural system, based on a network of water collectors, terraces, and cisterns, provided the resilience needed to absorb the impact of drought as well as to make the most of limited available moisture. Nabataean agricultural activities concentrated on *critical zones*—the wadi bottoms—that had high potential in nature as traps for energy, soil, and water. The system was also based on *capturing wastes*, in this case the rainfall that fell on steep slopes and the thin soils that covered them, and transferring those wastes to sites where they could be concentrated and used more effectively. Water that was surplus on one terrace, rather than being wasted by being retained there, became an input as run-on moisture for the next terrace in the system. Managed systems that use the "waste" from one sector as the "input" for another sector are extremely efficient and stable enterprises.

These gains were not accomplished without exporting costs onto other parts of the larger ecumene. Some of these costs were borne by the central administration, whether Nabataean, Roman, or Byzantine. In each instance, the costs were apparently regarded as justified, since without them the frontier could not be effectively controlled, caravans could not operate, border guards could not be supported, and defense buffers could not be maintained. Other costs were inflicted on nomadic pastoralists, whose best pastures were converted to cereal crops and whose rough hillside pastures were denuded (where anything other than ephemeral vegetation existed) to promote the movement of water and soil to lowland sites. The response of these more mobile elements of the population to such losses and costs was a factor in the instability that followed the Arab conquest and that undermined the viability of settled life in the northern Negev. But the denouement of this remarkable settlement system should not obscure its importance as an example of creative destruction. Slope areas were sacrificed (degraded) for the increased productivity of selected valley bottoms. These same lowland fields, even after widespread nomadic use of the region replaced settled agriculture, continued to grow the grain that was essential to the Bedouin economy. While less intensively used, less extensive spatially, and with far fewer labor inputs for maintenance, the same fields continued to be productive for nomadic herders as they had been for Nabataean, Roman, and Byzantine farmers.

Irrigation in the Mesopotamian Lowland

Mesopotamia is the alluvial lowland of Iraq that lies between the Tigris and Euphrates rivers (figure 2.4). It is an exceedingly flat, generally featureless expanse of land that is crossed by two large rivers, the Tigris and the Euphrates, which drain the rugged highlands of Anatolia and the Zagros Mountains of Iran. The hydrologic regime of each of the twin rivers is quite different, a product of their respective source areas. The Tigris is the shorter of the two rivers but, because it is joined by a series of tributary streams that drain high elevations in western Iran and eastern Iraq with significant winter precipitation resources, its discharge is close to twice that of the Euphrates (al-Khashab 1958, 41). Rising in the less elevated central and eastern districts of Turkey, the Euphrates travels a much longer distance to the Gulf, has a lower average annual discharge, is joined by fewer tributaries—none of which are located within contemporary Iraq—and floods earlier in the year than the Tigris. Both rivers share two common characteristics: in their lower courses neither river receives significant moisture, and both rivers are characterized by highly variable interannual flow regimes. This extreme variability in the flow behavior of the Tigris and Euphrates introduces an element of unpredictability into the Mesopotamian environment that is quite different from the orderly behavior of the Nile in Egypt. In Mesopotamia, ancient cultures (Frankfort 1956, 53) were aware of the uncertainty of life and of their profound inability to control many of the events—including floods—that had enormous power over them.

Flowing across a broad and, to the outsider, largely featureless, nearly tabletop flat plain in southern Iraq, each river has a low gradient. Meandering stream courses characterize both rivers, and elongated meanders are prone to be cut off. In these circumstances, frequent changes of river course occur (Adams 1981, 9), with consequent complications for the human populations located along the cutoff stream channels. The old levee banks of these abandoned channels provide a microrelief that is not readily observable to the naked eye. Nonetheless, such meander scars play a significant role in the hydraulic regime of the flood plain. The barrier effect that these minute, convoluted, sinuous topographic ridges play in the region's surface drainage is reflected in the more moist soils and denser seasonal and perennial plant cover found on their upslope side.

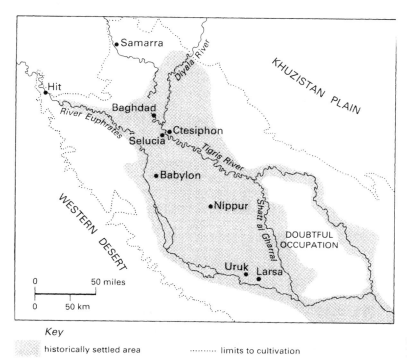

Figure 2.4. Mesopotamia and the Tigris and Euphrates floodplain. (From B. L. Turner II et al., eds., *The Earth as Transformed by Human Action*, 1991. Reprinted by permission of Cambridge University Press.)

Key

historically settled area ········· limits to cultivation

   Coping with the complications of a highly variable, unpredictable, and complexly interrelated river regime was essential to successful human use of the floodplain. Moreover, the alluvial floodplain of the two rivers south of contemporary Baghdad receives less than 200 mm (7.9 in) of precipitation. While the hills of northern and eastern Iraq receive enough precipitation to support nonirrigated field crops, the low rainfall in the south is unable to sustain dry farming. Without irrigation, settlement would be confined to the immediate banks of the two rivers and only small population totals could be sustained. It is ironic that one of the Earth's greatest early civilizations arose in an inhospitable desert plain with few indigenous resources other than the water of the two great rivers that crossed it, the muddy alluvial soils that the streams deposited, and the date palms and reeds that grew in abundance along the region's constantly shifting water courses and in its marshes.

   This riverine, desertic lowland is the setting for the second of our historical examples of land degradation. In contrast to the Negev, Mesopotamia illustrates the operation of *destructive creation*, the obverse of creative destruction. The many accomplishments of the Sumerians, who developed the world's first urban civilization over five thousand years ago (Kramer 1963, Adams 1966; Redman 1978), and their successor cultures not withstanding, irrigated agriculture in Mesopotamia failed to maintain human populations and environmental resources in a sustained fashion. Two and a half cycles of population growth and decline have occurred in Mesopotamia (figure 2.5), an oscillating pattern that is due in large part to the internal inconsistencies and instabilities that are deeply embedded in the

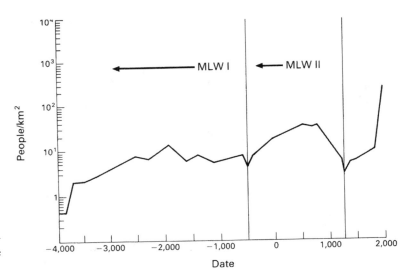

Figure 2.5. Cycles of population growth and decline in Mesopotamia.

structure of the region's irrigation system (Adams 1965, 1981; Adams and Nissen 1972; Johnson and Whitmore 1987; Whitmore et al. 1990). While evidence exists for a weak correspondence between lower stream flow and land degradation problems such as salinization (Kay and Johnson 1981), social rather than environmental forcing variables appear to be the major factors influencing land management and population dynamics in the region.

By *destructive creation* we mean a resource system that contains critical flaws internal to the system that undermine its viability. The flaws that promote destructive creation are often insidious, masked by shorter-term surficial signs of success. These flaws operate either to erode the system's sustainability in a direct fashion—for example, the deposit of sediments that clog water delivery canals—or to create such a delicately posed system that it is vulnerable to sudden, often irreversible changes of state when confronted with unexpected events. Irrigation systems are vulnerable to such catastrophic collapses whenever nature, managerial mistakes, or military conflict lead to an alteration in the main stream courses that provide water to the agricultural operation. This situation is different from the creative destruction that produced the run-on farming systems of the Negev because the factors that promote collapse in Mesopotamia are internal to the agricultural regime. In the Negev, external support was essential to the full flourishing of the Nabataean system, and its loss after the Arab conquest resulted in the decay of the Nabataean agricultural enterprise. But thirteen centuries later the same intrastructure, operating on the same management principles, worked as well as it ever had when it was restored. The same cannot be said for Mesopotamia.

The irrigation system of the Mesopotamian lowland exists historically in two states, disaggregated and integrated. The disaggregated system is the oldest irrigation configuration, and its appearance is associated with the spread of agriculture in the lowland. The hunter-gatherer communities of the floodplain were primarily confined to the vicinity of the region's major rivers, their many branches and subsidiary channels, and nearby swamps. The bulk of the floodplain was too dry to be of any real importance in terms of food production although seasonal grazing was possible. In the earliest stages of settlement, harvesting of wild plants and animals was the dominant mode of production. However, agriculture proved to be a superior food production system in two ways: it provided larger yields for a growing (albeit at a very slow rate) population, and it produced surpluses that could be stored in order to make it easier for people to survive the hazards of the local environment.

Agriculture itself could be made more secure, and the yields from cultivated plots along the riverbanks and levee back slopes could be increased by digging small ditches through the levee. This permitted the earliest farmers to transport water from the river to their fields for a longer period and to reduce their total dependence on the seasonal flood to deliver all their irrigation water. The construction and maintenance of these small, exceedingly primitive delivery ditches depended only on the labor of local kinship groups. This meant that small communities of several hundred individuals could generate the labor and leadership needed to operate the system (Fernea 1970). Under disaggregated conditions, with the exception of small cuts in the levees, the region's meandering natural streams were the main components of the water delivery system. Provision of drainage by the farming community was not considered, and excess water flowed from the fields down the floodplain's gentle gradient until it had fully infiltrated into the soil outside the cultivated zone or encountered the barrier of a nearby levee or meander scar. Here excess moisture accumulations favored the development of local marshes.

Because the disaggregated system existed at a small, local scale, such swamps were limited in size. Individual communities remained vulnerable to large fluctuations in the environment. Floods might wipe out an individual community, or changes in a river course due to a meander cutoff might destroy the productive habitat of a local population unit, but the survival of the floodplain population as a whole was never in doubt. Few, if any, of the driving forces that promoted change in local nature-society relations were under the direct control of the local population, and the system exhibited great stability over time.

These conditions began to change dramatically when the irrigation system evolved from a disaggregated state in which isolated settlement beads on a riverine string were scattered along the region's winding natural watercourses into an integrated system of water distribution and management. This change is associated with the rise of larger urban agglomerations and the development of a series of small city-states (Wooley 1965; Adams 1966; Redman 1978; Nissen 1988). The change is reflected in the architecture of the region's watercourses, which no longer meander randomly across the countryside but rather slash in straight lines through the landscape, delivering water from streambed to farm fields. Natural streams increasingly are replaced by straight canals. Evidence also exists in the form of documentary data recovered from an ancient irrigation archive dating to about 1990 BC, nearly a century before the famous ruler, Hammurabi of Babylon, linked the entire southern alluvial lowland into the first large-scale state structure (Walters 1970). This archive contains documents that make possible a reconstruction of the irrigation bureaucracy of the period, as well as time and effort calculations for canal construction. For example, eighteen hundred working days were needed to dig a 2.4 km (1.5 miles) long canal that was 1.5 m (5 ft) deep and 1.5 m wide (Walters 1970, xix). In short, by 5000 BC there was in existence in Mesopotamia a centrally managed irrigation system, the operations of which appear to be analogous to those of a contemporary Corps of Engineers!

The ancient irrigation system that developed in Mesopotamia was an exceedingly sophisticated structure in its aggregated state. For long periods this system was able to maintain large populations in a condition of relative prosperity and security. However, this apparent success in maintaining irrigation system functions masked latent flaws that either emerged episodically, after lurking beneath the veneer of prosperity only to appear when conditions were appropriate, or imposed increasingly greater burdens on the long-term productivity of the system. One set of long-term factors was a consequence of the way in which the Mesopotamian irrigation system dealt with its sacrifice zones.

There were two sacrifice zones: the forested uplands and the lowland swamps. Lowland Mesopotamia is a singularly one-dimensional environment from a resource standpoint. Aside from its rich alluvial soils, it possesses few resources. The petroleum wealth that has fueled modern Iraqi development could not be exploited until the twentieth century. Sumerians, and all subsequent agriculture-based civilizations in Mesopotamia, lacked basic timber and stone materials for construction, and they were completely without mineral resources. These items could be found only in the adjacent highlands; their acquisition figured from an early date in the economic and imperial designs of the lowland states. The Gilgamesh epic (Kramer 1959,184; 1963, 191–93) describes the expedition of a mythical ruler to a distant place (Lebanon or Turkey?) where cedar trees, the abode of a fearsome guardian, are cut down. There is no indication that the timber was brought back to Sumeria, yet it is not impossible that the epic adventure of this Sumerian hero records a more prosaic functional activity. Although glazed mud bricks and palm logs were the prime lowland construction material, there was no substitute for the hardwoods of the upland forest in any major construction.

Economic penetration of the highlands for mineral products stimulated both settlement expansion in the uplands and the removal of forest cover. The result was a millennia-long process of deforestation in the highlands that contributed to soil loss from these environments. In reducing the extent of upland forest, the acquisitive demands of lowland irrigation communities were not the only culprit, for the expansion of rain-fed agriculture in the uplands also played the major role in opening up clearings in the highland forests and exposing bare soil to the weather (Wagstaff 1985, 46–47). In a real sense, the health of many upland areas was sacrificed for the overall benefit of lowland civilizations. The result of this process of highland landscape change was increased sediment generated by erosion and moved into the downstream portions of the river systems. This resulted in increased siltation rates in the low-gradient floodplain districts of Mesopotamia. The increase in siltation imposed a heavy maintenance burden upon the irrigation system, a factor that will be discussed in more detail below.

The conversion of the highland forest habitat into more open environments dominated by seasonal grasses created new opportunities for the region's nomadic pastoralists, who found enhanced grazing for their flocks in their seasonal movements to summer pasture in the highlands. This demonstrates the degree to which almost no land degradation occurs without some human group finding a way to benefit from the negative development.

The swamps of the alluvial lowlands were the second major sacrifice zone. In large part, the extensive marshes of the alluvial plain are natural habitats. They are produced by a combination of low gradient close to sea level, abundant seasonal stream flow, impeded drainage produced by the old levee banks and meander scars of abandoned stream channels, and the steady deposition of sediments by the region's rivers into the head of the Gulf (Munro and Touron 1997). The majority of these swamps are permanent (Larsen and Evans 1978), but about one-quarter are seasonal (al-Khashab 1958, 62), the product of the annual flood regimes of the Tigris and Euphrates. The irrigated areas in the settled zones are characterized by the existence of significant swamps that are collectors of wastewater from the irrigation system as well as sources for considerable water loss. Evapotranspiration by the vegetation and evaporation from the surface waters found in these swamps, as well as infiltration into the groundwater and return seepage into canals and stream channels, represent the major feedbacks from swampland into the hydrologic cycle (figure 2.6).

These swampy areas typically develop at the tails of the irrigation system. An Akkadian poet described the relationship over five thousand years ago in a creation myth (Kramer 1961, 123). The god Anu having first created the vault of heaven, heaven created earth and earth generated rivers. Rivers made canals, and canals called into being marshes. And in these wetlands dwelt worms and by implication other noxious creatures, the miasmic airs, and loathsome organisms that posed a threat to the health and well-being of humans engaged in irrigation.

Swamps are encouraged by the microtopography of the irrigated portion of the floodplain and by irrigation practice (Adams 1965, 8). The region's irrigation systems are gravity fed, with the high point in the distribution system being found along the levees of the natural river courses and main distribution canals. Flow is downslope and away from the main water sources across the backslopes of the levee system. Excess water accumulates at the tails of the distribution system at the lowest points in the irrigated landscape. Usually, these areas also are associated with old meander scars that impede further lateral surface water movement.

The growth of initially ephemeral, but over time more permanent, wetlands is aided by the tendency of all farmers in all epochs to overirrigate if water is available. The *Sumerian*

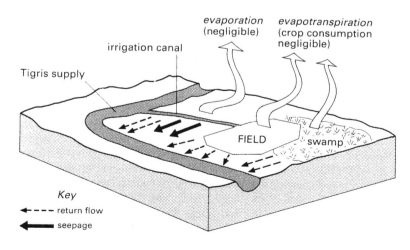

Figure 2.6. Water allocations of a hypothetical irrigation system unit. (Adapted from al-Khashab 1958, reprinted by permission of the University of Chicago Press.)

Farmer's Almanac, for example, urges its readers to water their barley at least three times after seed begins to germinate, but to throw on a fourth watering if possible in the interest of even better yields (Kramer 1963, 341). In an environment such as southern Mesopotamia, where the water table is close to the surface, frequent overirrigation inevitably encourages waterlogged soils, especially in microtopographic settings that promote water accumulation. This same tendency to overirrigate whenever water supplies are available is intensified by the difficulty of being sure of adequate water for the planting season, which occurs at the same time as the low-water stage in the twin rivers. Historically, this is an important contributor to the spread of land degradation in this area by promoting salinization (Adams 1978, 330) and swamp encroachment (Adams and Nissen 1972). These zones were sacrificed, from an agricultural standpoint, in favor of the better-drained soils in the prime farming zones. This is particularly true when, as is typical of ancient irrigation systems in Mesopotamia, no provision is made for draining either the irrigation plots or the swamps that develop at the tails of the system. That these swamps can also create ideal habitats for the snails and mosquitoes that serve as transmitters of schistosomiasis and malaria respectively also indicates that they can cause indirect negative impacts on the integrated irrigation systems that are their primary creators (McNeill 1976).

In south and southwest Mesopotamia, an important factor in the development of marshland was the failure to maintain crucial infrastructure. Late in the fifth and early in the seventh centuries AD, catastrophic floods broke through dikes whose maintenance had been neglected for many years. These floods inundated vast tracks of low-lying land. Although vigorous reclamation efforts partially rehabilitated lands flooded in the first disaster, the second epic flood's impact proved more permanent. Despite very strenuous efforts on the part of the Sassanian administration to reestablish the broken flood-control works and drain the waterlogged lands, little progress was made (Le Strange 1905). The lowered productivity of land resources damaged by both natural (floods) and human (neglect) factors undoubtedly contributed to the inability of the Sassanian state to resist the Muslim invasion and conquest in the second quarter of the seventh century AD (Lapidus 1981). While Arab administrators used much of the surviving Sassanian water-distribution and flood-control infrastructure (Moroney 1984)), agricultural recovery was not uniform in the Mesopotamian floodplain, particularly in the salinized and swampy south.

It is important to recognize that the impact of swamp expansion also has positive implications. In the far south of Mesopotamia, centered in the vicinity of Basra, entire cultures emerged that were based on fishing in the fresh, brackish, and saline swamps of the region (Adams 1981, 16). Analogous adaptation to the swamps is found in contemporary southern Iraq among the Marsh Arabs (Thesiger 1964). Although there are indications that fishing declined in importance after 2300 BC, if the number of fish species mentioned in ancient texts is any guide (Kramer 1963, 110), the existing evidence could also indicate a reduction in species diversity under the pressure of an increasingly saline environment. At the same time, an expansion of swamp vegetation would increase the habitat available for water birds of all types, and Kramer notes more than four dozen varieties of edible birds in the ancient texts. Moreover, swamp grasses constitute a valuable fodder source with which to support substantial domesticated animal populations. This resource constitutes a valuable foraging niche for the migratory herds of nomadic pastoralists, whose animals could combine freshwater wetland grazing in the winter with postharvest stubble resources and the semiarid steppe grazing of the nonirrigated floodplain districts. In sum, while swamp expansion was a negative consequence of irrigation system integration and expansion, for certain ethnic and economic groups the wetland environments that constituted an irrigation system sacrifice zone became a positive resource for their dependent herding and agricultural communities.

Today these marshy habitats are sacrifice zones in a different way. Massive development of overseason water storage capacity in the headwaters of the basin controlled by Turkey, Syria, and Iran have drastically reduced the amount of water carried by the Tigris and Euphrates and their tributaries (Beaumont 1996). Despite the widespread belief that punitive drainage projects launched by the Ba'ath Party regime of Saddam Hussain are responsible for the desiccation of the marshes of southern Iraq, there is good reason to believe that interannual storage and diversion of water in the Tigris and Euphrates headwaters are primarily to blame (Pearce 2001). Within Iraq itself, three major efforts with major ecological consequences occurred in the second half of the twentieth century. First, major barrages at Ramadi on the Euphrates and Samarra on the Tigris diverted seasonal floodwaters into the Abu Dibbis and Tharthar depressions. This water storage reduced the hazard of flood damage, provided increased water supplies for the irrigated districts, and diminished the supply of water that supplied the southern marshes. Second, a major collector drain was constructed from just south of Baghdad to carry waste water from the irrigation districts and carry it over 500 kilometers southward to empty into the Gulf. Running through the center of the Jazirah, this massive project and associated dikes provided regional scale drainage where none existed before and helped to starve the southern marshes of water. Third, the strenuous efforts of the Iraqi government in the 1990s to drain the major marshes around the confluence of the Tigris and Euphrates rivers began to massively affect the size of these landscape features. The impact of these drainage efforts is so extensive that hundreds of thousands of local residents have been disposed, important wildlife habitats have been altered, and desiccated wastelands have been created. Through these changes marshes now cover less than 20 percent of their former extent. The impact of their loss has aroused international concern amid fears that insufficient water can ever be provided to rehabilitate even a portion of the lost habitat (Partow 2001).

The critical zones for floodplain irrigation systems to function are the perennial irrigation zone along the natural river channels and the high potential floodplain soils commanded by the artificial canals of the integrated, centrally managed irrigation bureaucracy. The best drained soils on the river and main canal levee banks are vitally important resources, since these districts are the only ones in the Mesopotamian floodplain that historically could be cultivated on a year-round basis. The use of simple lift devices to raise water to the fields, the application of manure (often provided by nomadic herds grazing on postharvest stubble), and the application of considerable labor are the basic inputs needed to keep these soils in constant production. Alluvial floodplain soils require the application of a local genius loci principle in order to remain productive. This principle is based on two factors: alternate-year fallow of cultivated fields (Gibson 1974, 15–16) and the survival of salt-tolerant perennial bushes in the cultivated fields (Adams 1965, 5, 18). A related factor in maintaining the critical, high-potential zones is a competent managerial system.

Alternate-year fallow is an integral part of the traditional Mesopotamian agricultural regime. Only the limited, multicropped, perennial production areas along the rivers and distribution canals where low-flow water is available are an exception to the alternate-year fallow principle. Leaving alluvial soils fallow every other year is essential to productivity because drainage is not a part of the ancient Mesopotamian irrigation system. Without the wetting provided by irrigation, soils drain adequately after flooding and dry in their surface layers. The water table declines to sufficient depth to enable the next production year's application of irrigation water to wash any accumulated salts out of the surface soil and away from the root zone of cultivated plants, thus preventing salinization. Reduction in the frequency of fallow would encourage a rise in the water table and a movement of salts toward the soil surface. Any increase in salt content in the root zone is immediately reflected in a lowering of yields from the cultivated crops. Historically, salinization appeared suddenly near the end of the third millennium in Mesopotamia, at a time when irrigation systems were aggressively expanded by both local and regional administrations. These increases in soil salinity are reflected initially in a shift away from cultivating wheat in favor of the more salt tolerant barley, a threefold decrease in soil fertility over a seven-century period as measured by an increase in barley yields, and the appearance of surface patches of salt in temple fields (Jacobsen and Adams 1958; Helbaek 1960, 194–96; Walters 1970, 160–61). Explanations for the appearance of salinization are linked to the availability of more irrigation water, as local rulers dug canals that tapped previously unused Tigris River water (Jacobsen and Adams 1958, 2) to supplement the much more extensively developed canal distribution network based on Euphrates River water (Jacobsen 1960). The presence of greater water supplies available for longer periods throughout the year had the potential to encourage more intensive irrigation of the critical central floodplain cultivated areas than could be attained using simple lift devices. The spread of perennial irrigation to a greater area based on the larger and more reliable water resources provided by the integrated irrigation system had a high potential to become an inadvertent "engineered disaster" (Gibson 1974, 15). Also implicated are the demands of rulers for a greater return via taxes from the peasant farmers under their control (Gibson 1974, 16). In many cases the immediate returns to the central government were greater from large landholders, with whom autocratic governments were often linked politically and socially and who they naturally tended to favor, than from small landowners organized in a more tribal system. Favoritism of the landed rural elite, often resident in urban areas, debased the peasantry, reduced the availability of rural labor as sharecroppers were driven out of operation, and initiated a downward spiral in rural productivity and irrigation system maintenance that was difficult to arrest (Gibson 1974, 16–17). Whenever taxes increased as a means of concentrating rural surpluses in cities in support of centralized military and managerial bureaucracies, the farmer had little recourse other than to violate the alternate-year fallow cycle. Once initiated, only a short time (often less than a decade) was needed to bring the water table sufficiently close to the surface to begin to affect production negatively.

Perennial shrubs play an important role in support of the alternate-year fallow practice. Two shrubs in particular, *shauk* (*Procopis farcta*) and *agul* (*Alhagi maurorum*, camel thorn), are legumes that assume critical importance in the traditional agricultural system (Adams 1965, 5, 18). Each plant has a taproot that reaches deep into the ground. Evapotranspiration from their leaves, in the absence of irrigation, lowers the water table. In effect, shauk and agul function as natural pumps, lowering the water table in a natural way without requiring the expenditure of human or animal energy. Since both plants are legumes, they contribute to nitrogen fixation. Their deep roots help to loosen the soil and encourage greater water infiltration. Their leaves constitute a nutritious source of fodder once residual, postevapotranspiration salt has been washed off, and their branches can be collected for firewood. For such reasons, these fallow plants were a critical component contributing to the well-being of the traditional irrigated farming system. Until modern times, farmers left these "weeds" unhindered in their fields, a coexistence that was favored by the limited soil-turning ability of the traditional plow, which was unable to root out the shrubs.

Rather than competing with the shallow-rooted local cereal crops for water and nutrients, shauk and agul tapped deeper resources and recycled them to the surface. In an irrigation system in which systematic planned drainage ditches did not exist as a way to lower water tables and reduce soil salinity, these deep-rooted natural pumps were vital ingredients in maintaining overall system stability. While none of the primary or secondary sources suggest that removal of these useful shrubs was carried out systematically, it is not difficult to imagine pressures for increased perennial production contributing to their removal. The reduced presence or absence of these useful plants from irrigated fields would have created a much more delicately balanced field environment that demanded increasingly sophisticated micromanagement in order to avoid disaster.

As the irrigation system became increasingly integrated, it depended more and more on the managerial skills and competence of the irrigation bureaucracy. In the middle of the third millennium BC, apparent incision of streams and shifts in the course of the Euphrates (Adams and Nissen 1972; Nissen 1988, 129–64) instituted stress that placed even greater demands on the stability and organizational skills of the irrigation managers as they competed, often in military conflicts, for access to secure water sources. City states unable to control reliable water supplies declined in size and prosperity, and the center of gravity of political, military, and economic power shifted northward in the floodplain.

Episodes of civil war over water made an early appearance historically in Mesopotamia (Kramer 1959, 35-44), as did efforts to reform abuses of power and taxation (Kramer 1959, 45-50). Whenever these conflicts broke out for any prolonged period, the result was a decline in the viability of the irrigation system. The need for stability was substantial because the irrigation system required organized labor to counter the threat of siltation. Only a central administration could organize sufficient labor to keep the entire system clear of silt. Silt loads throughout the floodplain were always high, but environmental changes in the adjacent uplands increased the sediment loads of the river system. Only constant vigilance could keep the canals clear of silt and insure a secure supply of water for the fields (Jacobsen and Adams 1958, 7). Weaken this central administration, and rapid decline would follow.

However interested local communities might be in keeping their segment of canal functioning, success did not depend on their efforts alone. What happened upstream dramatically affected downstream prospects for prosperity. Without an effective administration to keep all units functioning in a coordinated fashion, the populations at the tails of the system experienced immediate negative impacts. Decline tended to spread from the insecure margins of the irrigation system back toward the centers of power in the more advantageously placed major cities. Greedy rulers promoted taxation and land management policies that precipitated mismanagement and a steady decline in population and productivity. Efforts to bring in slave labor in the Abbasid period, in order to replace a pauperized peasantry in carrying out land reclamation projects in the salinized and swampy south of the floodplain, produced a servile rebellion that nearly capsized the imperial regime. A neglect of investment in irrigation infrastructure, the loss of authority to an exploitative rural land-owning elite, the growth of rampant and corrupt tax farming, civil wars and conflicts with hostile neighbors, and the military inability of a weak central administration to provide security against the incursions of nomadic peoples led to the centuries-long decline of the Abbasid Caliphate (Lapidus 1988, 135–36).

Three centuries of progressive internal decline culminated in the Mongol coup de grâce when the sack of Baghdad in AD 1258 pushed Mesopotamia's second major population growth phase into total collapse and reduced the irrigation system once again to a disaggregated state. This story of an apparently successful system self-destructing is repeated episodically in Mesopotamian history, and is the bleak side of the creative destruction that generated the spectacular, albeit often unstable, efflorescences.

That Mesopotamian irrigation was able to cope with a host of factors over which it had little direct control suggests that the system, finely poised though it was, possessed considerable powers of resilience. Tectonic changes that controlled the advance and retreat of the head of the Gulf and shifts in stream channels were coped with through location shifts in the population, the development of new cultural adaptations to deal with a swampier environment, and the construction of major artificial waterways that linked much larger regional units together in one irrigation system. Efforts to expand into alluvial areas previously unaffected by salinization were attempted, often successfully. Changes in crops in favor of more salt-tolerant varieties, such as barley and the date palm, made it possible to deal very effectively with an increasingly saline environment. More use of previously underdeveloped Tigris River water through a complex system of regional canals characterized the Sassanian (Persian) pre-Islamic agricultural and population expansion. Yet none of these coping strategies could overcome the basic vulnerability of the Mesopotamian system to salinization, siltation, and mismanagement.

Summary

Creative destruction and *destructive creation* are opposite sides of the same coin. Each has operated throughout human existence, as individuals and communities have struggled to extract a living from the resources at their disposal. Whether by accident or by design, successful systems are those that are able to achieve sustainability for more than the transient lifetime of an individual. In the human use of environment, change—even destruction—in natural systems is inevitable. The slower that change, the more likely it is that adaptations can evolve that recover rapidly from disturbance and absorb change without a drastic alteration in system state. The process of creative destruction is able to accomplish this feat, replacing a less humanized natural system with a new structure that has long-term staying power. Destructive creation is the creative process gone awry in ways that are unviable.

Human efforts to survive in the Negev and in the Tigris and Euphrates illustrate these two contradictory processes. Both resource management systems developed in harsh, arid environments. Each system was built on a particular characteristic feature of the local resource base. In the Nabataean instance, it is the relative impermeability of local soils to infiltration that is at the basis of their ecotechnology. In Mesopotamia, it is the potential abundance of two exotic rivers that, if harnessed successfully, promises prosperity. In both instances, rainfall and river flow regimes are highly variable in time and space, creating great drought and flood hazards with which the resource user must cope.

Integral to a successful ecotechnology is the selection of sacrifice zones that can be destroyed in the interests of protecting and enhancing the zones that are critical to sustained use. In the Negev, the Nabataeans deliberately sacrificed crest and hillside habitats in order to encourage soil erosion, water movement, and water harvesting. Aided by the tendency of the region's soils to form a plastic crust that reduced infiltration, these upland water and soil losses were impounded behind low check dams and terrace walls in the nearby valley bottoms. Recognition of the soil characteristics of their habitat, as well as conceptualization of how water and soil—when concentrated in valley bottoms that were critical zones for agricultural production—could overcome the limitation of local climate, was a positive application of the genius loci principle. That the system created by the Nabataeans eventually declined when regional political conditions changed in the seventh century AD does not diminish the magnitude of their accomplishment. The same infrastructure, albeit on a reduced scale, continued to function right down to the present. By constructively using the waste products of one part of their habitat as an input for their critical agricultural zone, the Nabataeans turned a negative development, soil erosion, into a positive asset, a truly creative act. Contemporary Bedouin still reap benefits from the relics of the Nabataean system through the planting of crops on the former terraced fields.

In Mesopotamia, the Sumerians and their successors also developed a creative solution to coping with a desert environment. Unfortunately, their ecotechnology contained internal inconsistencies that undermined effective long-term system viability. Land facets at the tails of the irrigation system and in depressions were sacrificed because provision of drainage, which would have been difficult given the low gradient of the area, was not part of the overall design. Although both domestic livestock and wild animals could use some of these habitats, their overall contribution was negative. Disease vectors flourished in the swamps and wetlands created as an adjunct to irrigation, evaporation and transpiration contributed to an increasingly saline environment, and waterlogging encouraged a rise in the regional water table. With each downstream water transfer, water quality declined as surface water and ground water containing an elevated salt content returned to stream and canal flow, thus making the situation of downstream users more precarious. This failure to use wastes as a constructive input diminished the effectiveness of the Mesopotamian irrigation adaptation.

Over time, Mesopotamian irrigation managers, for all their evident sophistication, were unable to protect the levee back slope and rich alluvial soils from salinization. Violation of alternate-year fallow, the genius loci principle of the region, episodically promoted salinization and initiated a spiral of land degradation that was sudden in onset and rapidly produced negative impacts. Excessive siltation from the upland sacrifice zones created managerial problems for canal clearance that reached crisis proportions whenever civil conflict or invasions from neighboring states occurred. In sum, the Mesopotamian irrigation adaptation was a finely poised creation that operated close to the stability and resilience parameters that defined its existence. Destructive tendencies embedded within its social, political, and agronomic spheres promoted periodic land degradation and ultimate collapse.

The Physical Domain and Land Degradation

WHAT DRIVES land degradation? By our definition, decreases in an area's biomass productivity potential or utility due solely to natural processes are not considered land degradation. Yet in reality, many times it is difficult to determine the direct causes of land degradation. In the 1880s, cattle grazing reached a peak in the semiarid lands stretching from Wyoming to Texas. During this period, many of the grazed lands experienced a cycle of accelerated erosion that resulted in widespread valley trenching and gully formation. Thus, it is not surprising that much research indicates that the proximate cause of this accelerated soil erosion was overgrazing. Based on this belief, it would be classified as land degradation since it is the result of a human activity. Excluding or drastically limiting cattle-stocking levels can result in dramatic landscape changes, many of which provide increased habitat for wild animals (Bock and Bock 2000). But additional research has uncovered the fact that this period's climatic conditions also strongly contributed to the accelerated erosion. Thus the reality is that the accelerated erosion was as much due to natural factors as human. The trenching of valleys resulted from a coincidence of climatic conditions conducive to erosion and degradation of the vegetative cover by overuse (Leopold, Emmett, and Myrick 1966). What this one example illustrates is that, when examining land degradation, it is crucial to consider the natural and the human elements equally, without assuming that one or the other is dominant. It is wrong to neglect geophysical change and attribute all blame to human forces, as has been done in a significant part of the modern social science literature (Brookfield 1999).

Wide ranges of environmental settings exist on the Earth. Some places have rugged topography with cold dry climates and sparse plant cover; other locales could be areas of low relief and hot humid climates with dense vegetation. Because of the worldwide diversity of environmental conditions, intensities of natural processes and their impacts vary significantly throughout the world. One way in which these differences are manifested is in varied natural erosion and deposition rates throughout our planet. For example, the Thames River Basin in England, with a mean relief of 159 meters (521 ft), gentle slopes, and rainfall that usually is of a low intensity, experiences relatively low natural erosion throughout its area. In contrast, the Kander River Basin in Switzerland, with its greater mean relief (2,428 m, 7,965 ft), steeper slopes, and generally greater rainfall intensity, has high natural erosion (table 3.1). The widespread range of physical conditions found throughout the earth results in individual locations having different stability and resilient characteristics. With the variability of physical settings, the potential for land degradation varies from place to place. Terrains having the highest potential for degradation are called fragile lands.

The term *fragile* designates areas that, once disturbed by human activities, have a low resistance (resilience) to common natural events such as heavy rains and strong winds. Land degradation on fragile lands occurs in rich and poor nations, in areas using minimal technology or high technology, as well as in urban and rural areas. Furthermore, land degradation is not limited to fragile settings. Between 1946 and 1990, the uranium mining operations in Saxony and Thuringia (eastern Germany) resulted in enormous quantities of uranium production residue (tailings). The residues from the mining and milling released radioactive and other pollutants both on land and in the streams. The legacy left by the uranium production is a degraded landscape on lands that clearly would not be considered fragile.

While the majority of human activities are not as critical as those associated with radioactivity, the increase in the magnitude of many activities using contemporary technologies has increased the risk for severe land degradation. For example, because of the immense size of contemporary supertankers, there is the potential that their oil spills could create significant deterioration in regional environmental systems and affect thousands of square kilometers. This scale of potential impact is in sharp contrast to the more local potential of impacts in former days when tankers were much smaller.

The natural setting, including its energy properties (for instance, high relief and high rainfall or low relief and low rainfall) and earth materials, interacts with any given human activity. These energy and material factors incorporate both gravitational and biochemical attributes, climatological peculiarities, tectonic characteristics, soil properties, and rock types including their structural arrangement. Depending on the combination of the properties of these factors and the intensity of the processes found in an area, there would be a specific interaction between the natural and the human systems. For most activities, this interaction determines if an area's response to human utilization results in land degradation.

In this chapter, we examine different physical settings and their susceptibility to land degradation. Here we focus on the physical setting; we identify and examine critical aspects

Table 3.1 Relation of Relief to Denudation

River Basin	Mean Relief (meters)	Mean Slope Angle (sine)	Mean Denudation Rate (meters/1,000 years)
Thames River	159	0.02	0.02
Kander River	2,428	0.40	0.43

Source: Modified from Ahnert, 1970

of the geologic/geomorphologic domain, and consider how these properties affect a given area's response to human activities. In later chapters, when pertinent, we explore aspects of climate and vegetation and how they interact with human decisions with regard to land degradation. However, before we explore the geophysical domain, the concepts of frequency, magnitude, and variability will be briefly introduced with regard to physical phenomena, especially precipitation.

Frequency, Magnitude, and Variability

In daily weather reports, statements such as "today's temperature is 5 degrees above normal" or "the year's precipitation to date is 254 mm (10 in) below normal" are cited often. In fact, as expressed to the general public, weather is rarely normal as daily, monthly, and annual temperature and precipitation data are almost always either above or below normal in these reports. The reason for this "abnormality" is that the average is used as the expected (normal) in these reports for comparing current weather. The mean, being but one statistical measure of centrality, gives only a partial amount of information about any phenomena. This is especially true when dealing with physical phenomena.

First of all, for most physical phenomena, the most common population describing their attributes are log normal distributions. In the log normal distribution (right skewed), the majority of events are less than the mean (figure 3.1). That is, the majority (higher frequencies) of physical phenomenon is usually small or of a low magnitude—for example,

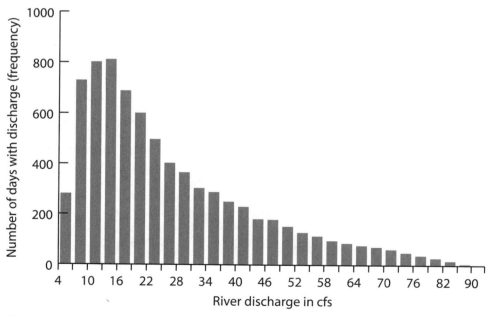

Figure 3.1.   Log normal distribution of the frequency and magnitude of a hypothetical river's flow. (Cartography by Anne Gibson, Clark Labs)

there are far more small earthquakes, volcanic eruptions, river flows, and rain events than their larger counterparts.

Because of the log normal frequency distribution of most physical phenomena, an interesting question is: What is more important with regard to land degradation, the high frequency low magnitude events or the low frequency high magnitude events? Soil loss, which contributes to both a decline in soil fertility and the soil's ability to hold moisture, must be prevented because it affects biomass production and hence land degradation. To accomplish this end, strategies must be developed that minimize the deleterious effects of both common weather events that contribute to soil loss (such as everyday rain events) as well as the extreme, high magnitude events that might occur only once every twenty years (such as a hurricane). Thus frequency and magnitude, which are clearly related to each other, need to be understood because of the role these properties play in land degradation. These two factors alone illustrate why the mean rarely is a sufficiently sensitive indicator to determine the susceptibility or likelihood that an area might degrade under different human land-use systems. Other information is required such as the range of magnitudes experienced and how often is it likely for a critical magnitude to occur. Variability measures add information that give further insights into the temporal, spatial, type, and intensity features of phenomena that a location is likely to experience.

Thus, variability is a property of physical phenomena that needs to be considered with regard to land degradation. Two common statistical measures used to assess some aspects of variability are the standard deviation and the coefficient of variation. (Detailed explanations of both of these may be found in any elementary statistics book.) Roughly, an area's variability characteristics describe how a property is distributed around a measure of central tendency (mean, median, and mode). As variability increases, central tendency measures (for example, mean) become a poorer description of the site's likely (expected) property. Precipitation characteristics will be utilized to illuminate characteristics of variability.

Every set of precipitation values is associated with a frequency of occurrence. For example, Ibadan, Nigeria has 900 mm (35 in) of annual rainfall, a 13 percent frequency of occurrence (figure 3.2). Ibadan's mean annual precipitation is 1,230 mm (48 in); its modal precipitation is 1,200 mm (47 in). Annual precipitation ranges have been experienced in western Nigeria from a low of 715 mm (28 in) to a high of 1,925 mm (75.8 in); that is, the absolute range in annual precipitation at Ibadan is 1,210 mm (47.6 in). Because of this precipitation variability, during dry years the area is clearly semiarid and moisture deficiencies are severe. In the wetter years, the climatic conditions are humid with many periods of moisture surplus. Relatively large precipitation variability is normal for locations such as Ibadan.

Precipitation variability is a particularly sensitive climatic attribute for subhumid and semiarid areas. The expected ranges in precipitation in these climatic zones often cross critical moisture boundaries. Precipitation variability is as much a critical climatic attribute as absolute values of precipitation in these zones. Temporal moisture variability tends to increase as precipitation decreases. At Ibadan, a location close to the annual moisture surplus/deficit boundary, management strategies must be developed to maximize agricultural outputs during both the wetter periods and the drier periods while protecting the environment. In our inquiry into land degradation and the environment, human systems must be sensitive to fluctuations in the physical domain.

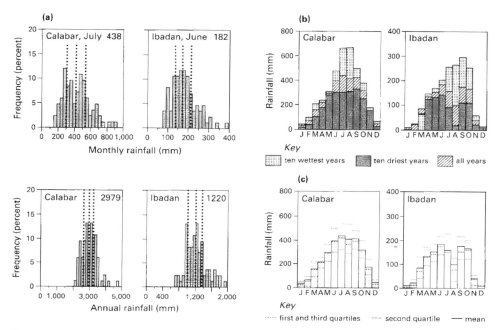

Figure 3.2.   The rainfall characteristics of Ibadan and Calabar, Nigeria. (Reproduced from the *Atlas of African Rainfall and Its Interannual Variability*, 1988)

Key: (a) Frequency distribution of rainfall: dotted lines indicate the first, second, and third quartiles; the mean (in mm) is given in the upper right-hand corner.
     (b) Monthly mean rainfall
     (c) Monthly rainfall statistics

Magnitude, frequency, and variability of natural phenomena need to be understood not only in terms of temporal patterns as illustrated in this discussion, but the spatial variability (areal discontinuity), as well as variability in types of phenomena experienced at a location need to be grasped. Spatial variability of precipitation is greater in drier climates than in humid ones. A reason for this is that there is a greater reservoir of moisture in the atmosphere under humid conditions. Thus a smaller volume of air supplies sufficient moisture for precipitation in humid compared to arid areas (figure 3.3). This greater moisture allows neighboring precipitation cells in humid areas to be closer to each other than in arid areas. The greater moisture supply also normally permits rain cells to persist longer in humid areas and allows a larger area to experience rainfall. Thus in convection precipitation, usually there is a greater spatial variability of rain events in arid compared to humid zones. Developing strategies to cope with this variability is required to prevent land degradation. Variability in types of phenomena also differs from place to place. Again with regard to precipitation, a humid middle latitude location such as New England in the northeastern United States experiences a wide range of precipitation types including snow, hail, sleet, drizzle, and thunderstorm downpours. In contrast, coastal central California normally experiences precipitation only as rainfall with a range of intensities.

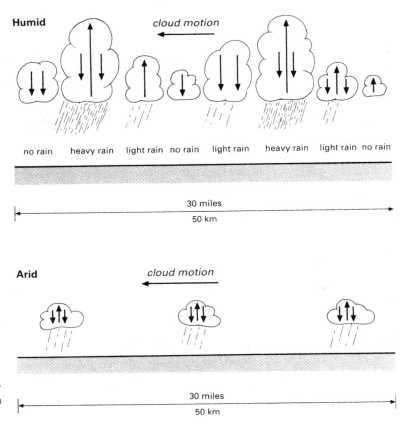

Figure 3.3. Spatial distribution of precipitation between humid and arid areas.

The Geologic/Geomorphologic Domain

The nature of the Earth's materials as well as the stability of the Earth at any location provide the setting in which human activities occur. Humans since the Agricultural Revolution make modifications, either accidental or purposeful, to natural settings as they pursue their livelihoods. In some places a specific activity/modification will result in land degradation. In another area, the identical human induced change will result in no significant environmental change. These differences, one site degrading while the other site does not, illustrate the basic fact that nature has not created equal physical settings. Some locations are extremely stable and resilient; others are not. Geological factors contribute to the sensitivity of an area's susceptibility to land degradation.

The Geologic Setting

The sudden and violent events resulting from internal forces within the Earth have been a component of Earth history almost since its formation (Huggett 1990). Recent manifestations of catastrophic tectonic activity having implications for land degradation are the volcanic eruptions of Mt. Pinatubo, Mt. Unzen, and Mt. Usu. While all three mountains are in highly active volcanic and seismic zones along plate tectonic boundaries, the volcanoes were considered dormant until their recent eruptions. During the years when the mountains were dormant, the inhabitants living in the surrounding areas developed intensive agricultural systems that took advantage of the fertile volcanic soils found on the slopes and gentle surrounding lands. The inhabitants utilized lands in the immediate area of the volcanoes as if the mountains would never erupt. On the Philippine farmlands surrounding Mt. Pinatubo, volcanic ash from the eruptions has buried the agricultural fields in this very fertile area of Luzon. Harvests were widely destroyed by the 1991 eruption. In the short-term, the eruption has necessitated the abandonment of a large proportion of this region's lowland farms. On the steeper slopes, the pyroclastic (volcanic ash) deposits interacting with the area's normal rainfall have resulted in landslides and mudflows that have degraded the affected hillslopes. Mudflows continued to take place during heavy rains and result in ongoing instability on the former relatively stable farmlands.

The 1990–1994 volcanic eruptions, following 198 years of dormancy for Mt. Unzen in Nagasaki Prefecture, Japan, damaged over eight hundred houses by its ash flows as well as covering lands utilized by a number of human activities. In both Japan and the Philippines, the utilization of the formerly productive lands affected by the eruptions requires major investments in both capital and labor before land-use systems will attain their previous levels of productivity and utility. Furthermore, the large amounts of unconsolidated deposits resulting from the recent volcanic eruptions make these areas more susceptible to mass movements than the preeruption conditions. The new physical attributes of these areas increase the likelihood of land degradation if former agricultural land use becomes reestablished in these areas. The area has become more fragile and for a number of years high frequency events will have greater impacts on the environment than in the preeruption period.

Likewise the March 2000 eruption of Mt. Usu on Japan's Hokkaido Island resulted in large unconsolidated deposits of pyroclastic materials. This created an extremely fragile setting. If these deposits are disturbed by human activities or heavy precipitation, landslides or mudflows could be set into motion. It will take years before these deposits, through erosion and depositional processes, become stable. Thus human occupancy of these areas in the immediate future is risky.

These recent volcanic eruptions illustrate an impact of ordinary tectonic processes on the status of the land resource. Volcanic eruptions occur erratically, and often during long periods of dormancy humans will develop and utilize the lands without concern to the volcanic hazard. Yet because periods of volcanic activity will occur, even if the specific time cannot be predicted, areas in which tectonic plate collision and continental-type materials are juxtaposed are susceptible to catastrophic damage. Pompeii is an example from the ancient world; Pinatubo, Unzen, and Usu are contemporary examples. The Pacific Northwest, for example, in the vicinity of Mt. Rainier, Washington, with a large population and the presence of large dams, is a likely site for future degradation due to the interaction of human activities with a highly active tectonic zone. Small tremors in this area during 2004 are clear indications of potential future eruptions in the vicinity of Rainier. Recognizing the high risk potential in the Mt. Rainier vicinity, communities have installed early warning systems to evacuate potentially affected areas downstream from dams if an eruption occurs.

Not all volcanic eruptions have the potential to cause catastrophic destruction. Volcanic eruptions associated with plate divergence and hot spots are generally far less explosive than those occurring in areas of plate convergence with continental materials. In areas of plate spreading, ash deposits normally comprise a smaller proportion of the volcanic deposits. Nevertheless, destruction of the land resource is still possible in these settings. Lava and tephra from the eruption of Eldfell in 1973 on the Island of Heimaey, off the southern coast of central Iceland, covered seventy houses and farms. The formerly productive pastures that provided grazing for livestock were replaced by freshly deposited basaltic materials. Bare rock has replaced soil as the surface material on the former pastures and farmland. Depending on one's perspective, the 1973 eruption can be viewed as either constructive or destructive. For farming, the eruption destroyed many productive lands. For the fishing industry, the largest employer on the island, the eruption improved the local environment because some of the lava deposits reached the ocean and improved the harbor, thereby giving additional protection to the island's fishing fleet during storms.

Attempts have been made to minimize the destructive effects of basaltic lava flows, the dominant materials associated with hot spots and plate divergence, by trying to alter the actual and potential routes of the lava flows. On Hawaii, earth embankments and trenches have been constructed to divert lava flows away from the town of Hilo and its surrounding housing subdivisions. On Heimaey, during the 1973 eruption cold seawater was pumped on the flowing lava to divert its flow away from the town.

From the geologic viewpoint, almost all volcanic activity is constructive, since it increases the amount of earth materials above sea level, which counters erosion processes. From the vantage point of this book, most volcanism results in at least a short-term decline in the utility of an area for most human activities. Not all tectonic activity with land degradation ramifications is catastrophic. One such example is the slow submergence of many coastlines at the rate of millimeters per year. This dynamic increases coastal flooding in many parts of the world, especially as sea levels are slowly rising too. Along the northeastern coast of the United States, tectonics result in coastlines slowly sinking at between 1 to 5 mm per year (Brown and Oliver 1976). To protect developed coastal areas, costly engineering works and maintenance are required. In spite of these attempts, when coastal storms and hurricanes occur the results are often flooding and large economic damage. Coastal zoning ordinances would likely be far more cost-effective than trying to minimize land degradation by attempting to offset coastal submergence through such engineering works as sea walls. For political reasons, often areas are rebuilt after coastal flooding, which encourages continued development of these risky areas. The result is an ongoing cycle of land degradation in which coastal storms occur continuously but are offset by economic subsidies due to the political pressure exerted by the area's inhabitants. The Town of Hull, located south of Boston, Massachusetts, is situated in a setting prone to coastal flooding where rebuilding has occurred due to government aid.

Gravity is ubiquitous both temporally and spatially. It is continually exerting a downward force on all earth materials and particularly is effective in moving unconsolidated surface materials. With the wide range in form and strength of materials found on or near the surface of the Earth, slopes react to the constant gravitational force in a variety of ways. Slopes can fail instantly as manifested in rock falls, landslides, and mudflows. At the opposite extreme, earth materials on slopes can move imperceptibly downslope at rates of only millimeters per year. This latter slow movement, known as soil creep, generally is unimportant in the context of land degradation. But, in some high-energy settings, it can be a factor when interacting with human activities. One such example occurred as a result of building a road through the Luquillo Mountains in eastern Puerto Rico, where steep slopes, high rainfall, and thick, weathered unconsolidated materials exist.

Road construction in Puerto Rico through El Yunque Rainforest has triggered slope slumps and landslides where previously the dominant process was slow soil creep. Slope failures continue today, generally during periods of exceptional rainfall. When these failures occur, they cover or break up the road and cause its temporary closing until the road is repaired. Biomass production in the affected portions of the hillslope is reduced as less fertile earth materials become exposed. The likely cause of the failures results from the cutting of a portion of the lower hillslopes for the road. This reduces the resistance of the hillslope and with this reduction in the slope resistance, gravitational forces become stronger than the internal cohesion of the slope. It is the existence of deeply weathered materials under the slope surfaces along with through flow and high water tables due to the humid tropical climatic setting that result in this instability. Once the slopes are disturbed by road construction, this sets into motion these accelerated mass movements. Once the slopes become unstable, they are difficult to stabilize without costly engineering works. In other less high-energy environments, road cuts also can set into motion accelerated mass movements. In central Massachusetts, during the building of the interstate highway system in the early 1970s, road cuts through some heterogeneous glacial deposits resulted in earth flows in places where the road cuts exposed clay lenses. The catalyst for slope instability was once again the removal of material from existing stable slopes as well as cut slopes that were too steep for clay. To stabilize these slopes so as to prevent earth materials from flowing onto the highway surface, the clay lenses were removed from the slopes once an earth flow occurred and gravel fill was substituted (figure 3.4). Whenever human activities alter topography, especially where surface materials are not consolidated, the potential increases for setting into motion slope instability. Sometimes, the instability clearly results in land degradation such as the closing of roads and the resulting economic costs and exposure of infertile soils. In other areas, it is possible to cut into consolidated materials leaving nearly vertical slopes (figure 3.5), which remain stable but can occasionally result in rock falls.

Lithological factors also need to be considered in the context of land degradation. Environmental conditions affect the resistance of rock types. As but two examples, granites are resistant in the cool humid climates of New England, but generally are nonresistant in the hot humid tropics of rainforests. Likewise limestones are nonresistant in humid areas while resistant in arid areas. One example of land degradation resulting from lithological factors occurred as a result of an urban area trying to meet its water needs. In 1928, to meet Los Angeles's ever-growing need for a reliable water supply, a dam was built in the San Francisquito Canyon. It was a poor choice for locating a dam as the canyon walls were comprised of mica shale. As the water rose in the reservoir behind the dam, the mica shale began to absorb the water and gradually softened, becoming less resistant. On March 12, toward

Figure 3.4. An unstable road cut through unconsolidated materials along Interstate 290, Worcester, Massachusetts. (Photograph by L. Lewis, spring 2005)

the end of the rainy season, the rock's threshold of resistance was surpassed and the dam failed. The dam's initial surge of water was over 61 meters (200 ft) high as it flowed down the canyon into the Santa Clara River. Besides the loss of life and damage to buildings, topsoil from 3,240 ha (8,000 acres) of farmland was gone (Reisner 1993, 99). The erosion of these soils converted former highly productive farmlands into nonproductive areas. This clear example of land degradation resulted from the interaction of the local area's rock properties (lithologic) with the human activities of dam building and farming.

The Geomorphologic Setting

Until rather recently, geomorphic investigations largely ignored human activities as a significant geomorphic variable. The roles of wind, water, and ice, the natural geomorphic agents altering the geological setting, have been and continue to be stressed in geomorphic investigations. Illustrating the minimal stress placed on humans as geomorphic agents, Chorley, Schumm, and Sugden (1984) devote only twenty pages in the appendix out of 607 pages in the total text to the examination of human activities. Bloom (1997) and Esterbrook (1999) don't even examine humans as geomorphic agents in their texts. The avoidance of human activities in geomorphic texts and research is far more common than their inclusion.

Figure 3.5.   A stable road cut through bedrock along Interstate 290 in Northborough, Massachusetts. (Photograph by L. Lewis, summer 1994)

Yet today humans are the dominant geomorphic agent or a very important one in many locales. For example, contemporary warfare with the power of modern armaments alters local landscapes. During the fighting against the Taliban in Afghanistan, in 2001–2002, the dropping of 15,000 pound bombs clearly created large impact craters. In addition, the impact of bombing on caves in the Tora Bora region has altered the area's topography. These types of activities clearly affect both landforms (geomorphology) and land degradation. In the Netherlands, over 30 percent of its national territory is the direct result of public work construction. Large areas of Holland have been reclaimed from the sea. The building of extensive engineering works has replaced the salty Zuider Zee with the freshwater Lake IJssel. Along much of the U.S. northeastern seaboard, as much if not more material is eroded or deposited by earth-moving machines as by the area's rivers. Sanitary landfills have created hills where once depressions existed (figure 3.6), and highway construction has cut through hills and filled valley bottoms to alter terrain gradients totally to meet the needs of automobile and truck transport. Geomorphic impacts of road construction are almost ubiquitous.

The flow of the Chicago River no longer is directed into Lake Michigan, its natural route, but has been reversed by dredging its channel. Instead its waters eventually reach the Mississippi River as its discharge today is completely controlled by human manipulation. In Egypt the damming of the Nile is resulting in significant erosion in its delta. The Colorado River rarely flows into the Gulf of California anymore as its waters are usually completely diverted to meet the needs of either irrigation or urban centers (particularly Los

Figure 3.6.   Landfill hill on Interstate 495 near Wrentham, Massachusetts. (Photograph by L. Lewis, spring 2005)

Angeles and Phoenix). In Turkey, through a series of dams, the hydrology of the Euphrates River system and increasingly that of the Tigris is being altered. To understand land degradation, there is a need to consider the interactions of human activities with tectonic (geologic) and natural surface processes (geomorphic).

It has been estimated that natural erosion results in 10 billion tons of sediment annually reaching the oceans (Judson 1981). Existing agriculture, grazing practices, and activities associated with urbanization result in oceanic sediment deposit estimates between 25 to 50 billion tons per year (Brown and Wolf 1984). Farming activities interacting with natural factors account for actual soil loss in agricultural areas (Lewis, Verstraeten, and Zhu 2005; Verstraeten and Poesen 2001). Thus human activities have escalated the erosion capabilities of the natural geomorphic agents of wind and water. For example, the U.S. Corps of Engineers spent years and millions of dollars straightening river channels to improve drainage and sometimes navigation, which also increased, in many cases, erosion and destroyed ecosystems. Today many of these straight river channels are being reengineered to approximate their former natural channels to reverse some of the environmental damage resulting from channel straightening. The largest such project is in central Florida. In an attempt to save the Everglades in southern Florida, drainage is being altered from the straight canals that permitted development in the area. In the following sections we explore some aspects of the conditions that lead to accelerated erosion and mass movements. A few common links between land alterations resulting from human actions and the increase in erosion rates are presented. The resulting surface modifications contribute to the pattern of worldwide land degradation.

Fluvial erosion results from a complex set of relations among precipitation, gravity, vegetation cover, and the erodibility (resistance) of the soil to erosion. These variables vary widely from place to place and the relations between them are complex. Because of this complexity, along with economic constraints and in many areas a poor database, it is difficult to develop strategies that can be implemented to minimize soil loss. For example, soil losses remain high on many American farms in spite of government conservation programs, educated farmers, and the use of advance technology. Some of the critical erosion aspects of precipitation, gravity, groundcover, and erodibility are examined in the following paragraphs. These are critical physical variables affecting erosion and mass movements. Following this chapter, the human constraints contributing to ongoing soil losses, such as labor shortages and economic factors, are presented in the remaining chapters.

Erosivity is an erosional force, which is associated with precipitation, that initiates both soil detachment and transport, is linked with rainfall intensity and amount, and varies widely on our planet. High erosivity areas are associated with high erosion potential. Humans must adapt different strategies for specific climates because of the wide range of erosivity from place to place. Figure 3.7 shows that the range of erosivity in the lower forty-eight United States ranges from less than twenty in the drylands of the West to over 550 in the South. Higher

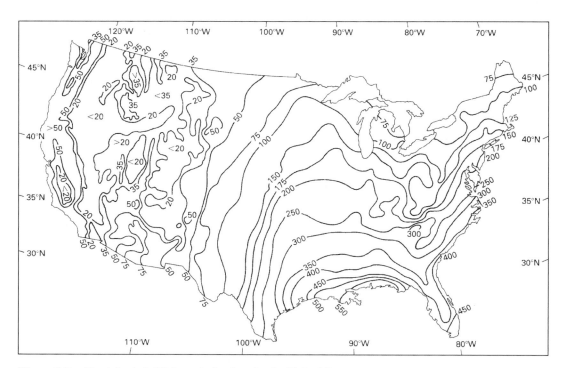

Figure 3.7.   Erosivity (rainfall intensity) values for the United States.

erosivity values are found in some humid tropical areas due to both high rainfall amounts and very intense rains. Tactics need to be in place to neutralize areas having high precipitation–erosional potential whenever humans disturb lands that experience high erosivity.

The importance of gravity with regard to erosion is relatively obvious. Slope affects both the direction and the velocity of surface flows. On flat and very gentle gradients, water will be stagnant or sluggish and have little or no erosion potential. As slopes increase, the directions and zones of concentrated water flows are determined, and the potential for higher velocity water flows and hence erosion increases. The erosional role of gravity is largely determined by an area's topographic characteristics. Steep highlands or lowlands have greater erosional potential than flat highlands or lowlands. When steep lands are disturbed, they have a greater chance of experiencing increased soil loss than does a gentler area. The role of gravity alone largely accounts for steep lands falling within the category of fragile land. Hence steep slopes have a greater proclivity toward land degradation when disturbed by human activities.

The nature of a surface's groundcover plays a critical role in either curtailing or accelerating soil loss. Human activities, as they alter the vegetation cover, are probably the greatest cause of increased worldwide erosion rates. Vegetation has numerous properties that reduce erosion.

1. The interception of rainfall reduces the total amount of water reaching the ground. For instance, broad banana leaves actually trap small amounts of rainfall, and a percentage of this trapped precipitation directly evaporates without ever reaching the ground. Hence this intercepted precipitation is unavailable to cause any soil loss.
2. As rain strikes plants, they absorb a proportion of the falling drop's energy (erosivity). This reduces the precipitation energy available to initiate soil movement.
3. Plant matter (living and decomposing) that completely covers the ground protects the soil and prevents rainsplash, a major catalyst of soil loss.
4 Groundcover increases surface roughness and thus lowers the velocity of water flowing over the ground.
5. Plant roots increase the resistance of the soil as well as remove moisture from the soil. This latter property means a greater amount of rainfall is needed to saturate soils. This attribute is important, as saturated soil is prone to erosion.
6. Soil infiltration is often increased by vegetation, which acts to reduce surface flows.

In summary, as vegetation cover decreases, the erosional potential generally increases. For example, construction is a human activity associated with large areas of disturbed bare ground. Unless conservation strategies are implemented at the time of construction, the disturbed areas are almost always major areas of sediment production (erosion).

Erodibility—the resistance of the soil to particle detachment and transport—is primarily determined by the soil's texture (particle sizes: clay, silt, sand, and so on), chemical properties, organic matter, and soil structure. All of these properties are dynamic and change in response to both natural and human phenomena. For example, soil texture can be altered due to vegetation clearing. This exposes the bare soil, and wind or surface water flows can remove the finer soil components resulting in an overall coarser texture. During a rain event, erodibility may increase due to surface seals or changes in particle orientation (Lal 1988). Plowing by breaking up the soil can destroy soil structure and make it more susceptible to erosion.

Soils vary widely in their susceptibility to erosion. Highly erodible soils have a potential to erode that is about ten times greater than the most resistant soils. As the erodibility of a soil increases, greater care must be employed to prevent degradation. Some soils, such as those composed of fine sand and silt-sized materials, are extremely vulnerable to erosion. The best strategy for areas with highly erodible soils is to develop land activities that minimize disturbance of the soil resource. Particular goals in such areas should be permanent, complete groundcover and activities that reduce surface water runoff. Where agriculture is needed, drilling the soil instead of plowing and planting perennial crops rather than annual crops need to be considered.

Wind erosion does not occur when soils are protected by a groundcover. Hence, under natural conditions, it rarely occurs in humid areas except in some coastal areas having sandy beaches. Only during extreme storms, where winds are high, do particles coarser than sands have the potential to be eroded by wind. Usually it is only silts and clays that can become airborne. Even in these storms, wind's effectiveness will be limited if rainfall wets the surface. Thus as an erosional agent, the effects of wind are restricted to areas where dry surfaces exist and where there is absence of vegetation cover. Under natural conditions, wind erosion is restricted to a number of limited settings within arid or coastal areas. When human activities reduce vegetation cover, such as in agriculture, overgrazing, industrial pollution, and construction, eolian erosion becomes possible even in humid areas, especially during dry periods. With regard to land degradation, it is especially in the semiarid zones and climatic areas with pronounced dry seasons, such as in the Southwest United States or Sahalian Africa, that wind erosion becomes a significant factor once vegetation cover is degraded by human activities.

Wind deposits exist in all climatic settings. In humid areas, these deposits often occurred under drier or colder climatic conditions than exist today. For example, in central Indiana in the humid American Midwest, forested hills of naturally stabilized sand dunes occur. Formed during the colder periglacial conditions that existed in the region toward the end of the Pleistocene, these hills are fixed in location due to the protection of the vegetation cover that evolved in the warmer humid conditions of contemporary Indiana. Even if the vegetation cover was disturbed on these sandy hills, due to sufficient ground moisture that reflects contemporary humid conditions, wind erosion potential on these relic dunes would be minimal during most of the climatic conditions experienced in this area.

In contrast, in semiarid southern Mauritania, because of overgrazing, fixed sand dunes have become mobilized as the vegetation cover degraded. These dunes have migrated southward and covered most of the country's better agricultural lands (Lewis and Berry 1988; Thiam 1998). Because this area is dry and minimal soil moisture exists most of the time, vegetation can reestablish itself only with great difficulty once destroyed. Once stable conditions are upset in such dry climatic settings, it is almost impossible to stabilize areas characterized with fine surface materials unless there is an available water supply to irrigate newly planted vegetation until it becomes established. In areas of marginal moisture surplus (semiarid areas or areas having very permeable soils), it is extremely important to maintain a good groundcover. Once disturbed, these lands are difficult to stabilize if wind erosion occurs since the winds remove the finer, fertile components of the soil. Loss of this fine material reduces the soil's ability to hold moisture. Equally serious, the movement of the soil materials prevents seedlings from becoming established without major human intervention.

The importance of wind is continuously increasing both in magnitude and areal extent due to misuse of land. Occasional dust storms occur in the American Midwest/eastern Great Plains because of the exposure of vast tracts of soil to wind during periods of plowing and sowing. When farm units were smaller, fields were smaller and windbreaks of planted trees minimized the effectiveness of wind. Today many of the windbreaks have been removed to increase the efficiency of farms using more powerful and larger equipment. To date, much of the deleterious effects of wind erosion in this area have been countered through fertilizer applications and improvements in seed quality.

Erosion, unlike mass movements, requires a transporting medium such as water, air, or a dump truck. Mass movements are the detachment and downslope transport of earth materials under the direct influence of gravity. Mass movements include a wide range of types such as rapidly moving landslides, fast-flowing mudflows, and extremely slow soil creep. All mass movements occur when stresses acting on a slope exceed the subsurface's material strength. Land degradation is usually associated with the more rapid family of mass movements. When farm fields suddenly slide down a slope, exposing infertile materials, or when hillslopes suddenly fail with resulting landslides covering houses, roads, or emptying reservoirs, not only is the degradation obvious, but it is dangerous as well. Engineers usually evaluate the likelihood of rapid mass movements using the concept of the safety factor. As the safety factor decreases, the probability of mass movements increases. When the safety factor value is less than one, mass movements become inevitable. Usually an external source of energy, such as an earthquake or heavy rain, triggers a rapid mass movement.

Slopes are always being altered either through natural erosional/depositional processes, such as a stream eroding the base of a hillslope, or by human actions, such as terracing, road cuts, or lot preparation for construction. These changes alter the shear stresses acting on the slope. Slope shear resistance likewise is dynamic. Earth material resistance generally decreases as you get closer to the surface. Chemical weathering processes account for this pattern. In addition, water saturation from precipitation, heavy traffic, the removal of vegetation, and the decay of root systems result in lowering the shear resistance of earth materials.

Permafrost areas, such as parts of Alaska, northern Canada, and Siberia, are particularly fragile settings regarding mass movements. Any disturbance in the tundra vegetation almost always lowers the reflectance (albedo) of incoming solar energy. During warm periods, increased local thawing in these disturbed areas results from this change in the albedo. The result is that mass movements are accelerated. When extensive engineering works are built in permafrost areas, mass movements triggered by changes in the local heat budget result (Ferrians, Kachadoorian, and Greene 1969). A hummocky landscape, thermokarst, can result. The resulting features upset local ecosystems as they alter the environment. As human incursions increase into tundra/permafrost areas, the potential of important changes in the topography may result due to an array of accelerated mass movements. These potential changes clearly need to be considered when oil extraction, such as in northern Alaska, is proposed or undertaken.

Regardless of location, it is almost universal that the earth's surface is being continuously modified by human activities. Steep lands are terraced to facilitate cultivation. Hillslopes are cut and valley bottoms are filled to construct highways. Reservoir and dam construction, which spread throughout the world during the twentieth century, have greatly modified landscapes. Reservoirs are one of the few surface changes initiated by humans that can easily be identified from space. All surface modifications set into motion complex changes that potentially impact the geomorphologic stability of areas. Terracing is just one example of a morphological surface change that alters the stability of the affected area.

In areas where population pressure is high, steep lands are often cleared for cultivation. Terracing is a favorite strategy to attempt to make the farmlands in this fragile setting sustainable. In many cases, with proper construction and continuous maintenance, terraces are a valid conservation tactic (Shaxson et al. 1989). Like any terrain modification, because of complex feedbacks between processes and responses in both physical and human domains, the effectiveness of terracing to control environmental damage on steep lands is often ephemeral. The following brief summary illustrates two terracing examples. The first example shows how changes in economic opportunities for the local population resulted in land degradation. The second example shows the need to construct terraces in accordance with local conditions. When this does not occur, as in our second example in Central Province, Kenya, it is likely that the intended conservation strategy of terracing will fail and land degradation will occur.

A primary intent of terracing is to minimize erosion on steep lands by controlling surface water runoff. Terrace construction reduces the slope angle of the areas where crops are planted (the bench portion of the terrace). The terrace walls between the benches are steeper than the original hillslope and require ongoing maintenance to remain stable. In both Spain and Italy, this has required farmers to repair their terrace walls over the centuries. For hundreds of years farming on terraced lands in many parts of Italy and Spain remained in continuous operation in what appeared to be very sustainable, stable conditions. With the formation of the European Economic Community (EEC)—now the European Union (EU)—economic opportunities in other EEC countries became available for many of the inhabitants in these areas of terraced farming. The greater economic opportunity resulted in an out-migration of many farmers. Abandoned terraced farms became widespread in many districts. Without the continuous maintenance of the terraces, erosional processes were no longer countered and terrace destruction became widespread as the terrace walls gradually degraded. Today terrace destruction continues as a constant phenomenon in many of these formally stable areas. Hillslopes are slowly readjusting from their human altered morphology (terraces) to a non-terraced slope, the natural equilibrium. In the process, a significant amount of the soils found on the terrace benches have eroded from the slopes, lowering their agricultural potential.

In many highland areas of central Kenya, forest clearing and terrace construction marked agricultural expansion on steeper slopes. In most cases, the terraces accomplish their goal of allowing a sustainable agriculture to exist by both conserving water and minimizing soil loss (Lewis 1985). But in some areas, the terraced lands were underlain with a layer of marl, which is impermeable. The decrease in slope on the terrace benches increased infiltration of rainfall. The change in vegetation cover from forest to tea lowered evapotranspiration losses. The result was that the hill's substratum became saturated as a perched water table developed above the layer of marl. The combination of the saturated ground and the wet marl surface lowered friction at the marl interface. The result was that landslides occurred (figure 3.8). This type of slope failure, and hence land degradation, would have been prevented if there had been proper terrace drainage to stop excessive infiltration. Because subsurface conditions vary widely in this Kenyan highland area, it is difficult if not impossible to develop site-specific terracing strategies within existing financial constraints. The net result in the area where the subsurface marl exists is that terraces, which were to contribute to a sustainable agriculture, resulted in destructive land degradation.

Illustrated in the previous portions of this chapter, the laws that govern physical processes operating in natural systems need to be obeyed to prevent environmental problems. Most often, land degradation is initiated when, intentionally or accidentally, people disturb the existing balances that exist in natural landscapes. Depending on the physical, chemical, and vegetational attributes of a given setting, alterations in land systems to meet human needs, such as terracing and groundcover changes, may set into motion dynamics that result in instability. Thus it is imperative to develop land-use practices that sustain or

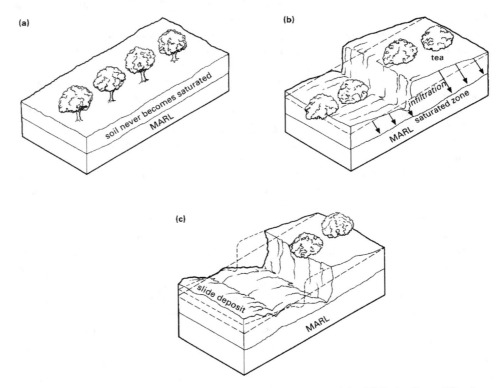

Figure 3.8.   The interaction between land clearance, terracing, and landslides in Central Province, Kenya: (a) under tree cover, (b) cleared of trees, planted in tea, and terraced, and (c) landslides due to greater infiltration of water on the terrace bench and saturation of the subsurface. (From Lewis and Berry 1988, by permission of L. Lewis.)

restore balances in critical components of the land system. To work in harmony with the cause/effect relations among the human and physical systems, these relations must be both identified and understood. It is the aim of the remaining portion of this chapter to examine the important catalysts of change in the physical domain. Human actions must work in harmony with these catalysts, "Not to do so is to court waste, damage (land degradation), sometimes even injury and death" (Robinson and Speiker 1978, 90).

Catalysts of Change in the Physical Domain: Agriculture

From the land degradation perspective, one of the most important land resource components is soil. When misused, it not only has direct impacts on the affected area, such as decreased fertility and lower crop yields, but also it can have severe offsite ramifications, often distant from where the accelerated erosion occurs. Accelerated soil loss on farm fields due to the lower cover of crops in protecting the soil from rainfall often results in the rapid siltation of reservoirs. Where irrigation systems exist, accelerated soil loss can destroy the delivery systems of water to the fields through their rapid siltation. Poor agricultural practices and the misuse of soils have been documented for millennia dating back to the misuse of lands in ancient times (Jacobsen and Adams 1958). Less obvious links to environmental problems associated with land cover change, agriculture, and resulting increased soil erosion also exist. Off the coast of Kenya, dying coral has been connected to accelerated soil losses occurring hundreds of kilometers distant from the reefs. With the change in land cover from forest to agricultural crops, increased settlement, and road building in the central highlands north of Nairobi, sediment loads increased in the streams draining the area. Eventually a proportion of these materials, transported in river flows, reached the Indian Ocean. Then offshore currents and waves distributed these sediments over wide zones of active coral growth. The coral could not survive in the turbid water produced by the increased sediments transported from the distant central highlands. As the coral reefs degrade, coastlines lose the protective shield that absorbed a large percentage of incoming storm wave energy. Two results of the degrading coral are less protected shorelines that suffer increased coastal erosion and a continuing decline in other components of the coral reef ecosystem.

Because of increased soil loss and declining fertility due to poor management practices, soil resource generally is deteriorating worldwide. The decline is of such a magnitude that it represents a serious global pattern. The overwhelming cause of most soil degradation is an array of human activities. Specific linkages between these activities and resulting land degradation are examined in following chapters.

Under natural conditions, the loss of soil fertility due to leaching (removal of soil nutrients by water in solution as rainwater or irrigated water percolates downward through the soil) and surface erosion is countered by:

1. the continuous recycling of nutrients through the decay of vegetal and animal matter and then their incorporation in living biomass; and
2. the addition of new nutrients into the soil system due to the natural breakdown of soil parent material by weathering processes as well as, in some river valley bottoms, the deposition of alluvial materials.

In contrast to the natural recycling of nutrients through vegetation and new nutrients entering the soil either through weathering or deposition, agricultural and modern forestry systems curtail nutrient recycling. In both agricultural and modern forest systems, harvests generally are transported to markets not in the immediate area. The nutrients in these exported farm stuffs are in the exported biomass. Hence, they are removed from local soil systems. For example, tomatoes grown in the industrial agriculture of southern Florida are widely marketed in the northern United States during the winter months. Their export stops the natural recycling of the nutrients in the biomass from ever entering the local soils where these products grew. Thus even without any soil erosion, nutrients are continuously being depleted in most commercial agricultural areas. To maintain soil fertility in most cases, supplemental fertilizers are needed to offset the loss of the soluble minerals incorporated in the exported farm products. Where this is not possible, soils likely will slowly decline in fertility.

Another contributing factor leading to the decline of soil resources is the accelerated soil loss associated with most modern farming systems (figure 3.9). Fields planted in annual crops go through an annual groundcover cycle. During the planting season when the soil is plowed, the soil is disturbed for planting and groundcover is minimal. Under these conditions, the potential for soil erosion is high. Groundcover gradually increases during the growing season as the plants develop from seedling to mature crop. As the groundcover increases,

Figure 3.9.   Ephemeral gully on farmland in central Belgium. (Photograph by L. Lewis, March 2002)

the erosional potential decreases. Finally after harvest, groundcover decreases and soil-loss potential once again increases. Soil resources have been completely destroyed in some areas due to poor farming methods. Within a national context, Haiti, through land misuse, has experienced massive soil losses over the last two hundred years. Today in parts of the country, bare rock is exposed where forests and thick soils once existed (figure 3.10). The widespread destruction of the soil resource through accelerated soil erosion has decreased the nation's agricultural potential to such a degree that destructive land degradation is widespread throughout much of the countryside. No longer are vast areas within the country able to support agricultural or forestry activities.

Widespread accelerated soil loss associated with contemporary agricultural and forestry practices is being curtained by implementing different strategies. For example, instead of clear-cutting a forest, which exposes large tracts of bare ground with a high potential for soil erosion, selective cutting strategies can be substituted to leave some trees standing, which adds partial groundcover. Depending on the density of the remaining trees, soil losses can be greatly reduced. The problem is that these selective, more environmentally benign harvest methods usually are associated with higher short-term costs than clear-cutting. In some agricultural areas, instead of plowing fields prior to planting, seeds are now planted using seed drilling strategies that disturb only a small portion of the field. This allows a groundcover to remain during planting and curtails soil erosion during the critical planting period. Again, there can be short-term economic disadvantages associated with these land preparation

Figure 3.10.   Soil degradation on steep slopes north of Le Caye in southwestern Haiti. (Photograph by L. Lewis, winter 1984)

methods. For example, groundcover can increase the risks of plant disease requiring more fungicides to be applied. Similarly, lower yields might result because the groundcover uses some of the soil moisture that otherwise would solely be utilized by the planted crops.

While clearly important, nutrient loss/decline in soil fertility and accelerated soil erosion are only two factors causing widespread soil degradation. Soil capacity to hold water, the infiltration rate, aeration, and pH (acid, neutral, alkaline) are other critical properties affecting biotic activity and hence crop yields. Soil porosity is particularly important in drier climates where growing seasons are often longer than the rainy season. If soil storage capability or impermeability is decreased due to soil loss, soil compaction, or other causes, a smaller proportion of an area's precipitation will be available for plant growth and yields will decrease. Increases in soil pH due to poor irrigation practices—or conversely decreasing pH due to acid rains—degrade soils and can either cause lands to be abandoned or agricultural yields to decrease.

Synthesis

The risk of land degradation from any specific activity varies greatly from place to place. In this chapter some of the critical factors in the physical domain that affect this risk were introduced. An area's stability and resilience properties are determined by complex relations among its geomorphic, climatological, and groundcover attributes. Depending on the local characteristics of these variables, there will be a specific interaction between the natural and human systems. If critical thresholds are reached, land degradation will occur.

While generalizations about the risk of land degradation in a given setting for specific activities are possible, they should never be considered more than a first approximation. For example, with regard to agriculture, high-energy settings, such as areas of steep slopes, would appear to have a higher erosional risk than low-energy areas such as plains. However, because land degradation results from the interaction of numerous physical associations, emphasis on any single property of the environment provides only a cursory insight into the physical setting's relative susceptibility to this problem. For instance, many low-energy areas also have high erosion potential. Semiarid plains, which are low-energy zones both with regard to gravity (low slope) and climate (low precipitation), are often as fragile for agriculture as are the high-energy environments of steep lands.

With any change in the land situation, a set of complex interactions usually occurs among the physical variables. Some of these changes may be acute, that is rapid and immediately obvious. The formation of a gully in response to a single storm due to the removal of the groundcover on a slope of highly erodible soils illustrates this aspect. Other changes may occur slowly over a longer time and not be immediately apparent. These chronic changes are often equally as destructive as acute ones. The slow rise in the local groundwater level due to the presence of a new dam and its reservoir eventually can contribute to an increase in mass movements. These movements result from increasing hydrostatic pressures on slopes (magnitude of stress) at the same time that there is a lowering of the internal friction of the slopes' regolith (decreasing the shear resistance of the slope materials). The 1963 Vaiont landslide in northern Italy exemplifies a chronic change in the physical domain that resulted in an acute case of land degradation: over $240 \times 10^6 m^3$ of material slid off a hillslope adjacent to the reservoir. This resulted in major destruction in the upper Piave River Valley three years after the reservoir was filled (Kiersch 1965). One final example is the slow removal of soil in water running off a field over time. This soil erosion, albeit very slow, will degrade an area by decreasing both the fertility of the land as well as lowering the soil's ability to store water. This subtle type of land degradation might only become apparent through lower agricultural yields or increased costs in fertilizers needed to offset the decrease in natural fertility. Many areas of the world have experienced slow, often undetected soil erosion. In some areas, this chronic change in the soil has necessitated the abandonment of agricultural activities, a clear indicator of land degradation. The wide range of physical properties found on this planet and the interactions among them indicate that a wide diversity of environmental conditions exist solely from the physical perspective with regard to land degradation potential. When human interactions are superimposed over physical interactions, the environmental responses to change become ever more complex. While all parcels of land are susceptible to land degradation processes, because of their physical diversity some areas are more prone than others to this problem.

While clearly not inclusive, the properties of precipitation and temperature that affect erosivity and moisture balance play a crucial role in many degradation processes. Although not discussed in detail in this chapter, the protection that plants provide against soil erosion is very important; when the groundcover is disturbed, land degradation potential increases. In the geomorphic realm, areas situated in active tectonic zones, locations underlain with highly erodible materials, or places situated in high-energy environments are settings where land degradation potential is large. The material presented in this chapter provides a base for our examination of land degradation. In the remaining chapters of this book, the various dimensions of the land degradation problem, especially interactions between the physical and human spheres, are explored.

Human Causes of
Land Degradation

IN ALMOST any climatic and topographic setting where people are present, regardless of the prevailing political and demographic situation, some land degradation will take place. This widespread occurrence of land degradation not only threatens local quality of life, but in some cases also actually threatens the viability of the affected areas. The existence of extensive land degradation cannot be explained solely by examining the physical, chemical, and biological processes and their interactions that are factors in land degradation. These natural processes are both reasonably understood and widely described. Furthermore, the technological interventions required to prevent or reverse the process of land degradation already exist for many land management problems. From a technological perspective, in agricultural, soil science, hydrological, engineering, and biological literatures, there are numerous well-documented and successful strategies and techniques in arresting land degradation in the land, water, and vegetation domains (OTA 1988; Pereira and Gowing 1998). Yet, worldwide, it is clear that destructive processes occur at a higher magnitude and rate than are restorative and conserving interventions (Brown 1992; 1998; 2001). This trend alone emphasizes the fact that land degradation is as much a human behavioral problem as it is a physical/technical one. Human factors that contribute to the degradation of our land, water, and plant resources under a wide range of environmental settings are the foci of this and the following two chapters.

Human Attempts to Stabilize a Dynamic Nature

Variability

Geologic, historic, and contemporary records all indicate that variability is a natural phenomenon in most Earth systems. From the planning and management perspective, variability creates numerous problems. For this reason, humans almost always try to minimize natural variability in most of the Earth systems they use. People do this in order to improve yield and efficiency in outcomes as diverse as food production, energy generation, transportation, and water supply.

Depending on the environmental, economic, and technological situation, to increase food production humans have invoked numerous strategies to cope with dryness, steep slopes, and variability in soil fertility. Improvements in transportation almost always result in the need to alter geomorphological phenomena to meet the specific needs of various modes of transport. Rivers, harbors, hillslopes, and valleys all have been, and continue to be, modified to meet the specific needs of transport systems. Any urban area requires a host of inputs and outputs to exist. These include basics such as food, water, and energy. Successful urban areas must minimize variability in the supply of these basic commodities. Furthermore, urban areas inherently modify the natural environment by their very existence. As but one example, pavement and other construction processes alter infiltration characteristics and hence change critical aspects of the hydrologic cycle. Urban areas, regardless of their location, alter the vegetational, geomorphological, hydrologic, and atmospheric realms both directly and indirectly.

Most human systems, regardless of the environmental, economic, and cultural setting, attempt to minimize the variability of natural phenomena as a strategy by which to improve their livelihood. Because of the feedback that exists among the components of natural and human systems, a change in any component to meet specific direct goals often triggers a number of other unintended adjustments. Unfortunately, many of these adjustments result in land degradation. All too often, strategies emphasize altering only the components of greatest interest. In most strategies it is either assumed that the other variables will remain relatively constant or that the components of the system that are not of direct concern can be just ignored. In other cases, humans simply neglect the ramifications of their environmental "control" because the changes they cause are inflicted on peoples and places outside their immediate sphere of interest.

For example, much of California experiences a highly seasonal pattern of precipitation. The summer months are generally dry, as precipitation is concentrated in the winter months. One strategy initiated to meet urban and rural water needs throughout the year was to build dams on upland headwaters. The water captured behind these barrages and stored in reservoirs during the periods of precipitation is then available for use during the drier months. In the process of capturing the rivers' waters, much of the river sediment that formerly reached the Pacific remains permanently trapped in reservoir sediments. Prior to dam construction, some of the sandy materials transported in the rivers reached the California coasts. This sediment was not only available for potential deposition, but actually comprised a significant supply of the materials deposited on the coastal beaches. With the deposition of these materials now in the reservoirs and not along the coast, beach deposition has been curtailed, but the erosional agents, such as waves, have remained constant. Without deposition to offset this erosion, many of California's coastal beaches have degraded. The previous relations between coastal deposition, which were partially dependent on the California river systems for material supply, and coastal erosion due to waves and currents, were upset due to water demands of the population and the human response to satisfying these needs.

Thus, the success of meeting a specific societal goal, in this case water demand, resulted in the unintentional degradation of some beaches quite distant from where the alteration in the fluvial system took place.

Environmental Deficiencies

For living organisms to exist, minimal energy, nutrient, and material requirements must be available in their environment. Different organisms have different requirements. Likewise, any environment possesses only a limited range of energy, nutrient, and material attributes. Humans have developed a multitude of strategies to overcome the natural constraints of any given area in order to meet human demands.

Humans have modified the natural world for a very long time in pursuit of their livelihood needs. In chapter 5 we discuss the replacement of forest cover by crops which, after the use of fire to hunt and modify landscape, is one of the earliest transformations that humans developed and systematically applied. Adaptations that had a major impact on local environments for thousands of years are numerous. In mountainous areas, farmers laboriously built terrace walls on steep slopes so they could create flat plots for their crops. In this way, they could take advantage of the higher rainfall of such uplands without seeing soil wash away. Domesticated animals kept close to family homes drastically increased the "hunting" success rate when meat was desired. Animal manure collected from stalls and pens was a valuable fertilizer. Only recently have large-scale animal factory farms begun to create solid and liquid waste streams that are impossible for nature to absorb. Subterranean dwellings in hot places where geology is suitable make it possible to keep cool without making substantial energy demands on the environment. What began as an adaptation to extremely harsh environmental conditions, making semidesert agricultural life more supportable, today attracts tourist income in a diversified local economy.

In the area of food production, various vegetation change and water transfer strategies have evolved for the wide range of environmental settings. Their successful application is reflected in the growth of human population. In the ten thousand years since the beginning of agriculture, the world's population has grown from a few million inhabitants to the billions of the contemporary world where today the majority live in urban settings.

On a global scale today, there is very little additional prime land that can be used to meet increasing food demands. The remaining nonagricultural lands, including abandoned agricultural lands, generally are not lands with naturally high agricultural potential. Most remaining forested lands are:

1. situated in fragile environments that have limited potential for crop production;
2. needed to meet our needs for wood production, recreation, or to satisfy our desire to maintain wildlife habitats; or
3. required to protect water supply for other urban and rural demands.

In moisture-constrained habitats, the overwhelming majority of the remaining nonirrigated drylands have either poor soil conditions that negate irrigation or are too distant from an adequate water supply.

An additional factor that limits areal expansion onto new lands to meet increased food demands is the reality that there is a continuing significant loss of land from rural activities due to urban expansion. Often, the lands being consumed for urban uses are potentially highly productive agricultural areas rather than marginal lands (Prostermann, Hanstad, and Ping 1996, 90). Preindustrial cities in particular are frequently located near productive agricultural land, and it takes a very strong planning function backed by strong police powers to prevent settlement encroachment on good agricultural land. Because of greater economic return in the urban sector, agricultural lands close to markets continue to be lost for nonagrarian uses. In some industrial economies, it is financially shrewd, but not necessarily ecologically astute, to sell agricultural land in urbanizing zones at urban house-lot prices and invest the financial gains in other land, potentially less productive, at a distance from the urban center. Using the funds gained in the land transaction from developers, the farmer can capitalize a new operation elsewhere at a much greater level and achieve economies of scale undreamed of in the old location (Hart 2001; 2003). To the extent that these highly capitalized operations involve dense concentrations of large numbers of animals in dairies, growing-out, and fattening facilities, the result can be serious environmental impacts, particularly on water resources (Howlett 2003). With existing land and environmental constraints, it is clear that future increases in agricultural and livestock production needed to meet increasing food, energy, and other material needs will largely have to be met through intensification of production on existing lands rather than by areal expansion.

Intensification strategies are numerous, and, although they can have significant impacts over a substantial territory, their most common goal is to overcome constraints and deficiencies at the local scale. Removing forest cover to promote crop growth and transferring water from surplus to deficit locations are two time-honored strategies that are discussed in the following chapters. In those locations where excessive moisture is a problem, wetlands have been drained and crops planted, always to the detriment of indigenous plants and animals. Steep slopes with thin soils have been modified to produce terrace "steps" that retain both deeper soil and greater soil moisture storage than nature would normally permit. Where soil is deficient, or repeated cultivation has diminished fertility, both organic and inorganic additions are made. Of these inorganic supplements, none is more important than nitrogen, and the industrial production of nitrogen fertilizer has had much to do with humankind's current ability (on average) to feed itself. The overly enthusiastic application of reactive nitrogen into soil and water systems has contributed to numerous environmental problems ranging from eutrophication to soil acidification to atmospheric ozone destruction (Smil 1997). Rotating crops to tap different arrays of soil nutrients, "fix" atmospheric elements such as nitrogen and carbon that have become deficient, and rest the soil by planting a useful forage crop that prevents erosion all constitute practices designed to overcome constraints and shortfalls while continuing to extract an increased output.

Seed, plant, and animal varieties used in any locality are an integral part of intensification strategies, since manipulation of genetic material to deal with limiting factors in the local environment has a long history. Breeding and biotechnology both develop hybrids that are intended to respond better to the human-controlled agroecology than their unaltered progenitors. In support of this effort, a variety of practices have evolved to alter the physical environment to make it better suited for the plant varieties and livestock desired for production. Examples of this approach include application of herbicides and insecticides to eliminate competition and predation, as well as efforts to create an ideal physical growing environment for plants through land leveling, drainage, and irrigation. While having beneficial effects in increasing food production, agricultural supplements are one of the major causes of water quality problems (James 1993).

The second major strategy is, through both plant and animal breeding or biotechnology, to develop varieties that can better adapt to the environmental constraints found in an area. Examples of this include the development of virus- and drought-resistant plant strains and the improvement of specific genetic aspects of animals, such as developing a larger goat or one that yields more milk. In the process of attempting to overcome specific environmental deficiencies affecting production, the strategies adopted often result in feedbacks into the environment that alter the status quo. In many places, the feedbacks resulting from the "improvements" that were intended to counter a specific location's limiting factors have resulted in land degradation. In the remainder of this chapter, the generic processes that originate in the human environment and result in degrading land resources are outlined. In most instances, these destructive processes originate in an effort to alter the Earth's landscapes to offset environmental deficiencies and variability while meeting human needs for agricultural and livestock products. In chapters 5 and 6, examples of these unintentional degradational developments are presented at several scales. Chapter 7 examines instances where changes in the environment have avoided destructive results and achieved sustainable development.

Falling Short: Human Factors Promoting Land Degradation

Land degradation is inherently a local process, at least in terms of its proximate causation, although local impacts are as likely to be driven by global forces such as climate change or international trade agreements as they are by local decisions involving stocking rates or cropping schedules (Reynolds and Stafford Smith 2002). Causation in land degradation is exceedingly complex and multiscaled, complicated by issues of perception, available technology, politics, land tenure, differing degrees of social vulnerability, institutional resilience, and ecosystem dynamics among others. Local resource users are more likely to be blindsided by the unexpected, often emanating from institutions and forces outside their influence and with unforeseen impacts on environment and sustainability, then they are to engage in deliberately wasteful exploitive practices. Current scholarship, particularly in arid ecosystems, at both the community (Robbins et al. 2002) and national scales (Batterbury et al. 2002), has moved away from simplistic models in which human population growth and pressure creates one-way impacts on and degradation of local resources. Instead, models of mutual interaction between livelihood systems and environment are envisaged ". . . where ecosystems respond to human actions while social communities adapt to changing ecologies" (Robbins et al. 2002, 326). In this mutually interactive dynamic, we associate land degradation with five generic failures to perceive, value, and respond to adverse environmental change. Each is considered briefly in this section.

Space: Unperceived Distant Impacts

Few humans are unaware of degrading conditions in their immediate habitat. Where these conditions go untreated locally, they are often symptomatic of deeply stressed, economically impoverished, institutionally bankrupt local communities. Most communities possess a wealth of adaptive strategies, functional technical knowledge, and sufficient institutional resilience to withstand extremely difficult conditions without initiating irreversible degradation (Mortimore 1998; Raynaut 1997). When serious degradation does occur locally, it often is a product of an out-of-sight, out-of-mind mentality. Such peripheral places, inhabited by marginal, politically powerless peoples, are impacted by decisions made elsewhere, but their local responses to changing economic and environmental conditions can accelerate existing trends in dramatic and often functionally irreversible fashion.

Thus changes that benefit some regions and populations, which from a national perspective constitute core constituencies, often export negative impacts onto distant places or numerically insignificant populations. Adverse environmental and social changes far from centers of power and influence seldom attract much attention. Where these negative developments are noted, they rarely are valued at the same level that they would be if they were affecting areas considered crucial to the viability and stability of national institutions, parties, and governments. Examples of this pattern are numerous. Parmentier (1996) describes how changes initiated elsewhere have impacted wetlands along the North African coast. Beginning under colonial administrations and continuing after independence, the primary pressure for change has come from agricultural development projects. These have involved stopping the seasonal flooding of both inland and coastal *merjas* (wetland pastures and meadows) by dams that store the water for release during the dry season to agricultural projects. Such irrigation projects, often linked to drainage schemes that complete the process of drying out wetlands, are based on a simple philosophy. Any water that flows unused to the ocean is wasted water! In the case of the Gharb Plain of the Sebou River in northern Morocco, the transformed water regime has created an agricultural district that is arguably the breadbasket of the country. But these gains have carried with them significant costs, often ones not calculated into the gains and losses associated with the development scheme. Five major off-project negative impacts exist. Herders who seasonally brought large herds of cattle to graze on luxuriant pasture no longer have access to that fodder resource. Hunters and gatherers from the surrounding settled population, who exploited seasonal plants and migratory waterfowl, are unable to support themselves from local resources. Migratory birds, using the marshes and coastal lagoons as stopover sites on the way from West Africa to Europe, find fewer resources to sustain their movements. Wetlands that were important breeding sites for freshwater and marine fish lose habitats critical for their life cycle. As Barbara Parmentier points out (1996, 165), agricultural managers have a set of objectives that involve intensified use of the landscapes entrusted to them. Their rewards come from the increased production generated by efficient use of land and water resources. Promotion does not come from increased numbers of birds counted in a coastal marsh as evidence of the improved viability of migratory waterfowl transiting their area to distant locales. Greater off take from marine fin fish spawned in coastal lagoons will not earn the agricultural managers of sugar cane fields or citrus groves an end-of-year bonus. These benefits of multiple purpose management are neither valued nor perceived as being important. The decline of production and viability in these areas and the impacts inflicted on other managers dependent on those resources may not always be distant spatially from those who initiate the transformations. But they are light years away conceptually, and it is this mental distance as much as physical distance that results in negative impacts that produce land degradation.

Another instance of unperceived impacts has occurred over the last two decades in the inland Gash Delta in eastern Sudan, where a multiplicity of external pressures have placed severe stresses upon local habitat (Kirkby 2001). The demand for charcoal generated by the quarter-million inhabitants of the nearby city of Kassala has placed woodland in and adjacent to the Delta under considerable pressure; current estimated rates of extraction exceed natural regeneration by ten times (Kirkby 2001, 228). Failure of the central government to invest in maintenance of the Delta's flush irrigation system has resulted in serious sedimentation and water-distribution problems. The expansion of rain-fed mechanized farming schemes on thousands of hectares in the rangelands outside the Delta have forced pastoralists, now excluded by the new cereal cultivation areas from traditional seasonal grazing resources, to concentrate year-round on Delta fodder resources. Degradation of these resources, formerly subjected to only seasonal use, is substantial. The gains in cereal production outside the Delta are not compared to the fodder production losses sustained within the Delta. No one associated with the mechanized farming operations acknowledges that the negative feedback loop of land degradation in a distant location is, in fact, a product of the tractor-dependent, fossil-fuel driven, large-scale cereal cultivation operations. In all such instances, the deterioration that occurs off the development project site in a distant place is not part of the "driver" population's mindset. Only local impacts directly connected to local practice are recognized; distant land degradation is simply not perceived. The norm is that environmental costs in nonproject or distant areas are irrelevant.

Time: Present Gain, Future Pain

Just as distant places are frequently ignored in the calculation of land degradation, so too are the future consequences of current actions often overlooked. In so doing, environmental costs are exported onto future generations. Concentrating the gains of resource use in specific units of space and maximizing the benefits of change for today while leaving one's descendants to deal with the environmental consequences are equally reprehensible practices. The pain caused by the need to make hard choices, the reduced biodiversity that constrains options, the burdens imposed by environmental remediation requirements all are avoided today by making them a cost for tomorrow. The shortsightedness of this search for immediate results and deliverable benefits while ignoring long-term consequences is exemplified in many ways.

One factor in failing to face future consequences is the desire of all political systems and leaders to deliver significant benefits for their citizens and adherents NOW. This drive to produce demonstrable results in a time frame that will redound to the credit of leader and party, as well as improve the material lot of the populace, is not necessarily bad. But it does generate a pressure for immediate results that is difficult to resist. This drive for immediate results runs counter to the *precautionary principle* (Jordan and O'Riordan 1999), which argues that any action likely to affect adversely either people or environment by instituting major, and potentially irreversible, changes must be approached with great caution. Only after very careful study and weighing of impacts in time and space should development proceed. Unfortunately, often the technology employed in a particular resource extraction activity is quite new and is used in settings where many parameters of the natural environment are not completely understood. The result is often a set of unexpected developments that occur years into the future as the impacts of the development activity slowly become recognized. In its demand that politicians, planners, and populace think long-term and that values other than economics factor into decision making, the precautionary principle challenges much that is taken for granted in status quo, commercial resource management.

Mining is a good example of an activity whose environmental impacts are concentrated in space. While local communities may benefit from the employment generated by the mine and the enterprises that arise to provide services to miners and their families, most of the profits from the mine operation are repatriated to distant locations. Environmental impacts, which can range from groundwater pollution, surface watercourse changes in both quality and quantity, increased atmospheric dust and land subsidence, appear primarily in the local habitat and are borne principally by the local community. In the case of gold mining near Fort Belknap, Montana (Emel and Krueger 2003), the land reclamation bond requirements placed on the mine owner as part of the initial development permit process were inadequate to withstand the company's eventual bankruptcy. This left the local community, the State of Montana, and the U.S. federal government to bear the cost of cleaning up the environmental damage, estimated to exceed $20 million, not covered by the reclamation bond. Other costs of mining operations are less easily quantified in monetary terms. In Butte, Montana, the end of mining operations has left older residential and commercial areas, traditionally constructed cheek by jowl to mine shafts and open pits, with a serious, long-term subsidence problem as well as intimate association with heavy metal tailings and arsenic-rich smelter soot (Baum 1997). In the last quarter of the nineteenth century, Butte boosters measured the economic prosperity of the local community by the thickness of the smoke (Wyckoff 1995, 482), either unaware or unheeding of the health consequences and environmental insults such contaminants represented for themselves and their children's children. The collection of polluted water in abandoned open pits, as well as the erosion of nearby slag heaps, poses a long-term problem to local groundwater quality, making Butte's Berkeley Pit the largest U.S. Superfund cleanup site (Wyckoff 1995, 494). This problem is fraught with irony and conflict. That the national government and its citizens, who benefited directly and indirectly from exploitation of Butte's copper mines, now have to pay a substantial portion of the remediation costs seems poetic justice. Local skepticism about the efficacy of such efforts in the face of the monumental effort required, thus leaving the resident population to cope with both costs and impacts, seems equally justified. Yet the very cleanup effort has spawned considerable local employment, and technological developments in processing polluted water for its heavy metal content raise the prospect of turning a huge environmental liability into a very considerable economic asset (Wyckoff 1995, 495). And at some level, left to its own devices and given enough time, at least on sites smaller than Butte, nature gradually manages to ameliorate the worst features of human

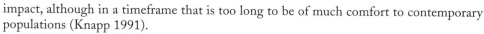

impact, although in a timeframe that is too long to be of much comfort to contemporary populations (Knapp 1991).

Exploiting groundwater can be a type of mining operation that imposes long-term burdens on future generations. This is particularly true when the groundwater is fossil in origin. Once withdrawn, this water resource cannot be replaced in anything other than a geologic time scale. Groundwater that is replenished by infiltration from rainfall and runoff is a renewable resource, one that can still be overdrawn and depleted but one that, in most cases, eventually nature will renew albeit sometimes requiring hundreds of years. The wise use of renewable groundwater for drinking purposes for humans and animals or to support intensive, closed environment agriculture where water losses are kept to a minimum is usually both reasonable and sustainable. Use of either renewable or fossil groundwater for open field agriculture, where evapotranspiration rates are high and water loss is excessive, almost never is justifiable. Yet this type of irrigation is the most common practice. Postel (1996, 43) has expressed concern that overdrawn aquifer layers might compact, filling the pore spaces that hold water and permanently reducing their water storage capacity. While the immediate users capture the benefits of a limited and precious resource, the depletion of the groundwater resource is a loss that seriously handicaps posterity.

An object lesson in the problems associated with intensive groundwater development for agricultural purposes is provided by recent experience in Saudi Arabia. Government agricultural policy in the early 1980s envisaged attaining agricultural self-sufficiency in critical foods as rapidly as possible (Brown 1998, 6). Central pivot irrigation based on groundwater development was a prime technology employed in pursuit of this goal. In less than ten years, grain production rose dramatically. From a base production of 260,000 tons in 1980, output soared to over 5 million tons a decade and a half later. Two factors intervened to initiate an abrupt decline in output. Very substantial subsidies for grain that were well above the global market price (Postel 1992, 32 estimated four times the global rate) had made irrigated grain cultivation and the investment in the technology to extract groundwater to sustain plant growth financially lucrative. The abrupt decrease in these subsidies in the mid-1990s undermined this monetary incentive.

An equally rapid decline in the groundwater table paralleled the drop in grain subsidies. Chasing water deeper into the ground raised drilling costs and increased the cost of domestic production. Concerns about the impact of fossil groundwater use were already expressed in 1991 (al-Ibrahim 1991). By that date a variety of ominous signs were appearing: water tables declined, some aquifers showed signs of compaction, ground-water quality decreased, particularly in coastal districts where subterranean salt water intrusion increased, land subsidence emerged in very severely affected areas, wells produced lower yields, pumping costs increased, well abandonments occurred with increasing frequency, and the need to sink increasingly deeper wells grew. The result of these emerging land-degrading trends was diminished groundwater availability, which shrunk the area that could be devoted to cereal cultivation. The result was a dramatic decline in yields and Saudi Arabia's return to its preexpansion status as a net importer of basic food stocks. With groundwater dramatically depleted, future development possibilities that require water—and in the desert what does not?—are now significantly reduced.

What Counts: Privileging Money Not Nature

Today the globe's dominant economic system, capitalism places primary value on the generation of capital, profits, and material goods. The bottom line controls the decision-making process of governments as well as corporations. Environmental costs that cannot readily be calculated in monetary terms are either ignored or assigned arbitrary values that usually underestimate the actual long-term costs involved. Insuring extra funds in the pockets of voters and bigger stock portfolios for favored special interests and campaign contributors is the primary objective. Identification of long-term costs and irreversible environmental impacts that might slow wealth accumulation is not viewed with favor. Politicians are often willing to sanction deficits in national budgets in pursuit of short-term political advantage, borrowing from the future in order to support the aspirations and lifestyle of the present. Lester Brown (2001, 21) maintains that in our shortsighted efforts to sustain the global economy, as currently structured, we are depleting natural capital. We spend a lot of time worrying about our economic deficits, but it is the ecological deficits that threaten our long-term economic future. Economic deficits are what we borrow from each other; ecological deficits are what we take from future generations. We engage in reckless ecological deficit spending because we refuse to recognize the price to be paid for borrowing from nature's capital and applying it solely to present benefits. Yet such behavior defies common sense and collective experience. Parents have no problem accepting the idea that they should send their offspring into a postcollege future free of accumulated economic debt. Everyone grasps the principle that you cannot assume more financial indebtedness than you can reasonably expect to repay in the future. No sane banker would lend money on excessively risky terms, although shortsighted credit card managers interested in drumming up immediate cash flow might! But the environment has no bankruptcy rules. Thus thirty-year home mortgages are acceptable, but crushing multiple credit card debts that might have to be repaid by others in the event of an unexpected demise are neither wise nor ethically defensible. Most individuals seem willing to behave responsibly in personal financial affairs, but are reluctant to project the same values into our collective ecological activities. The result is an increasing level of stress in the Earth's natural systems as a consequence of heedless human decisions about the long-term consequences.

As Jody Emel and Rob Krueger point out (2003, 11), employing the precautionary principle in natural-resource development requires weighting variables differently than is usually the case. Above all it requires a values shift from generating material benefits and capital accumulation at any cost toward a greater emphasis on environmental health and sustainability. Critics often challenge the utility of the precautionary principle, arguing that the concept is too vague in definition and its practitioners too sloppy in their use of evidence to make the principle a useful policy instrument. Indeed, some critics invert the uncertainty component of the precautionary principle. When adverse impacts cannot be demonstrated with hard observational data, direct causal linkages between development action and environmental change are contentious, and risks are at best only vaguely known. It is possible to contend that use of the concept results in the loss of development options that itself represents huge costs in foregone material gains. If the fears of adverse impacts prove unfounded or less serious than imagined, time, money, and opportunities to address social inequities may all have been lost. Moreover, efforts to direct money to dealing with potentially unknown and unquantifiable future costs may not turn out to be a justifiable expense (Jordan and O'Riordan 1999, 23).

It is this conundrum that underlies the debate over land degradation. On one side stand the advocates of growth and environmental change, armored in the certainty that massive material benefits will materialize from their development plans and convinced that concerns about future adverse impacts represent fuzzy science at best and unreasoned fear monger-ing at worst. In opposition are groups, fortified by moral righteousness in their defense of the environment, whose concerns about future habitat destruction receive frequent support from emerging evidence of environmental insults and deadly dangers unpredicted and unac-knowledged when projects were begun or new technology was introduced. Somewhere between the two extremes a balance must be found. As a brake on rampant development heedless of environmental consequences, the precautionary principle is a useful tool. It works only in a context of open information, particularly about the uncertainty that surrounds the future impacts of current activities that promise to initiate change in the natural world (Jordan and O'Riordan 1999, 19). Only in this way can the gains and losses that impact both nature and society be adequately, albeit often haltingly and unevenly, identified, addressed, and remediated.

Class: Identifying Gainers, Forgetting Losers

Much land degradation flows from a social bias to the way economic development and envi-ronmental change occurs. Essentially individuals and social groups with greater access to resources dominate the process. Change in the status quo is proposed in a top down fash-ion, with developers, politicians, planners, bureaucrats, and moneyed interests usually dri-ving the process. Consultation with less privileged segments of society seldom begins a proposal for change; nature and its advocates/defenders are invariably relegated to a reac-tive role. Even in contexts where public hearings are mandated or social impact assessment required, these process elements are customarily considered after biophysical, engineering, and financial feasibility studies are completed.

There are practical reasons for this delayed or deficient attention to socioeconomic and environmental concerns. Identifying the gains and the gainers from a proposed course of action is relatively easy. Tracing the more complicated causal connections that link losers in society and nature to change is often difficult and is an unpalatable task for the proposer of change. Designing development that takes account of the enormous variability and resilient capacity inherent in local human and environmental systems requires a grassroots approach. Others may not perceive one community's degradation as such (Warren 2002). An outsider's view of a local context may perceive rampant degradation via soil erosion, whereas many local inhabitants may find much less cause for concern, particularly in households where options other than agriculture, such as migrant labor or petty trading, exist (Batterbury 2001; Warren 2002). If the real cause of much local degradation is to be found in economic, social, and political forces operating outside the local context (Blaikie and Brookfield 1987; Sanders 2000), it is also equally undeniable that local populations are the ones most likely to bare the costs of such impacts. The perceptions of these local "losers," frequently located at a distance from the initiators of and beneficiaries from environmental change, are likely to be quite different from those of experts and managers operating at the national scale. Yet the views of adversely affected people are often not taken into account in meaningful ways. When efforts are made to incorporate local consensus into the development process, the concerns expressed and range of impacts recognized are often greater than those acknowledged by indigenous but nonlocal managers and outside experts and consultants (Sclove and Scammell 1999).

The Sardar Sarovar Dam on the Narmada River in India presents a classic illustration of these problems. First broached as an idea in 1863, the project came to fruition in the heady atmosphere of India's postindependence self-government (El-Bihbety and Lithwick 1998). The project is a multipurpose enterprise, designed to produce electricity (2,700 mw), irrigate in excess of six million hectares of presently cultivated and newly developed agricultural land, bring drinking water to 40 million people, and provide rural development options that would keep people on the land rather than have them move from presently drought-afflicted habitats to other locations (El-Bihbety and Lithwick 1998, 297). It also is a costly undertaking, requiring huge capital investments by the Indian government (US$ 1.3 billion in the 1985 plan) as well very large financial commitments by the World Bank (US$ 450 million) and other donor agencies, as well as a long time to build. As is the case in many development projects, calculating the economic benefits to be gained is easier to determine than the potential environmental and social costs. This is particularly the case with mega projects, which are so vast in scale and so complicated in character that it is difficult to identify all of the relevant variables and their interconnections let alone assemble data not readily available in standard census formats. Even identifying the number of people displaced by the Sardar Sarovar Dam and its reservoir proved a daunting task, an effort compounded by the Indian government's apparent disinterest in assembling data on populations displaced as a result of large dam projects (Roy 1999, 17). With twenty-nine other large dams, 135 medium dams, and three thousand small dams proposed for the Narmada and its tributaries, the problems occasioned by displaced populations are unlikely to disappear (Narmada Bachao Andolan 1998).

The fundamental issues involved in the construction of Sardar Sarovar and other large dams are at least fourfold:

1. What are the ecological and social impacts downstream from the dam?
2. What is the value of resources lost at the dam site and flooded by the reservoir impounded behind the dam?
3. What happens to the people displaced by the dam?
4. Do the benefits generated by the dam offset the costs imposed on the people and places adversely affected by the dam?

Located 170 miles from the sea, Sardar Sarovar will certainly have a significant effect on the landscape, ecology, and human and animal populations between the dam site and the Gulf of Khambhat. While the Narmada Valley Development Project's basic study documents identify downstream impacts to be important, no effort to estimate the cost of these impacts was made. Rather the World Bank in 1985 recommended examining those issues at a later date (El-Bihbety and Litwick 1998, 299). Project planners anticipated introducing fishing activities into the reservoir above the dam, and believed that these benefits would be considerable and would offset losses below the dam. But like potential losses, these gains were not calculated (El-Bihbety and Litwick 1998, 300). At worst, project developers assumed that downstream fisheries' losses would be offset by upstream gains. That the people likely to benefit from the new fisheries might be different from those who lost out below the dam seems not to have occurred to anyone. As a consequence of the failure to identify impacts and impacted populations in major economic sectors located downstream from the dam, it is impossible to know with certainty the level and spatial extent of potential environmental and social damage. Morse and Berger (1995, 375) expect these impacts to be significant, including the possible elimination of the last important *hilsa* (*Clupea ilisha*) fishery in western India. The hilsa is a salmon-like fish that seasonally ascends rivers in order to spawn. Not only would the Narmada River dam impede movement of the fish, but also changes in the flow, temperature, and turbidity characteristics of the stream are likely to have a major impact on aquatic habitats. In the absence of a specific impact study of the issue, all one can say, by analogy with other rivers whose waters have been impounded and diverted to other uses, is that the impacts promise to be considerable.

Significant negative impacts would occur when the Sardar Sarovar's reservoir begins to fill. Among those losses would be the local cultural landscape of sacred myth and mystery (Deegan 1995). For this loss it is difficult to establish a value, but only the most important material artifacts are likely to be moved away from flooded lands. Nonauthentic to some degree in their new locales, such artifacts are unlikely to command the same reverence that they did in original sites where gods literally walked on earth and established material realities. Although the dam's artificial reservoir is relatively small (37,000 ha, 91,000 acres), it will flood 245 villages inhabited by at least one hundred thousand, largely tribally organized people, inundate some of the richest farmland in India, and submerge important forest and biodiversity resources (El-Bihbety and Lithwick 1998). Clearly these displaced people, even if compensated by land elsewhere, are likely to be losers in the project's cosmic calculations. Even in Gujarat, where most of the benefits are likely to fall, Ashvan Shah (1995, 329) calculates that the richest 20 percent of the farming population will garner 57 percent of the project's benefits while the poorest 20 percent grasps only 3.7 percent of the gains. As befits a megascale project, these are nontrivial costs. One reason why it is difficult to estimate precisely what the costs of the project might be is the study process employed by the project managers. Rather than having all of the feasibility studies completed before the project was approved, the project was allowed to proceed with incomplete research and analysis. In violation of its own policy and rules (Morse and Berger 1995, 374), the World Bank committed itself to fund projects that lacked environmental impact studies. In 1987 the Indian government got around its own internal regulations by agreeing to undertake impact studies at the same time the dam was under construction. This astounding decision must have proceeded from a myopic conviction that project benefits were certain to outweigh any conceivable assemblage of adverse impacts. Thus the physical construction work could proceed in the certainty that research results would reward the project's true believers. Just in case the benefit/cost report turned out to be unfavorable, the concrete presence of the dam would make it impossible to stop the project's momentum. For both sides in the struggle over the Sardar Sarovar Dam and the larger Narmada River development project, the issue is fundamentally one of faith rather than fact.

Individuals and communities displaced by dam construction at Narmada were offered compensation in the form of access to land elsewhere. But this land did not always represent the same quality land as that submerged by the dam, and it frequently resulted in families and village communities being scattered in different locations. Resettlement was more complicated still. Most of the land commanded by the main irrigation water distribution canal benefited residents of Gujarat state, while those who lost land to the reservoir lived mainly along the river in Madhya Pradesh and Maharashtra. This uneven distribution of social and environmental benefits resulted in huge pro- and antidam rallies, on at least one occasion directly (although fortunately nonviolently) confrontational at the state frontier (Fisher 1995). Having to leave one state and move to another was a problem for many people displaced by the dam, as was the failure of the Indian government to provide compensation and access to land that would maintain predam living standards for at least 60 percent of the displaced population (Morse and Berger 1995, 373).

Do the benefits provided by a large dam project such as Sardar Sarovar, not to mention the complex of small-, medium-, and large-scale hydraulic projects envisioned for the region, outweigh the costs? This is an exceedingly difficult and complex question to answer, one that both national and international experts have often sought to avoid, at least in part. A number of particularly intractable (or potentially embarrassing) facets of the question were postponed or brushed aside as unimportant or unknowable and never entered the benefit/cost analysis. How one answered the question of the overall contribution of the dam depended on whether one's community stood to gain from the project. For proponents of the dam, not only expectations of material gain undergirded support. A distinct ideology of progress, of bringing the fruits of development and civilization to the impoverished, also contributed mightily (Narmada Planning Group 1989). Critics, usually advocating the rights of tribal communities slated to be displaced by the dam's reservoir (Dreze, Samson, and Singh 1997; Jain 2001; Sangvai 2002), feel equally passionately that fundamental values are at stake, that alternative solutions with less risk of major environmental disruption are available, and that important stakeholders are left out of the calculations in pursuit of poorly defined goals that benefit only special interests. Sober analysis based on sound time-series data is hard to find, but what does exist raises more questions about the viability of the project than are answered. These uncertainties suggest that the magnitude of the problems likely to arise and the adverse impacts predicted to result will not create pleasant reading and happy headlines in the years to come.

Magic Bullets: Imperfect Models and False Analogs

A "magic bullet" is an extraordinary artifact capable of doing things that ordinarily cannot be accomplished. We search for such special help when faced with a situation that is beyond our ordinary powers to control. For North Americans the "bullet" selected is usually a piece of technology. Often the choice of the particular item or practice we employ is ill founded because what works in our own experience and environmental context does not necessary work in other places. Frequently well-developed, if often unarticulated, beliefs are associated with the tools and techniques that we are convinced will solve a particularly nasty problem. If large dams or central pivot irrigation worked for us, we see no reason why the same solutions should not achieve the same results in other places. The bigger the scale of the proposed activity, the more likely we are to envisage its success. When less than perfect results occur, we seldom try to revise the method, and relive the same mistake repeatedly.

An example of the search for a magic bullet is found in the belief, widely held among water managers, planners, and engineers, that rainfall, runoff, and the floods that carry freshwater to the sea often constitute a wasted resource, lost to human use (Adams 1992). The goal of perfectionist practitioners of this perspective is to employ every drop of water usefully. For example, while Asit Biswas (1995) was willing to concede that some Nile water needed to flow to the Mediterranean for ecological reasons, particularly salt balance, he also accepted the goal of retaining as much water as possible and wringing the maximum agricultural benefit from it as being essentially correct. The utilitarian goal that undergirds this approach takes only humans as the objects of and beneficiaries from such intensified use. Improving on nature, overcoming its rough edges and damaging forces, squeezing every drop of material benefit from existing or potential resources, crafting a perfected nature is the goal. The best way to avoid such waste, advocates of rapid economic growth believe, is to employ the absolutely most modern technology to achieve resource management efficiency. One way to do this is to build large-scale dams, which are expected to achieve economies of scale in generating economic benefits.

Yet such projects invariably set in train consequences that have severe environmental and social impacts. As water is diverted from the dam's reservoir through distribution canals to irrigate areas that were either dependent on rain for cultivation or were not cultivated at all, water levels in the river downstream of the dam diminish. Farming becomes less secure as available water in the riverbed decreases to a trickle or stops altogether. In Pakistan, drastically decreased stream flow in the Indus during the last three decades has seen the sea advance in coastal areas to cover an estimated 486,000 ha (1.2 million acres) of formerly productive cropland (Eckholm 2003). Worse yet, people threatened with poverty have shifted into offshore fishing activities that have placed tremendous pressure on rich fishing resources. The diversion of freshwater to other places thus imperils the marine fishery, the coastal delta agricultural activities inundated by the sea, the freshwater fishery in the lower reaches of the Indus, and irrigated agriculture in southern Pakistan in general as dammed water is diverted from river flow to nourish agriculture in the northern and central zones of the country. In Southeast Asia, in the headwaters of river systems, displaced or colonizing farmers often clear steep-sloped land and generate soil erosion rates three orders of magnitude greater than those found on heavily vegetated plots (Douglas 1999, 227). The soil loosened by displaced farmers in the headwaters of a river system returns to haunt dam and irrigation system managers in the lower reaches of the stream. Increased rates of sedimentation produce clogged canals and reduced reservoir storage capacity. This degradation of infrastructure capacity and performance, in turn, reduces the life span of development projects and diminishes the rate of economic return on the capital resources that generated the project in the first place. Costly repairs and expensive investments in maintenance in order to remove silt may retard but can seldom prevent the slow slide of a system into decline.

Once advocated as a panacea capable of solving social and economic backwardness in a single bound, large-dam projects generate a host of unexpected difficulties and impacts including land degradation. Other magical solutions are often sought in institutions rather than technology. The ideologically driven forced collectivization of much of Eastern Europe after World War II produced large fields and converted pasture and woodland to cropland. A dramatic increase in soil erosion and sediment transport was the result. The collapse of collectives and state farms engenders—and the emergence of private tenure introduces—a new set of problems and opportunities in which 30 hectare (74 acre) fields would contribute the most to soil erosion control (Van Rompaey et al. 2003). In northern China, a similar change from collective to private land tenure in pastoral areas has produced more rather than less erosion. Land-use intensification and reduced animal mobility on private land holdings has promoted a significant reduction in vegetation cover and an expansion of sandy surfaces over the last thirty-five years (Jiang 2002; Wu and Long 2002). Thus utopian visions of dramatic material improvement in economic and social conditions generated a countercurrent of severe (but often unacknowledged) environmental degradation. Like many other examples of misplaced faith in the redemptive contribution of modern technology and institutional change, these innovations, particularly when designed on a grandiose scale, tend to cause more problems for the people aiming the magic bullet than they do for the specter they intend to slay.

Lessons to Learn

Humans are intimately associated with land degradation as both its authors and its victims. The Earth's normal variability also figures prominently as a driving factor, since extreme events often interact with land-use systems to produce land degradation. Efforts to reduce or eliminate this variability in the interests of securer livelihood conditions and dependable agricultural yields frequently unleash counterintuitive destructive forces that result in land degradation. The irony of this result is that fatal flaws in human values and behavior often lie at the base of such practices.

We identify five major failings that frequently short circuit efforts to promote development and produce creative, sustainable land-use practices. Failure to take account of impacts on distant areas, a strong tendency to export costs onto future generations, an inability to value nature adequately while emphasizing only immediate economic benefits, a preference for identifying gainers while ignoring losing habitats and inhabitants in environmental change, and an exaggerated faith in the ability of "magic bullets" to solve problems with minimum cost and adverse impact are major variables contributing to land degradation. The role of these variables is examined in the next three chapters.

5

Land Degradation at the Local Scale

EARTH SURFACE attributes reflect interactions among and between natural and human systems. Reflecting the dynamic nature of earth systems, changes occur in either system once critical thresholds are surpassed. To satisfy local needs, individuals usually must alter surface properties to permit the desired land-use activities. Urban, rural, agricultural, recreational, industrial, military, and transportation activities all have specific needs that require humans to modify natural conditions to meet the demands of these chosen endeavors. In this context, this chapter explores land-use change and land degradation at the local scale within the context of private and common property.

Following the Industrial Revolution of the late eighteenth century, nonagricultural impacts affecting land degradation increased in importance. Today, with new technologies resulting in more powerful and sophisticated machinery, the ability to alter the earth's surface significantly to satisfy modern needs is largely only constrained by economic considerations. The following are but two examples: (1) in Japan, with insufficient land available for a new airport, an offshore island was created to permit the construction of a new international airport. The magnitude of the rock materials utilized in the landfill for creating this island necessitated the removal of numerous hills on some small Japanese islands. These hills became sacrifice zones for the new land mass created for the airport complex; (2) in Boston, to depress a major artery that was both an eyesore as well as a barrier to pedestrian movements that negatively affected tourism, an expressway has been depressed below ground level. This created new open public space in the congested downtown area as well as new lands for commercial activity. The huge amounts of earth materials excavated in this public works project were utilized to fill in former quarries as well as to create a new island in Boston harbor where the city's sewage treatment plant is located. In today's world, it is economic and quality-of-life considerations, not technical ability, that act as the major brakes upon transforming the earth's surface to meet specific human needs. In altering the earth to facilitate satisfying the array of human desires, different sets of unintentional land degradation occur that were unknown or of minor magnitude prior to the Industrial Revolution.

Since the beginning of the Industrial Revolution, energy requirements and demand for raw and finished materials have continuously increased. As the industrial and now technology revolutions evolve, in order to satisfy these energy and material requirements, regions and countries are becoming less self-sufficient, until today phrases such as "the global market" are common. The complex movement of raw materials, semifinished and finished goods, and information from source and manufacturing areas to markets increases continually.

Improvements in and expansion of transportation systems, energy production, and mining have accelerated to meet the needs of modern societies. With their expansion and refinements, direct and indirect impacts on land resources are inevitable. The creation of one land mass and the degradation of another one to satisfy air transport needs in Japan is one manifestation of how human actions modify the earth to satisfy demands of the modern age. The Three Gorges Dam in China, which will generate electricity equal to 40 million tons of coal annually, is another. This new energy source will contribute to less air pollution than thermal plants that impact global warming. Yet in the creation of the reservoir, by changes in the flow regime of the river, and as a result of other direct impacts, ecosystems are destroyed and agricultural lands are lost (Chaudhuri 2003). With urbanization exploding, as more people now live in urban areas than rural areas, in tandem with improvements and expansion in the agricultural, industrial, and technological domains, another force has been added in altering the earth's surface. Common properties, under modern conditions, often have degraded as traditional constraints on their use are replaced by often less effective new controls. Governments, especially that of the United States, have encouraged the privatization of these properties for a multitude of reasons, including as a means of curtailing their environmental deterioration.

Unlike private ownership, common property is held and managed in common by groups of users. Common property resources were widespread prior to and during the Agricultural Revolution. While they represent a very ancient system of land use that predates the transition to the industrial age, common property resources still persist today. In the northeastern portion of the United States, most town centers in New England have commons dating back to the seventeenth and eighteenth centuries. Originally used as grazing grounds for the town's inhabitants' livestock, today most of these commons ban grazing and are used as town parks (figure 5.1). Some widespread resources that clearly remain common property in contemporary times are the oceans and the air we breathe. National, state, and local parks, likewise, possess many attributes of common property.

The strong emphasis on privatization of resources as a rational means of accelerating development is a powerful force that today places pressure on common properties. The relations between common properties and land degradation are the first focus in this chapter's inquiry on land degradation at the local scale.

Common Property Resources

The common property resources examined in this section contribute to worldwide food production. This production differs from the direct cultivation of crops, that is, the dominant

Figure 5.1.   Boston Common and Public Garden from space surrounded by the densely developed downtown. The Common is in the lower left-hand corner of the image. (IKONOS satellite image, permission granted by Space Imaging GeoEye, Dallas, VA)

food production arising from common properties usually involves the capture of animals and/or the exploitation of a resource that is fundamental to the production activity. For example, fishermen's activities are almost exclusively directed to the fish they catch; rarely is their attention concerned with the water medium in which the fish live. The water where the fish live is as much a common property as the fish themselves. In this sense, the common property resource of the water is an indirect, enabling resource rather than a direct object of human interest in the fishing ecosystem. In this respect, common property resources closely resemble primary extractive enterprises such as mining rather than an agricultural undertaking. For instance, fishing and the management of wild animal populations are characteristically capture ventures rather than the controlled agricultural activities in which humans directly manipulate domesticated livestock or plants. Today manipulation of common property activities is becoming a reality as governments attempt to manage some common assets to prevent their complete demise.

While the stocks of fish found offshore in the North Atlantic are still treated as a free good, their capture is restricted. Fishing along the east coast of the United States and Canada is controlled, not only in terms of the length and time of the fishing season, but also the types of fish allowed to be caught as well as their size. In addition, through the issuing of licenses, this common property can only be utilized by a limited number of people. The introduction of governmental controls, both U.S. and Canadian, on the utilization of the offshore fishing common property is a modern adoption of traditional controls that evolved with regard to most common property utilization in the preindustrial era. The pastures, marshes, forests, air, and water resources that are managed as communal property share with wild aquatic and terrestrial animals a common feature—they are in many places treated as a free good, open to use and abuse by all. The response of modern institutions to prevent the degradation of common properties remains an ongoing challenge. For example, the long-term interests of governments in maintaining a renewable resource, such as the fishing grounds of the Georges Bank (Canadian and U.S. federal governments) and the EU waters in Western Europe, through restricting commercial fishing are in conflict with the short-term interests of the fishermen. Needing to meet their immediate financial obligations, fishermen find that their livelihood is threatened by the major constraints imposed by governments, which want to bring back the fishing resource to prior levels. The shortened fishing seasons, as well as the placement of large areas of the common property off limits for any fishing, result in continuous disputes between the regulating government agencies and the fishermen. These conflicting interests between the interested parties result from changes in technology and political controls of offshore waters that allow overfishing. Thus managing of common resources under contemporary conditions often require a new set of rules in order to insure the environmental stability of the resource in question.

Characteristics of CPRs

Common Property Resources (CPRs) exhibit three important characteristics. First, most CPRs are controlled by a group of people for the individual and collective benefit of that group. The size of the controlling group varies from small, informal, unrecognized gangs to larger, more formalized clans or tribes, or even to nations. As an example of the latter case, introduced briefly in the previous paragraph, fishing off the eastern coast of North America is today controlled by two nations, Canada and the United States. The two states—as the representative voice of the people—claim the right to exercise authority over the "common" space (the Georges Bank fishing grounds) and to manage the resource in what the federal governments and their experts consider to be the best long-term interest of the countries. Often the national interest is in conflict with the short-term interests of the fishermen since restricting access to fishing results in the unemployment of fisherman. The federal governments' control of the common property—the fish—is viewed as crucial for the future sustainable productivity of the common property. National parks, national forests, state-managed rangelands, and coastal waters (the 320 km or 200 mile limit) are typical examples of nation-scale communal resources. The Maine lobster gangs described by Acheson (1988), who defend their economic interests by destroying the lobster gear of unauthorized interlopers, and whose management system has proven effective despite the worries of professional managers, are representative of the informal but persistent patterns of CPR management typical of local-scale resource use. The tradition of notching the tail of egg-bearing female lobsters is another example of management strategies that evolved to protect the common resource. When a female lobster that is carrying eggs is caught, most lobstermen notch the tail. As it is impossible to sell a lobster with a notched tail, this insures that an ample supply of lobsters will hatch for future harvesting. If this lobster is caught in a trap for the next two years (generally, this is as long as a notch remains visible) it will always be released. Notched lobsters have demonstrated they will contribute to the production of juvenile lobsters. This example demonstrates evolved strategies that protect the long-term viability of the common resource, but shows that these same practices may have a slight negative impact on the short-term (less income) economics of individual lobstermen even in the twentieth and twenty-first centuries.

Second, in most CPR systems it is difficult to restrict access to the resource. Local-scale social groups attempt to do this by carefully defining, usually on a genealogical or residence basis, who is part of the community that has a right to exploit the group's resource. "Outsiders" trying to gain access to the resource commonly cannot do so unless—by marriage, long apprenticeship, or contract—they are able to establish a working relationship with members of the group controlling the resource. But for members of the fortunate group, access to the resource is literally a birthright. As long as the individual possesses the right genealogical or residential qualifications, and has the requisite skills, technology, labor, and capital, barriers to using resources are few. In theory, this makes it difficult to avoid overexploiting the resource base in question. In practice, social controls often operate, together with capital and labor limitations, to reduce access to and abuse of common property resources to sustainable proportions.

A third feature of common property resources is that extraction of the resource by the individual inevitably means that the resource base available to others is reduced. Thus, the grass consumed by the animals of one herder is not available to nourish the animals of another herder. As long the intensity of use is low, this characteristic of "subtractability" (Berkes 1989) has no negative impact on either the environment or the group. When intensity of use increases, competition between individual resource users can produce both conflict and resource degradation. Usually, communities using CPRs have developed rules that promote reasonable equity in access to the resource while minimizing conflict and preventing degradation. Were groups not able to develop such institutions, it is unlikely that these communities would be able to survive. Since traditional CPR user groups have been able to maintain themselves for long periods, it follows that they have managed to create sustainable systems of common property management.

To a considerable extent, sustainable CPR systems strike a balance between communal responsibility and private advantage. Some scholars have stressed the difficulty of striking such a balance. Hardin (1968) argued that individuals will always be tempted to extract more from the common resource store than is their fair share, because to do so maximizes personal benefits while at the same time it shares costs among the entire community. Thus, when placed on common pasture, individually owned livestock will produce milk and other products for the owner but will consume grass that is potentially available to all. As more able and successful herders place more and more animals on the common pasture, increasing amounts of the resource are diverted to their gain. Any decrease in forage quantity and quality is a cost that is shared by all. According to Hardin, the result of this self-aggrandizing tendency is the inevitable deterioration of the resource base. Hardin called this triumph of greed and personal gain over communal good, environmental quality, and sustainable use the "tragedy of the commons."

He believed that rapid population growth was a major stimulus to adverse change in common property systems, and conspired with the structural problems indicated above to destroy prospects for sustainable resource use. Hardin's solution to the problem was either privatization of the common property resource, to make the user bear the real costs of excessive use, or the centralization of CPR management in the hands of governmental agencies with sufficient strength and wisdom to protect the public good. Whether through privatization or centralized management, the essential objective advocated by Hardin is to prevent open access to CPRs.

These three general features of common property resource systems, with respect to the environment and land degradation, are examined in the remainder of this section. Common property aspects of fishing, wild animal exploitation, wetlands, forests, grasslands, and water resources are introduced. They are considered in the context of whether they are large-scale or small-scale CPRs.

Management of Large-Scale CPRs

Large-scale CPRs frequently come closest to meeting Hardin's concept of a "tragedy of the commons." This is a consequence of the difficulties encountered in large-scale systems in restricting access to the resource. Some implications of this problem with regard to environmental deterioration are now considered.

The world's primary source of fish is the ocean, followed by freshwater lakes. In most cases, fish stocks that are located in either oceans or lakes are treated as common property resources and until the establishment of a 320 km (200 mile) economic exploitation zone controlled by riparian states, only limited and generally ineffective controls were established on the size of the catch from any particular fishery. The result of this strategy of generally unrestrained hunting of fish as the technology for catching fish improved has been a drastic and steady decline in global fish stocks. The path of degradation in the size of the fish population has followed a standard pattern, with overexploitation concentrating on the most desired species until their numbers drastically decline. Only then, and often very late, are efforts made to limit the size of the catch. With the decline in desired species, a switch follows where a second set of somewhat less desirable fish species are caught. Ultimately, fish such as shark and monkfish, which were once despised or limited to ethnic cuisine, become the gourmet choice of the elite. Eventually these "second choice" fish also begin to exhibit a dramatic decrease in size and availability, such as the contemporary reduction in the shark population. These trends are paralleled by technological innovations such as larger and more seaworthy boats, improved sonar to locate fish schools, and factory ships to process and preserve fish catches far from the fisherman's base. Such developments enable fishermen to travel further and to catch fish that their artisanal ancestors never would have had the ability to pursue. In the over fishing, the fish resource is drastically degraded. Eventually controls must be initiated to prevent the total collapse of the resource. In North America, to reduce the economic losses experienced by fishermen due to the quotas placed on the fishing tonnage to protect the common resource as well as areal restrictions to portions of the fishing grounds, the government has provided different types of financial aid. Nevertheless, fishermen have suffered financial losses due to the restrictions placed on fishing and their numbers are still decreasing.

Management of fish resources is difficult because territories often are hard to delineate and control, and because fish often migrate from one area to another over their life cycle. Thus access to the resource is difficult to control and, once caught, the fish catch benefits primarily those who caught it. Some examples of successful management of fish resources do exist. Ruddle (1989) demonstrates that Japanese coastal fishermen have a complex system of village tenure over and rights to nearby fisheries. This is accomplished in large part because the Japanese legal system makes no distinction between tenure on land and tenure rights at sea, and because modern national legislation controlling exploitation of the marine fisheries resources is based on customary practices. These practices guarantee sharing of sea space on the part of different communities, today organized as cooperatives, on the basis of different techniques employed to catch fish and of the right to take only certain types of fish. Outsiders granted access to tenured sea space are restricted in the timing of their access, are limited in the technology that they can employ, and are required to pay a fee for the fishing privilege. Like the lobster "gangs" along the Maine coast, such local-scale systems have proven capable of preserving the basic fish resource. Where local cooperative systems lose control of the CPR, as in the case of some of the coastal fisheries in Turkey (Berkes 1986), overexploitation by large-scale industrial fishing enterprises results in drastic declines in desired fish species.

Water Resource Degradation

The fish resource is only as viable as the water quality medium within which the fish live. Many examples exist in which the degradation of water quality and quantity negatively impact the aquatic environment. Except in some arid environments, where the principle of first use in time conveys a proprietary right in subsequent allocation of use, water is most commonly treated as an open access, common property resource. Overextraction of groundwater from the Ogallala Aquifer (discussed later in this chapter) is an example of how overzealous use of a common property resource creates difficulties for long-term sustainable use. Historically, as well as in many countries today, communities and factories located along their banks or shores used rivers and lakes as basic waste disposal systems.

Such was the case with the small tanning industries in the North Arcot district of India; the effluent from those factories polluted the region's water resources (Bowonder and Ramana 1986). Pollution is particularly threatening in this Indian region, as it is a low rainfall area. Some three hundred local tanning enterprises, limited in size by government legislation and local tradition, traditionally employ vegetable-tanning agents derived from forest vegetation. Deforestation drove up the cost of these agents and as a result the factories shifted to faster chemical-tanning agents. Chrome tanning became favored. This process uses over two hundred chemical agents including cadmium and arsenic (Bowonder and Ramana 1986, 3). Because of the low rainfall, stream flow is low during much of the year. During these dry periods, chemical agents become concentrated in the stream channels. Additionally, the region's soils are highly susceptible to infiltration, so during monsoon rains large quantities of contaminants percolate into the groundwater. Clean groundwater is an essential ingredient for the tanning industry in this area as well as the major water supply for domestic consumption. The unrestricted dumping of polluted effluent threatens an unregulated common property resource—the groundwater—as well as the viability of the basis of the local economy, the tanning industry, and the health of the local population.

Similar impacts on water quality characterize the industrial history of most "developed" countries. As an example, the waterpower sites along the Blackstone River in southeastern New England were privatized during the early stages of the Industrial Revolution and provided the basis for the early industrial growth in the United States. Resulting from this industrial activity, the Blackstone became one of the most severely polluted rivers in the country. This occurred because no control was exerted over the industrial effluent dumped into the river by each mill complex within the river basin. This pollution caused a heavy toll on marine resources and water quality of Narragansett Bay into which the Blackstone empties (Lewis and Brubaker 1989). With new water quality laws, municipalities have been required to build waste treatment plants. The intent is to insure that the water from the treatment facilities is of the same quality as the water removed from the system. The change from untreated to treated waste flowing into the Blackstone has succeeded in greatly reducing the discharge of suspended solids as well as other pollutants. In addition, with the changing economy of the area, most of the mills along the river have gone out of business, significantly reducing the flow of untreated waste into the waterway. Further improving the waterway was the creation of the Blackstone River Valley National Heritage Corridor, a new type of national park initiated with the intent to make the waterway into a tourist attraction that preserved the region's natural and historical heritage as suburban population expanded. Today the water quality of the river has greatly improved and boating is now possible. Yet, the historical past has a strong impact on the environmental situation. Heavy metals, including cadmium and copper, and PCBs remain trapped in the river sediments, especially in the former millponds. Despite the deindustrialization that has taken place with the collapse of the region's textile, electroplating, and shoe industries, their historic pollutants remain trapped in the sediments of the millponds. Major rain events have the potential to stir up these sediments. Thus a major focus in the basin has been to repair the dams at the abandoned mill factories to prevent a new cycle of scour that would send these "trapped" pollutants downstream and ultimately into the sea. The result is a legacy of degradation that has been inflicted on the current generation. This legacy imposes enormous costs on the entire river course if its discharge is to be cleaned up sufficiently to reach environmental standards approximating the water quality of the natural river system as it existed before industrialization. Yet the need to preserve the dams so as to keep the pollutants enclosed in the pond sediments prevents the Blackstone River Basin from obtaining the environmental characteristics that previously allowed many fish species to spawn in its waters. The dams present an impenetrable barrier to migratory fish. Atlantic salmon, American shad, alewife, and smelt—the anadromous fish once common in the river (Stolgitis 1991)—are today conspicuous by their absence. Nor will they return in the near future. Even if the Blackstone's water quality was improved to the point at which environmental conditions were suitable, the barrier effect of the dams prevent the migration of fish upstream from the oceans. To maintain the water quality of the Narragansett Bay, it is critical to keep the toxins trapped within the millpond sediments.

The Blackstone is a small stream, only 72 km (45 miles) long, the management of which is made more difficult by the problems of coordinating the activities of two state governments (Massachusetts and Rhode Island). The difficulties of managing a larger water body with multiple political jurisdictions such as the Mediterranean and Baltic seas can be enormous. The Baltic provides an example of the problems inherent in managing a large-scale CPR from an environmental perspective. In the not-so-distant past, nine sovereign nations surrounded this sea, making common policy on its management a difficult task. With the expansion of the EU, a more coordinated management policy is now possible for the region. Helping to coordinate decisions regarding the management of the Baltic fishery, the International Baltic Sea Fishery Commission (IBSFC) has been established. This institution includes all of the nine nations on the Baltic with Russia the only state not part of the EU. Fishing is an important economic activity in the Baltic, and industrial and agricultural changes in the countries bordering the sea have had a major impact on its marine environment. Some major changes occurring in the past sixty years are associated with eutrophication. Nitrogen, herbicides, and pesticide increases in surface water runoff (Fleischer et al. 1987), a fivefold increase in metal concentrations (Hallberg 1991), industrial wastes from the pulp and paper industry (Hansson 1987) among others, and locally significant nutrient discharges from a small but growing fish farming industry (Ackefors and Enell 1990) have created conditions for major change in the Baltic marine environment. Eutrophication has been the logical consequence of these new and increasing inputs to the sea, and many changes in the species composition of the flora and fauna have resulted.

The increased nutrient matter in the Baltic has promoted more primary production among the phytoplankton and zooplankton population. This increase in available food has sparked an apparent increase in the overall fish population. The increases in cod, herring, and sprat have benefited the local fishermen. In coastal areas, reports indicate that pikeperch's abundance have increased while whitefish have declined (Hansson and Rudstam 1990). Overall, despite some fluctuation in the species composition of fish population, the Baltic fish catch increased more than tenfold through the 1980s over fifty years (Elmgren 1989).

From the human perspective, with the increase in fishing, it would appear that eutrophication represents an environmental improvement. Specifically, eutrophication appears to be an act of *creative destruction* in which increased nutrient loading results in more primary production and larger fish stocks. But the picture is far more complicated than this optimistic assessment would suggest. Some of the increase in fish production is associated with more intensified fishing over the last century (Hansson and Rudstam 1990), coupled with limited knowledge of the size of the Baltic fish stocks earlier in the century. The lack of satisfactory time series data may mean that fish population increases attributed to the greater food stocks produced by eutrophication are illusory. In part, the greater success of human predation on fish is likely to be a result of the dramatic decline in the population of three species of seals and one small whale that were major competitors with humans for Baltic fish. These marine animals captured 5 percent of the primary production at the turn of the twentieth century, but today vestigial remnants of the seal population only account for 1 percent (Elmgren 1989). Moreover, it is now clear that fish ladders, once considered a successful device to insure the survival of Atlantic salmon (*Salmo salar*), have not overcome the obstacle posed by dams. By cutting off the spring freshwater pulse that signals the existence of food and nutrient conditions suitable for spawning, the hydroelectric industry has had a substantial negative impact on salmon. Combined with intensified fishing of salmon, the pulse-removal impact of dams has reduced the freshwater catch of migrating salmon between 50 to 90 percent compared to the 1950s (Jansson and Jansson 1988, 134). Boating, ship, and ferry movement along navigational corridors and the construction of harbor and marina facilities also have a significant impact on coastal aquatic vegetation. Since these areas are important sites for spawning, larval survival, shelter, and feeding possibilities for many species, population declines are common; but at the same time, other species appear to respond favorably to these same altered conditions (Sandström et al. 2005). Eutrophication, by creating favorable conditions for heightening the growth of algae and other organisms, decreases the oxygen supply. This, in turn, can have serious impacts on the composition of marine species, the availability of food supply for animals at higher trophic levels in local food chains, and the conditions needed for successful reproduction. Cod in particular are susceptible to these changes (Hansson and Rudstam 1990, 125), but other commercially valuable species such as herring are also likely to experience negative impacts. Thus, the balance between positive impacts that promote increased primary production and negative conditions that lead to sudden decline is often delicately poised. The result can be a sudden shift of the system into a *destructive creation* mode or environmental deterioration, with an abrupt collapse in the common property resource being managed. The difficulty of managing this complicated system in the face of scientific uncertainty has led Elmgren (2001) to call for a management strategy that constantly monitors environmental parameters and readjusts management decisions to take account of new data and nature's evolving responses to human impacts.

The Colorado River CPR illustrates some unforeseen consequences of trying to manage a river system while meeting the demands of all the states within its basin. Water supply for urban areas and agricultural irrigation, hydroelectric power generation, and recreation demands have all resulted in dam building on the Colorado. Since the construction of the dams, a series of environmental consequences have occurred. There has been a loss of sand along the channel banks as most is trapped behind the dams; changes in the flow regime and water temperature of the river have resulted in the extinction of some native fish species and the invasion of exotic fish and plant species (National Research Council 1996). In an attempt to restore dammed river systems to more natural conditions, new adaptive management strategies are implemented in a number of river systems in the United States. One strategy to approximate the natural regime of the now controlled rivers is to release large quantities of water to simulate natural flood conditions. The erosion and deposition associated with these high flows partially restores some predam environmental conditions along the river channel (Blakeslee 2002). Of course, the release of these waters reduces the water supply for other demands such as electricity generation, while the sediment supply is still restricted since sediments once deposited in the reservoirs behind the dams remain trapped. The result is that land degradation always is associated with the CPR management of rivers once dams are constructed. To date, strategies to mitigate the negative impacts are only partially successful.

Wild Animals

In most countries, wild animals are a common property resource. In many industrialized countries those wild animals valued for recreational hunting purposes are managed relatively carefully with a view to controlling total population numbers and insuring adequate hunting during the designated hunting season. Deer populations are continually monitored in the United States with hunting seasons being lengthened or shortened occasionally to keep the population within the desired magnitude. However, most wild animal populations in most places are not managed by human regulations.

Elephant hunting in the Congo exemplifies an animal resource that has undergone several stages of exploitation (Kisangani 1986). Before European colonization, elephants were a meat resource for the BaMbuti (pygmy) forest people. They captured the animals using either nets or bows and arrows. There were minimal competing groups since other inhabitants were not familiar with the forest environment and they feared becoming lost if they went deep into the trees. Technological limitations prevented the BaMbuti from overexploiting the elephants. During the Belgian colonial era overexploitation was avoided by Draconian measures—colonial authorities prevented natives from hunting in national parks and reserved lands and successfully limited the number of licensed hunters. Hunting was restricted to elephants older than fifty-five years. Given the long gestation period of the elephant and the eight to ten years required before reaching sexual maturity, restricted hunting of younger age cohorts preserved the reproductive structure of the elephant population. Increased off-take from younger age cohorts only began when economic recovery after World War II and late 1950s independence processes relaxed government control. Postindependence efforts to establish small indigenous cooperative hunting groups failed as the demand for ivory in the international marketplace increased. African elephants were particularly vulnerable to ivory hunters because both males and females have tusks, thus exposing the mature breeding population to disastrous pressure. Elephants were killed indiscriminately by poisoned water and fruit, regardless of age and sex, in many parts of the Congo and East Africa, as opposed to the selective reliance on firearms to kill mature males during the colonial period. At independence, the Congo had about 150,000 elephants throughout the country (Kisangani 1986, 156). Contemporary herd figures do not exist for the Congo, but except in a few isolated parts of the country they have almost completely disappeared. This rapid decline in Congo's elephant population mirrors the situation in much of sub-Saharan Africa.

A contrasting example of an animal that has received minimal management attention is provided by coral reefs. Marine animals are common property resources, and coral reefs are a vital element in many tropical coastal ecosystems. Coral minimizes coastal erosion in many tropical areas. Storm waves expend most of their destructive energy on the coral reef and not the shoreline. When portions of the coral reef are destroyed by a storm, over time, the coral slowly grows and repairs the damage. Once corals are destroyed, coasts experience accelerated erosion, a direct form of land degradation. Corals are also important marine ecosystems because of the diversity of species they help to sustain (Kuhlmann 1988; McClanahan 2002). In southeastern Asia, major portions of human protein intake come from the marine species that live among the reef. The algae produced in coral reefs sustains many economically important fish, pearl mussels are often harbored by coral reefs, and tourists are attracted to the aesthetics and biodiversity of coral habitats.

Yet corals are highly sensitive to fluctuations in temperature and to changes in water quality, especially turbidity. Corals throughout the world are at risk. Deforestation in nearby terrestrial river basins increases sediment delivery to coral areas, which reduces sunlight reaching the coral. Dredging operations around harbors or for mineral extraction likewise makes coastal waters increasingly cloudy. Without adequate sunlight, reef building cannot continue. Pesticides and herbicides from agricultural operations, industrial effluent, insect control spraying, blast fishing, physical damage and oil discharge by passing ships, increased water temperatures from water releases of coastal power-generating plants, and removal of coral for souvenirs or the growing hobby of marine aquariums all have negative impacts on coral survival (Ward 1990). The impact is analogous to poaching of wild animals whereby the short-term users benefit while the larger community suffers.

Management of Small-Scale CPRs

Small-scale CPRs have a long history in much of the world. They have been the primary way in which common property resources have been managed over time. It is a fundamental difference in scale, and therefore in the immediacy of local control, that distinguishes small-scale CPRs from the larger-scale CPRs. Small-scale CPRs are generally managed successfully because the community that manages the resource and the beneficiaries are synonymous. Any violations of use rules and any negative change in the quality of the resource being exploited are observable to everyone. Action can be taken quickly either to exclude those who mismanage from access to the resource or to lighten the pressure being placed on the resource by establishing temporal or quantity limits on use of the resource.

Many examples of small-scale CPRs exist (National Research Council 1986; Berkes 1989). Draz (1990) examined the *hema* (protection) concept as applied to grazing, fodder, and forest resources in the Arabian Peninsula and found approximately three thousand contemporary or historical examples in Saudi Arabia. Most of these hema zones were organized at the village or tribal level and relied on local sanctions and leadership to enforce compliance with community rules. Hobbs (1989) reports the existence of a strong environmental ethic among the Khushmaan Bedouin of the Egyptian Red Sea Hills that has developed out of long familiarity with—and sometimes abuse of—their native habitat. In Syria the traditional concept of protection has been resurrected after decades of neglect to form the basis of contemporary rangeland management (Shoup 1990).

In Morocco the *agdal* regulates pasture access. An *agdal* is a pasture area that is subject to explicit rules governing access to the resource. The community group controlling the agdal establishes who can graze, determines when grazing can begin and when it must end, and sets limits on how many animals may be placed on the pasture. In the western High

Atlas, at Oukaimedene, an agdal has been in operation since the seventeenth century under conditions of environmental stability and with remarkably little conflict (Gilles, Hammoudi, and Mahdi 1986). The agdal is especially important in controlling grazing in mountain environments. Both the traditional pattern of nomadic pastoralism and village-based transhumance depend on the existence of productive high-altitude pastures upon which herds can graze at some point during the summer dry season. Management of an agdal is by consensus and is exercised by a controlling council and one or two particularly knowledgeable local resource managers. As long as community solidarity is maintained, their decisions are easy to enforce. Whenever consensus breaks down, however, conflicts emerge that undermine the viability of the system (Artz, Norton, and O'Rourke 1986).

Often these disputes lead to the intervention of central governmental agencies. At other times these nonlocal institutions intervene in pursuit of a differing definition of the common good. For national institutions, the nation-state represents the best interests of all its citizens, and so should control the use and management of common property resources. In North Africa, this led colonial powers to replace tribal order with colonial disorder (Bedrani 1991) as traditional CPR institutions were undermined in favor of European settlers and local privileged groups (Bencherifa and Johnson 1991). Lands alienated from community control placed particularly severe pressure on those areas that continued to be managed as CPRs. Few local CPR groups were as successful in resisting the alienation of control over and access to local resources as the Bontoc and Kalinga rice farmers of northern Luzon, who managed to repel the efforts by the Philippines government to construct a series of hydroelectric dams in their territory that would have flooded their fields (Malayang 1991). Most local CPR groups have experienced difficulties because their tenure rights to resources were not formally documented, and so they had no legal status with the colonial or postindependence governments. Even where such rules were codified, as with the *dina* laws regulating the interactions of herders, agropasturalists, farmers, agrofishermen, and fishing groups in the inland delta of the Niger River in Mali, postindependence government intervention has undermined the traditional local authority system of masters of land and water who monitored use and settled disputes (Moorehead 1989). These management institutions were already under stress from more than a decade of drought, but government institutional intervention has contributed to substantial environmental decay. In all such situations in which a central government is unable to replace a functioning local CPR system successfully with a management regime of its own, the result is a scramble for access to the resource, which results in severe and rapid land degradation.

In India, disputes over the legal status of village forest councils have placed in jeopardy small-scale CPRs that were effectively managed. These local groups were successful because they were able to limit the amount of wood taken, restrict hunting and collecting in the forest to appropriate seasons, protect young animals and plants until they could reach harvestable size, forbid the cutting of "keystone resources" such as fig species that were valuable food sources to wildlife and domestic stock, and completely protect special places that could serve as genetic reservoirs for forest regeneration (Gadgil and Iyer 1989, 247). Income from user fees was reinvested in employing a watchman to protect the CPR. The role of spiritual values in nourishing successful management of CPRs in this and other instances is particularly important, since these values create a moral imperative that justifies individual sacrifice in favor of a common good. Gadgil and Iyer (1989, 241–42) argue that Indian caste society played a major role in successful CPR management too. These tightly knit groups were linked to the exploitation of particular ecological niches, yet were connected in multicaste villages into a network of symbiotic relationships. Each group was sufficiently small to achieve consensus in the use of its particular CPR, but could not survive if it failed to respect the CPRs of other groups whose specialized production was essential to community well-being.

Acid Rain and Forest Preservation

Located just east of the German-Czech border is the city of Most, the site of a large open-pit coal mine. Some of the coal from this area is exported to Germany. In addition, coal from this border area in the western portion of the Czech Republic is utilized in the Czech Republic to produce electricity. Much of this electrical energy is then exported to Germany. A short distance west of Most, just over the border in eastern Saxony, is the Erzgebirge (Ore Mountains). The forests that cover these mountains are primarily evergreen, especially along the higher elevations. The Saxony Forest Service (Sächsische Landesanstalt für Forsten 1994) owns and manages most of the forested land in these mountains. As such they are a CPR.

The environmental dynamics occurring in this portion of eastern Germany and the western Czech Republic illustrates how economic factors contribute to a degrading state forest (CPR). The critical environmental period for the forest is during the winter months when the Erzgebirge, especially in the higher elevations, is often covered by fog. The German state of Saxony imports large amounts of electricity from the Czech Republic. This electricity is generated by fossil fuel-burning power plants in the vicinity of Most. The direct costs of importing this power are less than the costs of generating electricity in Germany. Thus, it appears to be a good economic decision for Germany to import this power from the Czech Republic. But the economics of this decision become more complex if the increased environmental costs of managing the forests in this border area are considered.

"The dieback of entire forests in the recreational Oberes Erzgebirge area has greatly reduced its attraction to tourists" (Sächsische Landesanstalt für Forsten 1994). The causes of this devastation, acid condensation and bark beetle infestations, are directly linked to anthropogenic factors associated with the Czech Republic's production of electricity, a large proportion which is exported to Saxony. During the winter months, winds blowing from the east are polluted with the acidic exhausts from the power plants. The pollution into the atmospheric CPR is treated as a noncost in the production and export of the electricity. As the winds rise over the Erzgebirge, condensation in the form of fog often occurs in the winter months. These acidic fogs bath the evergreen forests. Over time, the soils in the area become more acidic and the trees covered by the acidic fogs for days are weakened and eventually die. The weakening of the trees makes them more susceptible to the bark beetle. In an attempt to control this pest, the Saxony forest service identifies all diseased trees and cuts them down. The result is that today large areas of the forests in these mountains have degraded and a large proportion of the forest is now characterized with an open canopy. In an attempt to save the forest, the Saxony forest service at its experimental tree nursery are raising plants for planting in the declined forest areas in the Erzgebirge (Sächsische Landesanstalt für Forsten 1994, 23). In addition during the summer months in an attempt to lower the acidity of the forest's soils, protective lime applications are applied to the forests thought airdrops by both helicopters and airplanes (figure 5.2). If the costs of the damage to the CPR forests of Saxony along with the costs of tree planting, tree clearing, liming, and other expenses resulting from the acid condensation due to the power generation were incorporated into the costs of the electricity, it is likely that the importing of electricity from the Czech Republic would not be the economical bargain that it appears to be when only direct costs are considered. Because the forests are CPRs, the Saxony government picks up the costs of combating land degradation through the use of general tax revenues. Even with the German concern/love for their forests and the massive governmental interventions attempting to offset the negative effects of the acid fogs, the forests in the Czech-German borderlands continue to degrade. This example illustrates how the preservation of this CPR is threatened by activities both internal and external to the country. Interestingly, the forests on the Czech side of the border are likewise suffering degradation for the same reasons. But the Czech government is not able to allocate the same amount of resources to try to offset the damage, as is the case in Germany. Thus the economic benefits derived from the sale of electricity are offset to a degree by the environmental damage to their forests. An approach to incorporate the costs of environmental damage to these CPRs in electricity costs needs to be initiated if these forests are to be preserved. If such costs were incorporated into the generation of the electricity, it is likely that pollution controls would be initiated at the power

Figure 5.2. An airplane loading lime in Saxony, Germany, to aerially distribute it over the forest to counter the effects of acid rain. (Photograph by L. Lewis, August 1996)

plants, which would curtail the damaging acidic fogs that are the primary factor degrading these forests.

Mismanaging the Range

Animal husbandry is an important component of agricultural systems. Not only do domesticated animals provide important food resources, but also they supply fiber, leather, tools, fertilizer, transportation, and traction power to the human communities that utilize them. Animals are kept in either of two modes: as a component of the crop production system of largely sedentary groups, or as the primary focus of activity of more mobile herding communities. Our concern here is with nomadic pastoralists who exploit the grass resources of rangelands using extensive techniques. In extensive systems, it is the animals that move to the fodder resources, thus expending energy to reach and convert plant material into products that are valued by and useful to humans. While efforts to raise animals more intensively are ancient, the costs and transportation problems created by attempts to bring food to stationary animals made large-scale intensification exceedingly difficult to achieve. Only in the past two centuries has significant intensification proven possible, and in many instances these efforts to achieve concentrated intensive production have generated significant environmental problems.

Both creative destruction and destructive creation processes have operated in animal husbandry. Creative destruction has been the least common occurrence in recent times, but it has taken place in several settings in which traditional pastoralists have altered environmental conditions in their favor over long periods of time. Pastoralists have accomplished this by promoting the growth of grass and particularly valuable fodder trees and shrubs at the expense of less desirable woody vegetation. Similarly, sedentary farmers have created sustainable agricultural systems based on creative destruction principles in which animal manure was a vital input that maintained long-term soil productivity. Unfortunately, destructive creation is the process frequently encountered in animal husbandry. This land-degrading tendency is often associated with futile efforts to intensify production. These intensification efforts may arise spontaneously from within the animal-keeping community, but more commonly they flow from land-degradation pressures set in motion by the misguided efforts of distant groups to enhance production (Grainger 1990, 76). In many instances, these intensification pressures convert *critical zones* for the herder, usually the dry season pasture resources, into *sacrifice zones* that are extremely difficult to rehabilitate. We examine several examples of destructive creation in both highly industrialized and modestly industrialized countries in this section.

Pastoral nomads practice an extensive mode of animal husbandry. Animals and herders move together between seasonal pastures in order to find adequate grass and water. Most often the lands that pastoral nomads use are too risky for settled communities to exploit via crop production. Either water is too scarce, rainfall is too scanty and variable, growing seasons are too short, slopes are too steep, or soils are too infertile to make farming a viable enterprise. Under such restrictive conditions, specialized herders of animals are often able to extract a living from such agriculturally marginal environments by moving their herds frequently and flexibly from one pasture to another (Johnson 1969).

By engaging in frequent seasonal location shifts, herders are able to avoid prolonged impact on critical zones. By far the most important zones for nomadic pastoralists are the dry or winter season pastures that nourish their herds when natural fodder is limited. Equally important are the watering points in these dry season zones, and the grazing areas within reasonable distance of each well, spring, perennial stream, or cistern. For "vertical" nomads, who rely on movement up and down slope to escape seasonally dry conditions, the critical zones are the high mountain pastures where temperatures are cool, grass is relatively abundant, and water can be found. These pastures set an upper limit on the number of animals that the community can keep. Good quality dry season pastures insure that newborn animals will likely survive and will be available to exploit the wet season pastures in the lower elevations at other times of the year. Institutions have evolved to control the number of animals that can be kept in these critical pastures (Bencherifa and Johnson 1991), and pastures can be closed by tribal elders if grazing conditions are too poor to sustain the herd in a given year in order to insure adequate grazing in subsequent years. Often the animals moved from lowlands to highlands are special breeds. Barth (1961, 6) reports that when the Basseri tribal migrations were halted in Iran, massive mortality in the tribe's sheep herds occurred. The sheep were incapable of withstanding either the winter cold of the uplands or the summer heat of the lowlands. Movement between the two areas was essential for the well-being of the herds. In effect, the animals and the herding system had coevolved into a structure that required movement for survival.

Activities that expanded the amount of grassland at the expense of forest in such uplands can be regarded as an act of creative destruction as long as the impact favored more grass for more (but not too many) sheep and goats, without setting in motion negative feedbacks such as soil erosion that destroyed the growth medium for the desired pasture grasses. Equally important were the institutions and values that promoted responsible use of the environment, protected useful fodder trees, and established guardians to control use of limited resources (Hobbs 1989).

In "horizontal" nomadic systems the critical resources are not alpine pastures but rather the grazing land around wells and the pastures in wetlands and on floodplains of permanent streams. Tribal groups return to these areas after the rainy season is over, floodwaters have retreated, farmers' crop cycles are complete, and fields are in fallow. Because farmers cannot keep enough animals to use all of the potential dry season fodder, and because only one crop can be grown each year due to climatic factors, a niche exists that pastoral nomads traditionally exploited. Protecting access to that niche was of vital importance to the herder. In the traditional system, access was never threatened to these lands during the postharvest period because the farmer needed the dairy and meat products and manure of the migratory herds to enhance human nutrition and maintain soil fertility. In the inland delta of the Niger River in Mali, these relationships were codified into a legal system that protected the rights of all parties and formalized the timing of their respective periods of use (Gallais 1972; Bremen and de Wit 1983). In eastern Libya, formalized understandings existed between noble and client lineages about who had to leave drought-stressed critical zones adjacent to permanent watering holes in order to insure that there always was a balance between human population, animal numbers, biomass carrying capacity, and water availability (Peters 1968). Even more elaborate stratified social structures among the Touareg had in part similar objectives (Bernus 1990, 152–55). The result of these devices was to protect the critical dry season pastures from severe pressure in stress periods, and to reduce herd sizes to levels that would permit vegetation regeneration when a cycle of better conditions returned.

Good conditions during the rainy season were handled in just the opposite fashion. When pasture was rich in any one tribe's or clan's traditional territory, other herding groups were allowed to enter the district and help consume the bounty. Access to high-quality pasture, particularly in the spring when lambs and calves are growing out, is essential to a nomadic pastoral operation. But the sporadic spatial distribution of rainfall in drylands insures that some herders will lack grazing in their territory in any given year. Thus, allowing other groups access to an abundant seasonal and temporal resource when and where it exists enables a group to claim access to the resources of its neighbors at a future time when its local resources are scanty. The result is the open grazing system described by Perevolotsky (1987) in the South Sinai, one that is replicated in most traditional grazing areas throughout the world.

In many traditional nomadic pastoral systems fire played a major role in environmental management. Burning the previous year's unconsumed dry growth of perennial grasses just before the next rainy season is a common practice in much of sub-Saharan Africa (Carr 1977, 17–19). Although poorly timed and too frequently repeated burning can destroy vegetation cover and promote soil erosion (Goldammer 1993), the wise use of fire has many advantages (Stocks and Trollope 1993, 320–21). It stimulates perennial grasses to produce new shoots, thus jump-starting growth. Fire also discourages the growth of shrubs and trees in a grassland since most woody plants lack a bark that is thick enough to insulate the plant from the heat of the fire. By destroying trees and seedlings, fire reduces the moist, shady habitat that is most conducive to tsetse flies. It is in these tsetse-free areas, particularly in regions of higher rainfall, that domestic livestock are concentrated (Bourn 1978). When fire-influenced grasslands become less hospitable to tsetse flies, the vector for the transmission of trypanosomiasis (sleeping sickness), humans and animals benefit (Lewis and Berry 1988, 76). Large tracks of African environment have been cleared of tsetse habitat by the use of fire. The result is an act of creative destruction in which overall woody biomass is reduced by the destruction of trees and shrubs to promote the growth of grassy habitat that is more useful to animal herders. Over time, grassland plants, herd animals, and herders increase at the expense of other species. They coevolve through a frequently repeated series of destructive acts which, provided that they are neither too intense nor accompanied by too rapid an increase in herd size, produce a sustainable pastoral system. The artificial nature of this balance between grass, trees, herds, and humans is illustrated by the changes that occur whenever the pressure of fire is removed from the system. Lewis and Berry (1988, 373) report that the diversion of manpower from the herding economy to the mines in Zambia resulted in reduced frequency of fire. Without fire to sustain the grassland, shrubs and trees regenerated rapidly, tsetse-borne sleeping sickness expanded, and domesticated livestock experienced a sharp decline in numbers. Efforts to use chemicals as an alternative means by which to destroy tsetse habitat have had limited success, and pose serious risks for the environment (Linear 1985).

Unfortunately, destructive creation is a more common process in dryland pastoral systems than creative destruction. This trend accelerated in the twentieth century as the balance of political and economic power shifted away from pastoral groups in favor of settled farming and urban populations (Johnson 1993b). There are many reasons why degradation has become the dominant trend in pastoral rangeland environments.

The most important factor is the loss of critical zones in the traditional pastoral system to agricultural development. These zones do not have to be large in area to have a major impact on the viability of pastoral communities. If they are important areas of dry season pasture, the loss of even a small critical zone can have far-reaching impact. The conversion of river floodplains from an annual cropping system to a multicrop regime is a case in point. This most often occurs when the desire to intensify agricultural production leads to the construction of a dam to store water for year-round irrigation. Not only is former dry season pastureland lost beneath the reservoir, but also conversion of farmland below the dam from basin or recessional cultivation to perennial irrigation makes it impossible for migratory herders to graze on seasonally fallow land or postharvest crop residues. Any additional recessional grazing created along the banks of the reservoir is inadequate compensation for the lost seasonal floodplain pasture and is usually accompanied by greater exposure to waterborne disease (White 1988). Pastoralists are then thrown back into year-round dependence on rangeland resources that historically have only been asked to support them for part of the year. The inevitable degradation that results makes pastoralists and their habitat the sacrifice zone for the production gains that take place in the irrigated agricultural sector and/or the generation of hydroelectric power sent to distant locations to satisfy a multitude of urban and industrial demands as well as power generation for irrigation pumps.

The expansion of dryland farmers into pastoral zones is another example of this process. Because rain-dependent farmers do not have access to enough surface water for irrigation, they are limited to zones in which they can hope to receive enough rainfall to support a crop in a majority of years. These areas may be marginal from the standpoint of cereal crop cultivation, but they are often the best, most critical zones for pastoral activities (Sollod 1990). Their loss has contributed in large measure to land degradation in northern Niger (Bernus 1980), the lowland fringes of the Kenyan Highlands (Campbell 1981, 1986; Little 1987), the High Plateau of Algeria (Bedrani 1991), Xinjiang in western China (Banks 1997), and this story can be repeated in many of the Third World's semiarid rangeland provinces where farming and herding can both be practiced.

Changes in pastoral practice and technology that encourage excessive concentration of people and animals also promote grassland degradation. The most common culprit in such instances is the introduction of bore wells, which are able to tap groundwater resources unreachable by traditional well techniques. These bore wells replace shallow traditional wells constructed by hand. Overgrazing always is a problem around any well, but the limited water supply produced by traditional well technology discouraged the concentration of too many animals at any one well (figure 5.3). Also, local herders maintained control over access to their wells. The introduction of government-funded wells was accompanied by loss of control by local herders over who could bring their animals to the well, and over how much livestock could be watered.

Permanent settlement near traditional dry season grazing and water also was very limited in the past. The development of secure, good-quality water resources attracted settlement by herders and the creation of infrastructure (schools, health clinics, police posts,

Figure 5.3.   Land degradation around a bore well in Kenya's Central Province on the edge of the Rift Valley. (Photograph by L. Lewis, February 1983)

vehicle repair facilities, and much more) to serve the increasingly permanent population. Few former nomadic pastoralists were willing to give up their animals when they settled if they could possibly avoid it. Therefore, grazing pressure around wells increased dramatically, producing a small pimple of desertification potential in the midst of a larger, unaffected area. Frequently the herds circulating around these deep-well sites were owned by merchants who preferred to invest capital in traditional means of production such as animals. Often masked by apparently decent grazing during periods of adequate rainfall, these splotches of degradation burst into full view whenever drought diminished annual fodder production and caught large herds of water-demanding sheep and cattle in exposed locations (Bernus 1980). In the Sahelian drought of the early 1970s, the results proved disastrous for many herds, as well as very costly for the environment. There are reports from the central Sudan that, despite the occurrence of better rains in the late 1980s, some of the most desirable fodder grasses have yet to reappear around important well and village sites (Khogali 1991, 89). In contrast, in other parts of central Sudan a lightening of pressure has resulted in at least partial recovery of the vegetation (Olsson 1985).

The pressures that encourage excessive concentration of people and animals near spatially limited resources are not restricted to developing countries. They are found in the rangelands of industrialized states as well. In such situations, the problem of degradation is increased by the commercial context in which the herder must operate. In the Gascoyne Basin of western Australia, sheep ranchers must generate income from sales of sheep and wool. Whenever prices drop for wool or mutton, off-station sales are difficult to justify economically. When depressed prices coincide with drought, the station operator is encouraged to keep more animals on the station than can be sustained without degradation. Owners are reluctant to destock their operation because they hope to bring sufficient animals through the bad period in order to capitalize on what they anticipate will be a future rise in price. Thus, animals are retained for economic reasons rather than reduced based on ecological criteria. These animals are not kept on the poorest pastures, but rather are held on the land with the highest potential to produce fodder. The ironic result is that degradation is concentrated on the best and most productive land, while the least productive areas remain relatively unscathed (Walls 1980), a cycle that is typical of much of pastoral Australia (Williams 1978).

Contemporary range management postulates the need to balance livestock numbers in an area with its carrying capacity over a succession of years (Child et al. 1987). The basic idea is to optimize livestock and forage production. This goal is achieved by averaging forage production and the weight of livestock over time to produce a livestock unit per hectare figure that can be supported without degradation in most years. The approach is inherently a steady state model intended to insure production without degradation regardless of conditions in any year. Some fodder goes unconsumed in good years to retain forage capacity to sustain the herd in bad years. This controlled management approach constitutes the conventional wisdom of range management and contends that rational decisions can be made that achieve both profit and land preservation simultaneously. Drawing on decision support system models, practitioners of the conventional wisdom integrate a powerful array of tools for measuring and monitoring the state of range ecology and translate this into practical management decisions (Stuth and Lyons 1993).

Such approaches invariably assume a set of long-term equilibrium conditions that can be managed coherently to optimize the productivity of range for human purposes. This position is increasingly challenged by the "new ecologists," who argue that dryland grazing systems in particular are disequilibrium systems in their fundamental nature and must be understood as such if there is to be any possibility of successful management, a fact they maintain was grasped experientially by traditional pastoral range managers (Behnke, Scoones, and Kerven 1993). Alan Savory (1999) poses another interesting challenge to the conventional wisdom of range management. Like the practitioners of conventional wisdom, Savory insists that environments must be approached holistically rather than treated as a set of isolated components. In brittle environments such as rangelands, where moisture is in short supply for a substantial part of the year, what matters to Savory is how quickly vegetation breaks down and returns nutrients to the soil. Large masses of grazing animals, the buffalo in the Great Plains or the mixed herds of ungulates in sub-Saharan Africa, maintain ecosystem health by breaking up soil crusts and trampling unused fodder into the soil where it can decompose rather than desiccate. In this scheme, the critical issue is not the short-term disturbance introduced into the local habitat, but rather the length of time concentrated use continues. Savory views trampling by dense masses of animals as a form of plowing, which prepares the seedbed for future growth. Only when prolonged does trampling produce soil compaction, reduced infiltration, lower soil moisture, and create inhospitable conditions for germination and seedling establishment. Translated from wild herd animals to managed livestock conditions, Savory's method envisages a dramatic increase in stocking rates provided the livestock spends very short amounts of time in any one spot. Rapid movement is the key to increasing carrying capacity and enabling income-challenged farmers to boost their potential profits. Claims of doubled carrying capacity certainly capture the attention of financially strapped land managers, particularly in conditions where economic austerity and changes in agricultural policy are placing commercial herd managers under considerable stress (Archer 2001). Whether such prophecies of profitability will be fulfilled without significant negative impacts on the environment awaits future judgment. But certainly the practitioners of conventional wisdom are convinced that any short-term gains are guaranteed to be offset by future catastrophic collapse.

A third major factor promoting destructive creation in rangeland is loss of control over the range resources that most traditional pastoral communities have utilized. In many African rangelands, tribal lands were nationalized in the period immediately following independence from colonial rule. The common property of a community became the common property of the nation-state. Most governments have found it difficult to substitute a reasonable system of control for the former tribal system. For many pastoral groups, especially in the drier rangelands, trees are an important source of forage in dry seasons and in drought periods. Protecting trees from excessive cutting is essential to pastoral survival (Hobbs 1989, 104–6; Barrow 1990). Both because most modern pastoral development strategies have tended to ignore the importance of trees and because the loss of proprietary communal rights to rangeland removes traditional managerial constraints on the exploitation of rangeland, dryland arboreal resources have experienced considerable degradational pressure during the postindependence era, especially in Africa. This is particularly true in cases in which charcoal is a major source of domestic fuel, and drought offers no alternative for pastoralists but to consume their trees for nonpastoral purposes (Dahl 1991).

In many countries, as a consequence of new tenure arrangements following the assertion of national control over rangelands, blocks of land were set aside for group ranches or reassigned for commercial ranching schemes designed to produce meat for export. These ranches seldom worked well for three reasons. First, they removed important sections of range from use by traditional producers, thus forcing them to concentrate on producing the same number of animals on less land. This constituted an inappropriate form of intensification, since at best limited capital was invested in the traditional livestock sector. The resulting degradation in range quality outside the fence was usually attributed to nomadic pastoral mismanagement of a common property resource. In fact, it represented a sacrifice zone created by the ranching scheme favored by governmental action and support. Second, ranches were fixed entities with rigid boundaries in a highly variable and fluctuating environment. Whenever rain failed to fall within the fence, herds were trapped, unable to move legally to adjoining areas that had received better rainfall. The strategies developed for ranches generally ignored the *genius loci principle* by not taking into account the highly variable nature of precipitation in drylands. Degradation then often occurred within the fenced perimeter. In Syria, nationalization of pastoral areas set off a scramble for access to the land by private individuals without effective constraints on herd numbers. Only by reestablishing control through the mechanism of pastoral cooperatives based on the old tribal system and traditional concepts of environmental protection (*hima*), an acknowledgement of the *genius loci* principle, was further rangeland degradation arrested (Draz 1977, 1990; Shoup 1990). Privatized individual or group ranches not only fit poorly into the variable environment of traditional pastoral areas (Kipuri 1991), but also they set in motion significant social conflicts. In Kenya corrupt land control boards and unscrupulous individuals often cooperate to buy up the land titles of group ranch members for a fraction of their value, dispossessing the least powerful members of the community while promoting their own social and economic aggrandizement (Galaty 1999). For the well connected, the law has become the method of choice to advance individual advantage, often at the expense of sustainable land management.

Destructive creation in animal husbandry systems is the product of a failure to view rangeland development as part of a larger system. By treating development segmentally rather than holistically, changes in agricultural systems have cut traditional links to pastoral areas and turned them into sacrifice zones. When new technology, such as diesel pump-driven bore wells, has been introduced, it has often operated in a vacuum, divorced from meaningful association with local communities. The result has been the creation of local sacrifice zones adjacent to the new technology site and the initiation of severe degradational pressures. Stopping and reversing this degradation has proven to be very difficult, and only occurs in rare instances when developers and managers were willing to listen to the *genius loci* principle and base change upon time-honored precepts that are in tune with local society and environment. Whether the range management principles advocated by Alan Savory and his disciples will prove to be a stroke of genius loci brilliance or a misguided effort to enhance productivity that results in creative destruction remains to be seen.

Tropical Deforestation and Land Degradation

Approximately 1.6 billion people at least partially depend on forests for their livelihoods. For both this reason and the feedbacks of tropical forests with global climate, the state of the world's tropical forests was the source of great concern at the United Nations Conference on Environment and Development, held in Rio de Janeiro June 3–14, 1992. Management projects concerned with improving the status of tropical forests have been a major concern of international organizations and governments since the Rio conference. This concern is reflected by the number of projects funded with the intent of improving the environmental status of forests. Even with world attention and funding directed toward stabilizing tropical forests, "they continue to suffer unacceptably high rates of loss and degradation" (Cassells and MacKinnon 2002, 24).

Two major and divergent pathways of explanation for tropical deforestation, and hence environmental degradation, have been articulated for what is driving this ongoing process: single factor causation versus irreducible complexity (Geist and Lambin 2001). Shifting cultivation and population growth are identified as the two major primary causes for tropical deforestation (single factor causation). Complex socioeconomic processes in which single causes are impossible to identify is the general explanation offered by the other camp regarding why tropical deforestation continues unabated. According to Geist and Lambin, there are a multitude of reasons why tropical forests continue to be cleared and land degradation continues. For this reason, they conclude "that any universal policy or global attempt to control deforestation (e.g., through poverty alleviation) is doomed to failure" (2001, 97). Some of the factors/combinations examined in their book are expansion of cropped land and pasture, infrastructure expansion and agriculture, formal state policies associated with development, and population growth, poverty, and migration. As but two examples, we examine in the following paragraphs the unintentional environmentally destructive changes resulting from the "development" of the Amazon River Basin and Southeastern Yucatan (Mexico).

Land Degradation in the Amazon Basin

Since the turn of the nineteenth century, large-scale deforestation for agricultural development has been a recommended strategy for the Amazon. Attempts to clear large areas of rainforest in Amazonia, with the intent of replacing the sustainable, but low-intensity, slash and burn agriculture of the local population with pasture, plantations, or other forms of large-scale agricultural development generally have failed. The early optimistic view of the basin was that it had an unlimited potential once cleared of the dense, lush vegetation along the river's edges. Today it is a widespread perception that most agricultural ventures have partially failed because the "plateau" areas—98 percent of the area with altitudes up to 200 m (656 feet)—generally have poor soils deficient in plant nutrients, while the remaining 2 percent of lower, more fertile floodplains and swamps are too wet or susceptible to seasonal flooding, which curtail their agricultural potential (Salati et al. 1990). In addition, the potential for agriculture is not as great as one might expect given a twelve-month growing season because of cloud cover and factors related to the area's high temperatures. For example, evapotranspiration processes consume approximately 50 percent of the solar energy reaching the rainforest.

These high evapotranspiration rates are crucial in maintaining the precipitation patterns of the region. Calculations indicate that only 50 percent of the water vapor resulting in precipitation in Amazonia originates directly from moisture evaporated from the Atlantic. The remaining precipitation is from moisture recycled (evaporation and transpiration) within the basin. That is, the current humid climate found from the Atlantic in the east to the Andes in the west is a result of a complex interaction between climatic (especially winds and solar energy) factors and aspects of the rainforest (tree cover, soil infiltration, transpiration, evaporation, plant interception). While the rainforest vegetation is primarily the reflection of the climate, the rainforest feedbacks play a critical role in maintaining a humid climate, especially in the central portions of the region, which are distant from the Atlantic moisture supply and the topographic effects of the Andes (Salati et al. 1990). Thus, deforestation of the eastern portion of the rainforest has the potential to make central Amazonia's climate drier since surface water runoff will increase and transpiration and evaporation will decrease. These three changes in the local hydrologic cycle reduce the atmospheric potential of water vapor reaching the central portions of the basin.

The forest environment always has sufficient temperature and soil moisture to permit uninterrupted metabolism of the plant and animal domains. Soil moisture storage and vegetation cover play critical roles in maintaining sufficient soil moisture during the short dry season that exists in most areas of Amazonia (Richards 1952; Tricart 1965). With no unfavorable season affecting its growth, the forest—but not some individual trees—is evergreen.

Leaf fall is not determined by a seasonal attribute such as temperature in the middle latitudes or moisture deficit in grasslands; it is determined by specific attributes related to the growth pattern of the individual plants. With each plant having its own biological rhythm, there occurs a continuous supply of falling leaves to the forest floor. With the high temperatures and moist conditions found on the forest floor, plant litter supply and decomposition are in an approximate balance. The result is that ground litter is very thin throughout the area. Thus despite the high production of biomass, very little organic matter accumulates in the soils.

Species diversity is the rule throughout the rainforest. This vegetation diversity partially reflects microenvironmental variability in topography, soil, and water table (groundwater) conditions. The result is that in most areas there is no strong dominance by any one species. This creates a problem for the economic utilization of the forest for tree products. The low concentration of any single species under natural conditions increases the cost of bringing forest products to market. Another property of the rainforest is that because of the heights of the trees—over 40 meters (130 feet) for the tallest—the multistoried aspects of the forest canopy, and the very dense tree canopy, very little direct sunlight reaches the forest floor. A generally sparse underbrush results, except along wide rivers where ample sunlight reaches the ground through the opening provided by the rivers. This interception of direct sunlight by plant cover is crucial in preventing the soil surface from attaining high temperatures. This helps in the maintenance of high soil moisture during the drier season by lowering direct evaporation losses from the upper soil.

The primary matrix of stability in the Amazonian ecosystem is a delicate set of reciprocal interactions among the climate, vegetation, soil, and associated organisms (Committee on Selected Biological Problems in the Humid Tropics 1982). Any unilateral change in one of these factors without compensating changes to maintain the existing interactions usually results in deleterious conditions. The natural vegetation plays a critical protective role with regard to the soil by minimizing splash erosion, preventing the raindrops from compacting the soil, slowing runoff, and reducing the concentration of overland flow, and maintaining high soil moisture through the shade provided by the complete forest canopy. Once the forest is cleared and nontree crops are substituted, the soil climate is completely modified. With the removal of the forest canopy, sunlight directly reaches the soil, the soil surface dries out after a few rainless days, and soil temperatures rise. The cessation of leaf fall results in the supply of organic matter being curtailed and decomposition of the remaining organic matter occurs. On many rainforest soils these changes often result in a hard thin crust developing on the surface. All of these changes encourage increased runoff once a rain event occurs; thus, the potential for soil erosion increases. Even in gently sloping areas, because the weathered regolith is often very thick and rainfall is intense, rapid degradation of the soil can occur. Gully formation and increased mass movements all too often result (Tricart 1965).

Clearly, large-scale agricultural development of the rainforest is a risky strategy, especially when annual crops are substituted, because of the associated alterations in environmental conditions that accompany agriculture. In contrast, the traditional slash and burn agriculture practiced by the indigenous population and small traditional farmers minimized upsetting the existing relations among vegetation, soil, climate, and organisms over large areas. Their irregularly cleared small fields only permitted localized runoff. Furthermore, the surrounding forest provided enough shade to minimize temperature changes. Seed sources remained in close enough proximity to cleared sites to facilitate recolonization. Particularly critical was abandonment of the fields, usually taking place within a three-year period of use. Abandonment prevented significant alterations in the soil's textural and fertility properties. Minimal change in soil and temperature conditions allowed the forest to regenerate itself (Jordan 1989).

With the exception of the soils situated on the floodplain, the soils in Amazonia receive minimal new nutrients to offset any losses occurring due to poor farming practices. The widespread belief is that the low nutrient supply in the rainforest soils, due to the heavy leaching associated with the ample moisture supply and the high annual temperatures, is the reason why rapid declines occur in crop productivity. This is often cited as the reason why slash and burn agriculture and agricultural attempts to develop the area have resulted in land abandonment after a few years (Richards 1952; Tricart 1965; Meggers 1971). Yet on soils rich in nutrients throughout the humid tropics, such as recent volcanic-derived soils in Central America, the same drop-off in productivity often occurs. Thus, the causes of field abandonment must occur in response to other factors too. Studies indicate that nutrient stocks in fields at the time cultivation is abandoned, even on relatively infertile soils, are higher than in the undisturbed surrounding forest soils (Jordan 1989). Others have found little evidence of nutrient stress at the sites that they investigated prior to field abandonment (Proctor 1983). According to Jordan (1989), nutrient stress is only one factor to consider in explaining crop productivity declines in rainforest areas. Soil erosion, changes in pH, pests, and the invasion of weeds and shrubs are other critical factors contributing to crop productivity declines. After a few years of cultivation, weeds and other successional vegetation increase in density, and their success is at the expense of the planted crops. Thus, while crop productivity clearly decreases in the slash and burn agricultural system after a very brief period, this decline is offset by increased productivity of vegetation not coveted by the inhabitants. To counter this decline in crop yields requires more labor and capital than just clearing a new plot of land. Field abandonment allows the weeds and other non-desired plants to colonize the disturbed areas. Over time, the forest regenerates itself.

Contemporary major land clearing and the establishment of "permanent" pasture alter the environmental setting significantly beyond the minor perturbations associated with slash and burn agriculture. The resulting changes in the nutrient supply, soil texture, microclimate, and other factors result in significant changes in the environment that prevent regeneration of the rainforest (Hecht 1981). In other words, land degradation results. Activities in some parts of Amazonia appear to have reached this level.

With Brazil's population growth and a shortage of arable land for poor farmers in the heavily populated areas along the eastern periphery, the rainforest area was viewed by the national government in the 1960s as the logical area for both agricultural and economic development. President Castello Branco outlined a number of objectives in a 1966 speech to develop Amazonia agriculturally (Davis 1977). The goal was to establish an integrated development plan that would permit the Amazon Basin to become an integral contributor to the Brazilian economy, meet the food needs of the growing population, and serve as a new settlement area. One of the first steps of the Branco government was to authorize the building of the Transamazon Highway to improve the accessibility of the area. The strategies of the government failed to result in sustainable land use in the vast majority of the area for a multitude of reasons (Hecht and Cockburn 1989) that are beyond the scope of this book. But clearly one factor was the lack of realistic planning related to the environmental situation found in this humid rainforest area. In many areas, not only did the government's goal of economic development not take place, but also severe land degradation occurred that would limit future options.

Degradation in the basin resulted both from development of a basic infrastructure that was built to facilitate accessibility into previously remote areas as well as the agricultural practices of the new settlers. Paralleling the Amazon River, construction of the Transamazon Highway linking Recife and the northeast to the Peruvian border and the Brasilia-Cuiabá-Porto Velho highway became roads leading to disaster (figure 5.4). In particular, the Brasilia-Porto Velho road construction initiated major ecological problems, both direct and indirect,

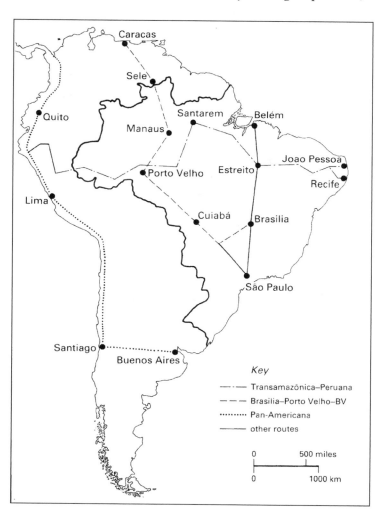

Figure 5.4. The Transamazon Highway and a network of regional routes are being constructed by the Brazilian government to open up the region to development. (From B. L. Turner II et al., eds., *The Earth as Transformed by Human Action*, 1991. Reprinted by permission of Cambridge University Press.)

because the road opened up vast new areas to migration and settlement. A poor database for the basin meant that road design was not tailored to the environmental settings found throughout the basin. In the process of bulldozing the land for the roads, the civil engineers learned that the topography was not as flat as their planning strategies had assumed (Smith 1982). The result was that exposed hillsides suffered extensive erosion to such a degree that vegetation was prevented from reestablishing itself in many cases. Also bridge construction resulted in the destabilization of many stream channels. The lack of reliable hydrologic data resulted in inadequate bridge designs and many bridges were destroyed due to unexpected high river flows.

Government policies encouraged agricultural, cattle ranching, mining, and logging projects in the areas opened up by the new roads. Implementation of these policies resulted in over 40,000 square kilometers (15,000 sq miles) being deforested during the 1980s in the single state of Rondnia (Salati et al. 1990). Over 30 percent of Amazonia, by the mid-1980s, was estimated to be damaged in varying degrees of magnitude as a result of these policies (Malingreau and Tucker 1987). A multitude of problems have been documented, ranging from mercury pollution associated with mining to land degradation resulting from ranching. Because of the magnitude of the area cleared for agriculture and ranching, these activities have been particularly harmful. According to Denevan (1981), the greatest threat to the rainforest is not agriculture but ranching. Because of the limited fertility of much of the forest's soils, pastures established after land clearing begin to decrease in production after only a few years. Unlike the traditional slash and burn agricultural system in which the land is abandoned when yields decline and the rainforest reestablishes itself over a number of years, ranchers adopted another strategy. They often burnt the old pasture to obtain a new flush of nutrients for the soil. This permitted new grasses to sprout with sufficient nutrients for livestock. Yet each burn resulted in slightly less nutritious grasses. After a ten-year period, the nutrient level in many of the areas had decreased to such a low level that not only must the pastures be abandoned, but also it is highly questionable if the rainforest will ever be able to recover naturally (Jordan 1989). Clearly, land degradation has resulted from this short-term strategy in many parts of the basin. On many lands opened up for development in Amazonia, the region's environmental constraints were not taken into consideration. The result was the sacrifice of a viable ecosystem with potential for other opportunities—many not yet discovered—for a limited number of exceedingly short-term gains.

Land Degradation in Southeastern Yucatan

Southeastern Yucatan, like Amazonia, is a tropical forested area. However, the area is characterized by a semideciduous or seasonal (wet-dry) tropical forest due to seasonal water deficits common from December through April. This drier type of tropical forest compared to the Amazonia rainforest reflects both the region's lower annual precipitation of 950 to 1,300 mm (37–51 in) and the seasonality of the precipitation. Additionally the region's underlying limestone allows water to infiltrate easily into the substrata, which exacerbates the moisture deficits during the dry season (Turner, Geohegan, and Foster 2003). The result of the moisture deficit is that between 25 and 50 percent of the trees lose leaves in the dry season. The drier the location, the greater is the loss of leaves (Klepeis 2000).

Beginning with the Mayan occupation of the area (3000 BC), the Yucatan has been exposed to human alteration of the forest and various intensities of deforestation. With the decline of the Maya civilization, forest regeneration occurred. In the late 1800s, the region was once again exposed to important human impacts and deforestation began due to logging of tropical hardwoods and the extraction of chicle. Logging activities resulted in a major decline in precious hardwoods such as mahogany. Beginning in 1967 the highway from Mexico City to Mérida and Chetumal was completed. Just as in the Amazon, road building opened up the area to migrants and land clearing expanded for both agricultural and livestock activities. In the 1970s major land clearing occurred in response to government-directed colonization and large-scale agricultural and livestock projects. This resulted in major land cover changes (deforestation) but little sustainable or successful agriculture. The distance from market (central Mexico), the shortage of water, and the generally poor soil quality contributed to this lack of success. Nevertheless, because of land scarcity in other parts of Mexico and the improved access to the Yucatan provided by road building, migration has continued into the area (Klepeis 2000). The result is increasing numbers of individual smallholders planting pasture and agricultural crops. Chile peppers, oranges, bananas, and tomatoes are the commercial crops, while corn (maize) and beans are the dominant staple crops. The type of agriculture/livestock that has grown in the area clearly is one of extensification compared to intensification. Soils have degraded, especially in terms of available nutrients. Fertilizers to restore soil fertility are too expensive for general use by most of the inhabitants. The overall result is that land covers have largely changed from a mature forest to a complex array of low intensity agricultural and pasture lands. "The number of options available to the government for obtaining economic benefit from the land have become fewer" (Klepeis 2000, 236). The establishment of the Calakmul Biosphere Reserve in the area is recognition of the difficulty of developing the area's dry tropical forest into a sustainable economic agricultural or livestock system. The goal now is to utilize a large proportion of the remaining forest cover to turn this southern Yucatan area into a tourism center with the forest itself and the Maya ruins within it as the attractions.

Conclusions

Common property systems work best when they are modest in scale and are managed by local groups who directly benefit from the resources exploited and are able to enforce compliance with management rules. Once systems increase in scale and management passes out of local hands, pressures are usually placed on the resource base that are difficult to sustain. Land degradation is the result. But traditional land management systems do offer principles that are useful, and in many cases are critical, to the wise use of these resources. Their incorporation into new systems of use that are responsive to modern conditions is a prerequisite to successful management strategies.

Agricultural Production Systems

For living organisms to exist, minimal energy, nutrient, and material requirements must be available. Any environment possesses only a limited range of the required attributes. Humans have developed a multitude of strategies to overcome an area's natural constraints and meet human demands. One of the earliest modifications that continue to the present, as witnessed in the previous section dealing with the Amazon and Yucatan, is that forests are cleared and replaced with groundcovers that better meet the immediate needs of the population. The alteration in the forest cover results in the conversion of forest biomass which is largely indigestible into increased production of digestible foods such as fruits and grains. Another widely utilized strategy having ancient roots is the transfer of water onto drylands to ameliorate the moisture constraints of these areas. Originally these transfers were local and utilized local groundwater or exotic rivers, such as the Nile, flowing through an arid area. Today agricultural demands often result in water diversions from river basins hundreds of kilometers from the irrigated areas. These water transfers onto drylands supplement local water supplies and permit both crop substitution and establishment of highly productive pastures. Increased yields in agriculture and livestock have resulted from this tactic. Significant food production increases for human consumption from many of the world's semiarid and arid lands attest to the importance of these water diversions. Increasingly, supplemental water supplies are being utilized in humid areas to increase food production by minimizing ephemeral dry periods when plants are under moisture stress and yields suffer. Irrigation is a critical component of agriculture in the humid east coast of the United States from Florida through Massachusetts. The use of sprinkling systems is widespread on farms in the humid East and is one strategy for maximizing food production. The water supplied to the farm fields reduces or eliminates soil moisture stress which is common during short periods of the growing season.

Intensification strategies often only utilize a limited number of the components of a natural system. Today on most lands, farmers need to add supplements to the existing lands to meet their production requirements. Cultivating the same crop (monoculture) on the same land year after year is the common practice in most high intensity farming systems. This practice has been favored by mechanization, better crop varieties, and the development of chemicals that protect the crop from pests and weeds (Power and Follett 1987). Most monoculture places high demands on the soil. These demands are far in excess of what most soils can sustain, even in the short-term. Thus one common by-product of agricultural intensification is the need to add fertilizers, herbicides, and pesticides to counter the limiting constraints of nutrients in the soil and create an environment that prevents both pest and weed infestations. These supplements, while favorable to food production, are a major cause of water quality problems worldwide. Irrigation, animal husbandry, and rain-fed (nonirrigated) agriculture constitute prime examples of intensified resource use gone awry in many areas. The following cases represent stories in progress, with an uncertain outcome, rather than conclusive events with regard to the environment and land degradation.

Irrigation in Egypt

Irrigation agriculture is an example of the difficulties that arise when efforts at intensification of food production in dry environments result in negative environmental consequences when feedbacks associated with it are ignored. Two examples, Egyptian irrigation in the Nile Valley, which relies on surface water as its supply source, and North American High Plains

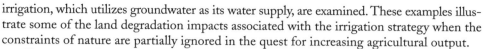

irrigation, which utilizes groundwater as its water supply, are examined. These examples illustrate some of the land degradation impacts associated with the irrigation strategy when the constraints of nature are partially ignored in the quest for increasing agricultural output.

From a population perspective, the Nile Valley is Egypt, since this is the area where almost all Egyptians live. As has been the case since biblical times, nearly all of Egypt's population is concentrated in approximately 3 percent of the national territory. Although the contemporary cultivated area is about 35,000 sq km (13,500 sq miles) (Baines and Malek 1980, 16; Fisher 1993, 360), a 30 percent increase over that available to the ancient Egyptians, the total dependence of Egypt on the waters of the Nile remains the same today as it was five thousand years ago. An elongated corridor of cultivated green in the delta and entrenched valley of the Nile stands in verdant contrast to the desolate desert, the expanse of which is only occasionally broken by flecks of oasis cultivation or mining operations. With existing technologies available in 1882, the 27,000 sq km (10,400 sq miles) area of the delta, valley, and Faiyum depression constituted the maximum possible extent of irrigated cultivation (Butzer 1976, 83). Approximately the identical area made up the Greco-Roman ecumene when the ancient population reached its maximum size. This land area probably reflects the maximum extent of agricultural land use that could be achieved with a non-fossil-fuel economy.

Before regulating the Nile's discharge through dam construction, its hydrologic regime was extremely regular. Summer rainfall in the Ethiopian and East African highlands and the Lake Victoria basin are the water sources of the Nile. The White Nile, originating in the Lake Victoria basin, is the source for the Nile's base flow, while the Blue Nile, with headwaters in Ethiopia, provides the bulk of the Nile's annual flood. The two Nile branches join at Khartoum, and for its northward course across the Sahara to the Mediterranean the Nile barely receives appreciable additions to its discharge. With no additional flows into the river during this northward flow, the river's discharge decreases due to both evaporation and channel losses to groundwater. The flood resulting from the Blue Nile waters reaches Cairo in August, peaks in September, and declines precipitously in October and November. From February to June, the Nile is naturally at low ebb, a mere trickle of its flood stage self. Ancient Egypt's political authority and social stability were directly tied to the regular and predictable occurrence of the Nile flood.

Until the nineteenth century, Egyptian life and livelihood was directly dependent on the annual, unregulated Nile flood (Collins 2002). Farmers utilized the floodwaters by means of a basin irrigation system (Willcocks 1889). In this system, the Nile functioned as both main water delivery canal and as principal drainage ditch. During each flood episode, the floodwaters breached or overflowed the river's natural levees and inundated the adjoining floodplain. Because the Nile floodplain is extremely narrow throughout most of the river's course in Egypt, in most years the flood covered the majority of the plain. At the height of the flood, only higher levee tops, village sites which were always situated on high ground to avoid floodwaters, and distant portions of the floodplain remained above water. The meander scars of abandoned former channels retained water for longer periods after the flood decreased. Farmers learned to augment this floodwater retention capability by erecting low, linear mounds that carved the floodplain up into a series of smaller basins. The increased water retained in the basins enhanced the amount of infiltration that occurred and increased the soil moisture store available for agriculture. Annual floods also distributed substantial amounts of silt on the fields. These silt deposits were important for agricultural sustainability by both maintaining soil texture and supporting soil fertility.

The entire agrarian system was simplicity personified. Little physical infrastructure was needed to bring water to or to drain water from the fields. When the flood crest passed, the river channel served as the main drain by which the receding waters were removed from the fields. The natural rhythm of riverine ebb and flow established the parameters of the region's single cropping season. For once the floodwaters receded, bringing supplemental moisture to fields was a monumental task, beyond the technological and organizational abilities of the inhabitants. During the low-flow periods, most fields were too high above the river level and too distant to make raising the water possible.

Yet the desire to intensify crop production and to improve on the productive capabilities of the natural regime was evident at an early date. Attention initially focused on developing double cropping along the levees adjacent to the river. The *shaduf*, a hand-operated lever action waterlifter, introduced sometime between 1800 and 1250 BC, enabled about 10 percent of the floodplain to be double cropped (Butzer 1976, 46). Adoption of the *sagiya*, an animal-driven waterwheel, in Ptolemaic Egypt allowed greater quantities of water to be lifted; another 10 percent of the floodplain could now be cropped during the low flow periods. Beyond these limits, areal expansion and intensification was unable to proceed until the nineteenth century. Nile silt deposits associated with the flood were sufficient to maintain soil fertility and texture in the areas inundated by the annual flood, especially if grains and legumes were rotated. Additional use of night soil, pigeon droppings, and fallow were essential to maintain fertility of the double-cropped areas close to the Nile. Additional gains in productivity depended on the manipulation of farming techniques and the introduction of new crops (Watson 1983) within the constraints of the time-honored and environmentally sound irrigation system.

Completion of the Aswan High Dam in 1967 was the last stage in a century and a half of effort to escape the constraints imposed on agricultural productivity by the natural river regime and the basin irrigation system. Population growth, particularly in the twentieth century, placed pressure on Egypt's limited agricultural lands, and made it easy to regard the floodwaters that escaped unutilized into the eastern Mediterranean as "wasted." Moreover, although the Nile River's flood regime was conspicuous in its regularity, low-flow episodes have occurred that resulted in food shortfalls (Bell 1975; Riehl and Meitin 1979). For periods as long as a century, lower Nile flows had devastating impacts on the integrity of Egyptian society and population (Butzer 1976).

Traditional basin irrigation was vulnerable to the impact of low river flow for two reasons. First, the high degree of reliability of normal magnitude floods discouraged investment in elaborate infrastructures of canals, diversion structures, and drainage ditches that would be needed to extract irrigation water by gravity flow from the Nile. Second, existing water-lifting technology was too limited in scope to irrigate a sufficient area to compensate for low-flood episodes. Storage of food reserves to offset lean years was the strategy adopted to cope with production shortfalls.

Efforts to increase the productivity of Egyptian agriculture began in the 1840s with the construction of a dam at the head of the delta. Other dams followed both in the delta and on the floodplain. The primary aim of these structures was to divert river water to the fields during the low-flow months. In accomplishing this, Egypt increased the areas where more than one crop each year could be planted. The basin irrigation system began to disappear and was replaced in large areas by a perennial irrigation regime. The Aswan High Dam was the last stage in a century-and-a-half long process. This dam went well beyond previous efforts to regulate the Nile and increase its agricultural productivity. Its aim was to store the floodwaters that previously had passed "unutilized" and "wasted" to the sea. By releasing this water throughout the year, the intention was to increase the multicropping of Egyptian farmland and to extend the perennial system to all existing cropland. In addition, new land, particularly to the west of the delta, was targeted for development with the newly saved Nile water. Drought and flood hazards were expected to be eliminated completely. Hydroelectric power generated at the dam dramatically increased the country's available power, and serves as a magnet of power-hungry fertilizer, chemical, and processing industries (Ibrahim and Ibrahim 2003).

Although largely foreseen, the environmental impacts of the Aswan High Dam are considerable. Examination of those impacts has dominated discourse, and the positive contributions of the dam to Egypt's industrial development have received less attention. Ideology, emotion, and the sheer magnitude and complexity of the problems involved undoubtedly contribute to this imbalance. Here we contribute to the one-sidedness of the debate by discussing only those adverse environmental impacts that result from the dam's construction. The land-degrading impacts of the High Dam are of two types: (1) those that result from the establishment of sacrifice zones as a consequence of the dam's construction; and (2) problems that have emerged to threaten both short- and medium-term sustainable agriculture in Egypt's critical agricultural zones.

Several sacrifice zones are associated with the High Dam. The most immediate of these is the portion of the Nile Valley now inundated by Lake Nasser (Lake Nubia in the Sudan). Rich riverine agricultural lands were flooded and over one hundred thousand Nubians were relocated (White 1988, 7). While health conditions in the new agricultural settlement schemes may have constituted an improvement, it is not clear that the productivity of the newly developed lands matched that of the drowned fields. Substantial change took place in the plankton and fish populations of the Nile once the dam was closed, with many species disappearing. In the short-term, the fish catch, largely *Tilapia* and *Alestes*, has exceeded expectations, but this has been at the cost of the diversity of the fish population, and may also be placing serious strains on the replacement capacity of the species most commonly exploited (White 1988, 9). In contrast, the seasonally fluctuating edge of Lake Nasser has created a new habitat for wildlife, and the same water and vegetation resources have attracted nomadic herders. In sum, lands sacrificed to the construction of the dam and its storage reservoir have had significant social and economic opportunity costs (particularly in moving or protecting the region's most significant archaeological monuments), but have been at least partially offset by gains elsewhere. This is not surprising, since the places and populations affected are relatively small in scale and planning for compensatory development was integral to the process of erecting the High Dam. As one moves further from the site of the dam, it becomes easier to sacrifice more distant populations and environments for the expedient interests of the beneficiaries of the project.

Still reasonably close to home, but more distant from the dam site and therefore more difficult to address, are the coastal erosion problems that emerged after the dam's construction. The coastal erosion threat was first articulated to a wide audience by Mohammed Kassas (1972), who contrasted the millennia of delta growth through deposition with the post-Aswan environment. Sediments once carried to the delta during flooding are now trapped behind the High Dam. This is starving the delta of essential building material. Without the annual supplement of silt and sediment to maintain land levels and to provide material for the coastal bars and beaches, the sea began to erode the delta and advance upstream (Hefny 1982; National Geographical Society 1992; Stanley and Warne 1993). A rising sea level due to global warming also appears to be implicated in the delta's problems (Milliman, Broadus, and Gable 1989). White (1988, 36) indicates that coastal erosion is largely concentrated in a few districts. However, when these menaced areas are the narrow barrier bars and spits that separate the delta's brackish lakes and lagoons from the open sea, not only is the loss of an important inland fishery threatened but also the breeding function of the coastal lakes for the marine fishery is lost. Moreover, conversion of the lakes and lagoons into open arms of the sea places at risk the low-lying reclaimed agricultural land at the northern fringe of the delta (Kassas 1972, 187). Today the northeastern portion of the delta is sinking at an annual rate of 0.5 cm (1/4 in). This subsidence has existed for thousands of years, but in the past it was countered by the deposition of Nile sediments. If current rates of deposition, coastal erosion, and subsidence continue, it is estimated that by the year 2100 the delta's coastal margins will retreat between 20 and 30 km (12–19 miles) inland from west of Alexandria to east of Port Said. Up to 26 percent of the most productive rural and urban areas within Egypt could be lost if present trends continue and massive engineering works are not undertaken to curtail the erosion and subsidence (Milliman, Broadus, and Gable 1989, 343b). The "emergence" of this sacrifice zone due to policies aimed at increasing agricultural production in the Nile Valley needs to be addressed (figure 5.5).

An example of more distant degradational impacts is that of the fishery losses that have occurred in the eastern Mediterranean. Annual fish and shrimp yields have declined drastically since the mid-1960s. These declines are attributed to the decrease in the Nile's discharge reaching the Mediterranean with the concomitant decrease in minerals and organic nutrients reaching the sea. This has starved marine food chains of fundamental resources. These decreases in marine life are an inevitable consequence of the dam. The now undiluted waters near the delta changed the habitat for many commercial species and encourage the invasion of more salt-tolerant Red Sea species, which have a much lower market demand. Because Egyptian fishermen were only a minor component of the Mediterranean fishery, non-Egyptians absorbed most of the losses experienced in the area. Because it was anticipated that declines in coastal lagoon and Mediterranean fish harvests would be offset by an increase in the freshwater fish catch in Lake Nasser, which the Egyptians largely control, it was easy to treat the eastern Mediterranean as a sacrifice zone of only minimal importance. In actual fact, the Lake Nasser fish catch has never lived up to its anticipated results, but improved fishing technology and the growth of a farm-raised fishing industry have largely compensated for the High Dam's initial adverse impacts on the open ocean and lagoonal fish catch (Ibrahim and Ibrahim 2003, 84–85).

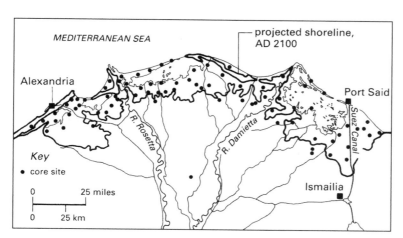

Figure 5.5. Construction of the High Dam at Aswan has reduced sediment river flows into the Delta. Coupled with rising sea level and a subsiding coastal landscape, this threatens the viability of the current shoreline.

A range of difficulties affects Egypt's efforts to enhance the amount of water available for consumptive use. In excess of 10 percent of the water stored in Lake Nasser is lost annually to evaporation and seepage into adjacent groundwater aquifers (White 1988, 8). This loss of water is nonproductive and constitutes a sacrifice imposed by the decision to build a dam in such an arid setting. The loss is especially significant because both Egypt and the Sudan are approaching full use of the Nile waters. Impending limitations in water supply have spurred plans to develop and control the water resources of the entire basin. Some of these plans envisage over-year storage of water in the lakes of highland eastern Africa (Tana, Victoria) and the reduction of water losses en route to Egypt. From Egypt's perspective, all of these plans create problems since the areas where they are to be implemented are beyond its national territory. All of the upstream countries likewise will need additional water in the future.

One large-scale effort that has been considered to increase water supply in the lower Nile is directed at the Sudd, a region of vast swamps in southern Sudan covering more than 5,000 square kilometers (1,930 sq miles) (figure 5.6). Sustenance for the marshes of the region, and of the agropastoral livelihood systems of the region's inhabitants, requires 14 billion cu m (18 billion cu yd) of White Nile water annually (Waterbury 1979). From the standpoint of Egyptian water managers, this is a wasteful use of water that could meet future water needs. Moreover, plans to store water in highland lakes make little sense if the water gained here is filtered through a giant swamp, which acts like a huge sponge soaking up water on its course to Egypt. The solution proposed is the Jonglei Canal, a bypass structure intended to divert water around the Sudd with minimal losses. The expectation is that if the project were completed, it would deliver 1.9 cu km (2,500 cu yd) of new water to Lake Nasser (Whittington and Guariso 1983, 449). To date, little progress has been made on the Jonglei due to political instability in the region and the opposition of the inhabitants living in southern Sudan. At least for the time being, this has prevented the conversion of the Sudd into a sacrifice zone. All of the proposed projects aimed at increasing water supply to Egypt will have severe environmental impacts. The Sudd project would destroy an ecosystem for

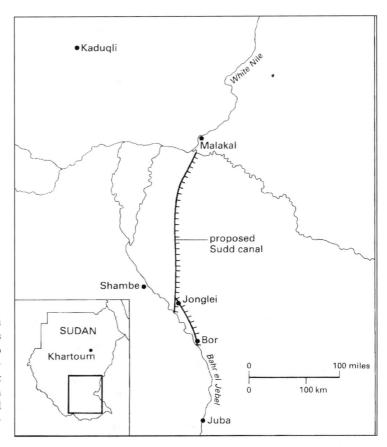

Figure 5.6. The proposed Jonglei Canal, whose construction has been interrupted by nearly two decades of civil war in the southern Sudan, was designed to divert Nile water from passing through the Sudd, a giant wetland, and transfer the water for use in irrigation systems farther north.

the sake of Egypt's water needs. All of the other proposed water diversions (the Machar Marshes, Bahr el-Ghazal) will also initiate major environmental changes too with resulting sacrifice zones distant from Egypt. Fortunately, all of the proposed projects will require international agreements as well as external financial assistance. This likely will impede any rapid decision and implementation on any of these water projects.

The critical zones that nourish Egypt's large and growing population have also experienced degradational pressure since the construction of the High Dam. In the effort to make the Nile's agricultural soils more productive, processes have begun that destabilize and degrade these environments. Lake Nasser acts as a huge sediment trap. The result is that the lower Nile water contains a lower level of solids and experiences a reduced turbidity than it did under unregulated flow (Whittington and Guariso 1983, 87). The changes in sediment, water regime, and water temperature have altered downstream channel characteristics as well as the number and types of aquatic organisms that can survive in these new managed conditions. Costly corrective measures have been required to protect irrigation intakes and downstream barrages due to the new dynamics of the Nile.

Within the floodplain irrigation districts (Grove 1982) and the desert reclamation areas to which Nile water is transferred, a litany of complications such as salinization, waterlogging, silting, increased bilharzia threat, and alkalinization imperil these critical zones (Ibrahim and Ibrahim 2003, 80-90). Kishk (1986, 228) estimates that from one-third to one-half of all irrigated land experiences salinization problems. Compounding these difficulties, the availability of water throughout the year leads inevitably to drainage complications. Efforts to extract multiple crops from the same field each year provide neither rest for the land nor opportunities for the land to dry out. In many instances not even the most careful management can prevent a rise in groundwater levels. In some areas of Upper Egypt, water table rises of almost 2 m (2 yds) have occurred (UNEP/GEMS 1991, 14). The delta has experienced lower increases in water tables but in this area the table is closer to the surface so that any rise creates environmental risks. Waterlogging and an elevated water table are closely linked to salinization, since as the subsurface soil becomes increasingly saturated the farmer has less ability to wash crop-threatening salts from the upper levels of the soil.

An additional threat to Egypt's critical zones comes from the linked changes that accompany the shift from seasonal basin to perennial irrigation. Without flooding, the annual layer of silt being deposited on the fields has ceased. The silts were essential to improving the tilth of riverine soils to dilute the windblown sandy material transported into the Nile Valley each year from the adjacent desert (Kishk 1986, 230) and to providing needed trace elements such as magnesium and zinc. The overall impression is that soils have declined in fertility due to the new management associated with the High Dam. Commercial fertilizers are now required in large areas while pesticides and insecticides are required to protect both the crops and humans due to the continuous humid conditions that now exist on many of the Nile Valley fields.

One result of these new inputs is a serious decline in water quality throughout the whole valley for agricultural, urban, and industrial purposes with particularly serious implications for water purification (White 1988, 336). The delta region has experienced the greatest decline in its hydrologic environment of all the areas within the valley. Because water used for irrigation in the delta is reused water from upstream agricultural and urban areas, contaminant concentrations are higher. The high nutrient loads in the water along with the lower river velocities due to the water removed from the channel for year-round irrigation create ideal conditions for eutrophication and the growth of macrophytes and algae (UNEP/GEMS 1991, 14). In effect, parts of the delta, a critical zone, are in danger of flip-flopping into a sacrifice zone as poor quality water is distributed in the area for agricultural purposes. The delta is experiencing problems analogous to a clogged toilet into which waste continues to be deposited. While treating the eastern Mediterranean as a septic tank into which contaminated water is discharged is hardly a long-term solution, the absence of the cleansing action of the annual flood is beginning to become increasingly apparent in perhaps the most sensitive environmental zone in Egyptian agriculture.

Two other land-degrading pressures are threatening Egypt's agricultural lands. These forces are only indirectly linked to the High Dam. The spread of water hyacinth (*Eichhornia crassipes*) in the canals threatens land productivity. This South American aquatic weed introduced as an ornamental and accidentally released into the wild is difficult to stop from spreading. It flourishes in the nutrient-rich waters of irrigation canals, basins, and reservoirs and consumes large amounts of water intended for irrigated crops. In an increasingly water-limited agricultural regime, the presence of water hyacinth is a serious threat to future productivity.

Urban expansion generates pressures that directly threaten agriculture's critical zones. This occurs in two ways in Egypt. One is direct physical expansion of urban settlement onto adjacent farmland. Cairo is the major, but by no means the only, urban center to alienate farmland for nonfood production uses. Kishk (1986, 229) estimated that an average of 120 sq km (46 sq miles) of land are lost to urban growth each year. From the standpoint of economic rent, such losses may represent a more productive use of the land. But from the perspective of sustainable agriculture, the constant loss of highly productive critical resources and the need to bring inherently less productive land into production as a replacement is tragic. Another urban pressure on agricultural lands is brick manufacture. Although illegal since 1985, mining of agricultural soils to a depth of 1 m (1 yd) for brickmaking removes 120 sq km (46 sq miles) annually for agricultural purposes (Kishk 1986, 229). There is every reason to assume that continued brick manufacturing losses still occur, albeit likely at a smaller scale.

In summary, the Aswan High Dam has had a profound impact on Egyptian economy and society. The power resources and industrial opportunities created by the High Dam have contributed greatly to the national economy. The over-year storage capability of the Nile's flood discharge in Lake Nasser has altered the natural regime and enabled year-round use of limited agricultural soils. Conversion to perennial irrigation in large parts of the valley has intensified the use of Egypt's limited land resources by as much as three times through multicropping. This has made it possible to keep pace with the nation's growing population, which continues today but at a slower rater (Ibrahim and Ibrahim 2003, 30).

However, there is a dark side to this story. Lands flooded by Lake Nasser and the fisheries of the eastern Mediterranean and the delta's coastal lagoons were consciously sacrificed. Coastal environments along the delta's shores as well as the agricultural zones of the delta itself are at risk to erosion, sea encroachment, and rising groundwater levels caused by the absence of flood-borne sediments to counter normal coastal erosion and delta subsidence. Water quality declines and groundwater table rises from excess accumulations of polluted water threaten the delta. Areas within both regions are on the verge of becoming sacrifice zones. Other sacrifice zones may well be created in Sudan, Ethiopia, Uganda, and the Congo in the future as a consequence of efforts to increase the supply of water reaching Egypt. And there is the prospect that Nile Valley countries, most notably Ethiopia (Collins 2002), unhappy with the privileged position accorded to Egypt in the division of the Nile's waters, may divert a larger share of the annual Nile flow and thus create at least a partial sacrifice zone in Egypt! Yet more alarming, because the impacts are more immediate in time and space, are the changes within the core of Egyptian geopolitical space that threaten critical zones—the Nile's vital agricultural soils—with salinization, waterlogging, nutrient and chemical pollution, and alienation for urban and industrial uses. The magnitude of the difficulties generated for sustainable agriculture and the scope of land-degrading forces unleashed by the High Dam are beginning to become ever more apparent. These threats are the malevolent aspects of the abundant opportunities created by the High Dam, and justify regarding the creation of the dam as an act of both actual and potential destructive creation.

Irrigation in the Great Plains

Compared to irrigation in the arid Nile Valley, Great Plains irrigation takes place in a sub-humid (precipitation = evapotranspiration) and semiarid grassland. The Great Plains is a vast tract of land with generally fertile soils extending from Texas in the south to Montana and the Dakotas in the north (figure 5.7). Drier along the western edge, the Great Plains become moister further eastward until they blend imperceptibly into the humid central lowlands. Typical of dry areas, this area experiences great interannual variability of precipitation. Gaines County in western Texas, citing but one example of this characteristic, experienced a low precipitation of 168 mm (6.6 in) in 1956 and a high of 960 mm (37.04 in) in 1941 (Sheridan 1981, 90). This general pattern of high moisture variability is constant throughout the period of historical record, as witnessed by data for Antelope County in northeastern Nebraska. In this county, 292 mm (11.48 in) are recorded for 1894 and 950 mm (37.04 in) in 1903 (Center for Rural Affairs 1988, 19). Permanent streams are infrequent in the region, and where they

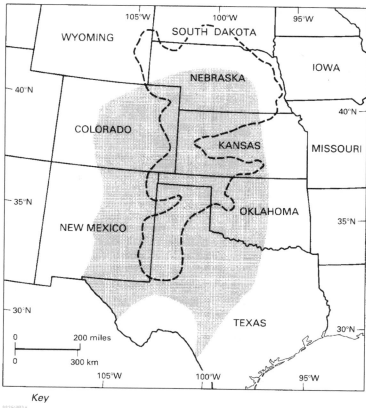

Key

area in which significant groundwater overdraft is occurring

boundary of Ogallala Aquifer

Figure 5.7. Declining groundwater in the Ogallala aquifer threatens the survival of irrigated agriculture in many parts of the Great Plains. (After the U.S. Geological Survey 1982)

do occur, they experience large seasonal variations in discharge. High stream flows are common primarily in the spring resulting from snowmelt. During the summer growing season little if any surplus precipitation occurs, and flows in the perennial rivers largely reflect the inflow of groundwater into stream channels.

In contrast to the limited presence of surface water, underground water resources were abundant in a large portion of the region prior to overpumping to support irrigated agriculture. The groundwater resources are part of the Ogallala Aquifer, a complex formation of sedimentary deposits derived from eroded materials associated with the Rocky Mountains. These formations stored vast amounts of water, which in scale exceed those of Lake Huron (Lewis 1990, 42b), and underlay much of the central and southern plains. In 1950 the aquifer was estimated to contain 1.7 quadrillion L (456 trillion gal) of water. While there was a vast amount of water in the aquifer, it is unevenly distributed. The average thickness of water-bearing layers is 61 m (200 ft), but it ranges from less than 15 m (50 ft) in Texas to nearly 366 m (1,200 ft) in Nebraska (Powers 1987, 2). Although some of the water in the Ogallala Aquifer is fossil water, much of it is renewable. The recharge is dependent on a small and variable rainfall as well as on snowmelt to produce sufficient moisture to infiltrate into the groundwater store. The recharge rates are very slow when compared to the continuing rate of withdrawal from the aquifer. Recharge is on a time scale of centuries with maximum rates in the highly permeable Sand Hills of Nebraska. Here 15 cm (6 in) of annual recharge occurs on average. In the southern portions of the aquifer the recharge is less than 1cm (½ in) (Powers 1987, 2); yet, due to the higher temperatures of the southern plains, this is the subregion with the highest water demands for meeting crop needs.

After several false starts in which farming systems inappropriate to long-term climate conditions were practiced (Bowden 1975), the region evolved a mixed system that combined animal husbandry with grain farming. Wheat was the crop in the drier west, corn in the moister eastern fringe, and beef cattle were raised everywhere. In the drier western and southern districts, and especially after the Dust Bowl experience of the 1930s, extensive mixed farming systems that left large areas of native grasses in place became the rule.

This system was technologically limited by the ability of farmers to extract groundwater (Glantz 1990, 16). The windmill had made permanent settlement possible on the plains, but the amount of water brought to the surface was limited. Only enough water for settlers and their animals could be extracted from near-surface aquifers. The deeper and more abundant Ogallala resources were not exploitable by wind-powered technology, and most farmers were unable to gain access to enough water to practice irrigation. Drought risk encouraged a flexible, diversified form of dryland agriculture, one that was sustainable and resisted excessive soil erosion as long as the conservation techniques and increased scale of farm operation mandated by the Dust Bowl experience remained in place.

Changes in technology, agricultural economics, and government tax policy encouraged massive exploitation of the Ogallala Aquifer after 1950. The introduction of more powerful diesel and gasoline-fuel-driven turbine pumps made it possible to extract water from deeper underground. In 1953 a new irrigation technology was developed to take advantage of the newly accessible groundwater resource. This was the center-pivot irrigation system, so called because it linked an overhead sprinkler boom to a central water source. The overhead boom was mounted upon a series of wheeled towers and was anchored at one end to

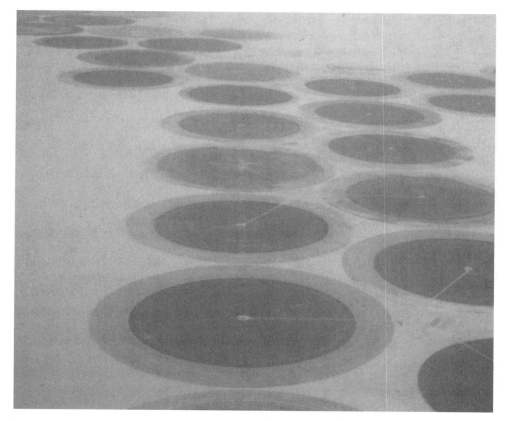

Figure 5.8. Central pivot irrigation on the outskirts of the Kufra Oasis, Libya. The technology, developed in the U.S. High Plains, made it possible to grow field crops in the center of the Sahara based on fossil groundwater. (Photograph by D. L. Johnson, March 1970)

the water source. The entire structure rotated around the well as a central pivot and was designed to irrigate approximately one-quarter section, or 65 ha (160 acres). In practice the corners of the quarter section could not be reached by the boom because the center-pivot irrigation system inscribed a perfect circle (figure 5.8). This left approximately 11 ha (27 acres) in each quarter section without the benefits of the system's water. The American penchant for giantism has led to the construction of center-pivot systems capable of irrigating 214 ha (530 acres). In the space of two decades, the rectilinear landscape of many parts of the American West has been transformed into a spectacular array of circles (Sutton 1977).

The major advantage of the center-pivot system was its great structural flexibility, which made it possible to irrigate tracts of land that otherwise were topographically unsuited for gravity irrigation farming (Center for Rural Affairs 1988, 2). No massive land leveling was needed to make this type of irrigation possible. This made it feasible to convert marginal range and cropland, without incurring large land-preparation costs, into more intensive uses that produced a higher value product. Lewis (1990, 43) suggests part of the economic incentive for the conversion when he claims that irrigation can produce yields six hundred to eight hundred times greater than dry-farming technology. In Nebraska and the moister eastern plains, land was converted to corn, while in the southern and drier plains cotton was the crop of choice. Another advantage of the system is its low labor requirement. Once established, it operates automatically with the aid of a computer to synchronize water application along the boom and to initiate movement to a new position on the circle. Expensive, time-consuming construction and maintenance of canals and ditches, or the physical movement of static pipe systems, are eliminated. Because the water is applied automatically and does not depend on rainfall, the farming system becomes drought-proof, guaranteeing good yields regardless of precipitation conditions. Moreover, fertilizers, herbicides, and pesticides can be applied as needed along with the irrigation water, thus further reducing labor costs.

The emergence of a sophisticated technology to exploit groundwater for irrigation purposes occurred at a favorable time. Although the capital costs of installing the irrigation system were considerable, the prices for crops were also high. Farmers, bankers, and agricultural extension agents all anticipated large profits for the foreseeable future. Government tax policy created a supportive financial climate by providing investment tax credits, favorable equipment depreciation schedules, tax concessions on land purchases, depletion allowances for groundwater withdrawals, and a generous capital gains policy. Crop price support programs, crop insurance, disaster relief payments, and low interest mortgage loans all favored farming over ranching. Widespread speculation in land became common, because simply plowing up land and planting it with wheat doubled the land's value (Huszar and Young 1984, 233). Banks jumped on the bandwagon by lending large amounts of money to farmers who wanted to intensify their production. In addition, the attitude toward regulation of groundwater withdrawals was sympathetic to rapid development. Common law treated groundwater use as a right belonging to the owner of the land under whose property the water was located. Any reasonable use of this water was permitted provided that the similar rights of neighbors were not infringed. Changes in this legal status for groundwater were slow to develop (Emel and Brooks 1988; Emel, Roberts, and Sauri 1992), were fiercely resisted in some areas, and only came about as depletion reached alarming proportions and conflicts over water use increased.

The boom time atmosphere that developed prompted rapid expansion in irrigated acreage as rangeland and marginal cropland were converted. In the space of three decades irrigated acreage increased fivefold, reaching over 6 million ha (15 million acres) in 1980. However, the stability of the irrigated acreage created is today threatened by a number of serious degradation impacts. By far the most serious is the dramatic reduction occurring in groundwater reserves, and the corresponding decline in the water table that accompanies aquifer depletion. Farmers who once obtained water from 15 m (50 ft) wells now must go to 75 m (250 ft), and as energy becomes more expensive, the costs of water extraction are not met by the resulting agricultural income. The result is that over 170,000 wells are no longer viable due to both economic and hydrologic factors (McKnight and Hess 2005, 258–59). In effect, farmers are mining the groundwater in the same way that coal miners mine coal. This is because the rate at which nature replaces withdrawn groundwater is far slower than the rate of depletion. The rate of withdrawal is more threatening in the southern plains where the amount of groundwater is more limited. Predictions that by the year 2020 the southern Ogallala will be reduced to one-third its former volume (Glantz 1990, 19) are reflected in decreased water pressure in existing wells and a decline in the water table by as much as 30 m (100 ft). Less drastic declines are reported for the northern High Plains. Were current rates of withdrawal and recharge to continue, Press and Siever (1974, 238) estimate that it would take several millennia before precenter-pivot irrigation conditions could be reestablished.

In this instance, a sacrifice zone is being created that will impact future generations more severely than present users because, while present users will be protected by social assistance programs from the worst effects of the impending impacts of destructive creation, subsequent inhabitants of the region will find their range of resource-use options drastically reduced. Dire predictions of complete collapse of the Ogallala are undoubtedly excessively apocalyptic, because as groundwater availability decreases, the cost of extracting the remaining resource increases. This encourages a search for new water supplies, all of which are currently much too costly to contemplate (Evans 1972; Powers 1987, 5) even if present users of distant surface water were willing to consider their release for transfer to the plains.

Increasingly constricted supply also encourages greater efforts at conservation and water-use efficiency. Solar-powered surge valves, runoff pits to collect and recycle irrigation and rainwater, and water applications that inject water directly into the soil rather than into the air are all reported (Lewis 1990, 43). Water losses from sprinkler irrigation have declined from 40 percent to 5 percent, while the estimated net depletion of the Ogallala Aquifer in at least one conservation district in Texas has declined from 2.78 billion cu m (2.25 million acre ft) in 1963 to about 3.70 million cu meters (300,000 acre ft) in 1985 (Wyatt 1988, 2). These measures promise to reduce the rate of Ogallala degradation, but they do not reverse the damage done to its ability to sustain irrigated agriculture and other livelihood options in the long-term.

Groundwater is connected to permanent surface water bodies such as permanent rivers, lakes, and reservoirs. The surface of these water bodies is generally the height of the water table. As water tables decline, less groundwater flows into these water bodies and they shrink. When water tables decline drastically, such as below the height of the river channel, river flows will cease and the stream will become ephemeral. Thus the lowering of groundwater by overpumping results in wetlands decreasing in areal extent, streams having lower discharges, and lake levels declining. All of these result in changing environmental conditions. Additionally, since groundwater is recharged by the infiltration of surface water, the quality of surface waters affects the quality of the groundwater. Pesticides, herbicides, and fertilizers are often lavishly applied in center-pivot irrigation. The use of these chemicals reflects the general tendency to push highly capitalized farms as close to their production limit as possible in order to generate income flows that will contribute to debt reduction or maximize profit. Monoculture and the reduction or elimination of crop rotations mandate the use of chemicals to compensate for the loss of natural regenerative processes. Nitrate fertilizers are a particularly prominent source of groundwater contaminants (Center for Rural Affairs 1988, 13, 17; Lewis 1990, 44). Degradation of groundwater quality seems to be a particularly dangerous example of biting the hand that feeds you.

Accelerated soil erosion is another negative response to center-pivot irrigation when it is poorly implemented. Many of the soils brought into intensive production were light or sandy soils readily susceptible to wind erosion. Often these soils were used as rangeland and protected from excessive wind erosion by a permanent grass cover. Conversion to irrigation increased their erosion risk by removing the groundcover. Especially after harvests, these soils are susceptible to wind erosion. With center-pivot systems capable of operating on 30 percent slopes, many areas that were topographically prone to erosion were brought into production. The large wheels necessary to carry the irrigation pipe-supporting towers across the landscape often created field scars that often evolved into gullies (Center for Rural Affairs 1988, 16; Piper 1989, 74). Although yield reductions due to soil erosion have been modest (table 5.1) due to the use of inorganic fertilizers, off-farm damage to recreation facilities, navigation, reservoir siltation, and residential areas has increased greatly (Colacicco, Osborn, and Alt 1989). Changes linked to the conversion process also increased the erosion threat. Planting of shelterbelts along field edges was a major strategy for reducing wind erosion vulnerability after the Dust Bowl (figure 5.9). The desire to build larger center-pivot systems made many of these shelterbelts obstacles to the farming operation and they were frequently removed. Their removal increases wind speeds and hence the potential for wind erosion.

Clearly, groundwater-based irrigation has exposed a critical zone, the region's agricultural soils, to both actual degradation and potential threat. Combined with the pressure placed upon the groundwater resources, the expansion of irrigated farming has been an act of destructive creation with limited prospects of developing a sustainable agricultural system. The most likely prospect for the region is a gradual reconversion of the agricultural habitat back to a much less intensive mixture of wheat cultivation and livestock rearing, a transition that will be handicapped by the diminished resource base of a mined groundwater table and varying degrees of an eroded and less fertile soil.

Table 5.1.   Yield Reductions in the Great Plains Due to Soil Erosion (percent)

Crop	Northern Plains	Southern Plains
Corn	1.9	0.8
Soybeans	2.4	—
Wheat	1.1	1.6
Legumes/hay	0.2	0.8
Cotton	—	6.4

Source: Colaciccio, Osborn, and Alt (1989, 36)

Dams and Urban Water Supply

During the past one-hundred-plus years the construction of dams has been an almost ubiquitous strategy for providing reliable water supply for urban and agricultural areas, the development of recreational areas, flood control, and electrical generation. The reservoirs formed behind the dams in many cases are so immense that they are one of the few human engineering works that is visible from space. Unless an urban area is situated adjacent to a large source of freshwater—such as Pittsburgh astride three rivers and Chicago adjacent to Lake Michigan—water transfers from areas of low water use to urban areas are a necessity. Contemporary urban areas require water supplies in excess of the potential of water production directly derived from the lands occupied by the urban area.

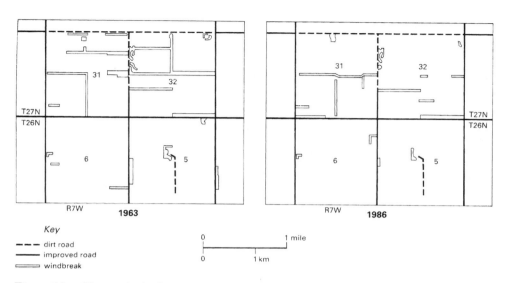

Figure 5.9.   Changes in the frequency of shelterbelts in Antelope County, Nebraska, between 1963 and 1986 reflect both changes in land management philosophy and the scale of farming technology.

While located in a humid climate, by the 1920s Boston was required to seek additional water supplies beyond its metropolitan area to meet the ever-greater demands for water resulting from its 60 percent population growth of the previous thirty years. The response to this need was the initiation of construction of the Quabbin Reservoir in 1928. This water project in rural west-central Massachusetts is approximately 161 km (100 miles) from the urban area that it was to supply. Without this water source severe water shortages would continuously exist, which would clearly impact on the economic well-being of the Boston metropolitan area. While local rivers exist in closer proximity to Boston than the distant Quabbin watershed, their ability to meet the water demands of the area was constrained by two factors. First, their water quality was already low in the 1920s due to pollution resulting primarily from industrial wastes. Second, it was impossible to build reservoirs on the rivers in the metropolitan area as densely populated urban communities already occupied the lowlands that would be flooded by the dam construction.

All reservoirs result in a sacrifice zone. The lands flooded by the reservoir for water storage represent an absolute loss of the land resource. This loss is further emphasized as often the valley bottoms that are flooded have very good agricultural soils. The sacrifice zone attribute is further amplified when reservoirs are utilized primarily as drinking water sources. In this case, multipurpose uses of the reservoir's waters, such as recreation, are often limited in many cases as a strategy to protect the water quality for the urban area. The Quabbin holds 1.559 trillion L (412 billion gal) of water which is used solely as a domestic water supply. Its construction required 311 sq km (120 sq miles) to be expropriated for its management, of which 101 sq km (39 sq miles) is flooded by the lake. Four small towns and several small villages as well as parts of a forest and agricultural ecosystem were destroyed as a result of the

Quabbin project (figure 5.10). All future options for land use of the flooded lands were removed for the benefit of a distant, more politically powerful metropolitan area.

The Los Angeles-southern California area further exemplifies the almost insatiable demands urban areas can place on regional water resources. This is an especially dramatic example because southern California has a dry climate with an explosive population growth. Seventy percent of all stream flows in California occur north of Sacramento, while 80 percent of California's population lives south of Sacramento. As early as 1905 plans were initiated to transfer waters from the Sierra Nevada (Owens Valley) to southern California to

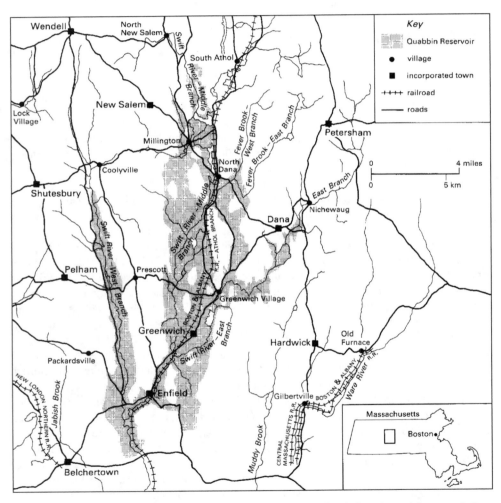

Figure 5.10.   The Quabbin Reservoir in central Massachusetts flooded a substantial area and eliminated several rural communities in order to store water for transport to metropolitan Boston. (Copyright © 1991 by the University of Massachusetts Press. Reproduced by permission.)

supplement the limited ground and surface waters captured in Los Angeles' local reservoirs. With southern California's continuous development, by the 1920s it was clear that if population growth were to continue in this region, additional water supplies would be needed. These water demands culminated in the construction of three major aqueducts to meet both the urban and agricultural needs of southern California. Today the resulting complex hydrological system transfers water from river systems over 320 km (200 miles) away. Runoff from snowfalls in the Sierra, and stream flow directly pumped from the Colorado River—river discharge derived primarily from snowfall and rain occurring in the Rocky Mountains over 1,000 km (600 miles) away—are diverted and transported by this aqueduct system to bring life and prosperity to arid southern California's thirsty farms and urban areas.

The Owens Valley, prior to the diversion of its waters, was once occupied by fertile verdant land fed by the meltwaters flowing off the nearby Sierra Nevada Mountains. Today it is a degraded, dry dust bowl (Sauder 1994). For all practical purposes, all of its waters have been funneled to meet the urban needs of Los Angeles. The contemporary conditions of the arid, degraded Owens Valley illustrate the concept of a sacrifice zone par excellence. This competition between traditional agricultural uses of water and growing urban and industrial demand for an increasingly scarce water resource (Englebert and Scheuring 1982) dominates discussion of water policy in the American West. Water-use competition is likely to force out of production those agricultural districts that are unable to achieve major water-use efficiencies (Englebert and Scheuring 1984), and thus increase the spatial extent of a type of land degradation analogous to that of the Owens Valley. Similarly, the diversion of the Colorado's waters to both California and Arizona is of such a magnitude that during most years it no longer has any significant freshwater discharges into the Gulf of California (Gulfo de Cortez). In fact, to meet treaty obligations, the United States desalinizes some of the Colorado's waters to meet the quota of freshwater that Mexico is entitled to receive. The result of these water diversions is that the Colorado Delta has shifted from a very productive ecosystem to a completely degraded unproductive state (Babbitt 1991).

The building of reservoirs to meet the water needs of urban areas always results in the degradation of the land in the areas directly impacted by the dam. Other changes, such as water table alterations and salinity changes in river flows, can result in an additional degrading land resource. Changes in the hydrologic balances associated with dam construction and water diversions are particularly prone to result in negative impacts in sensitive areas such as arid lands. Los Angeles and surrounding cities and towns exemplify some of the problems resulting from urban demands and the approaches that historically were developed to satisfy them (Hundley 1992).

In southern California, where local aquifers existed, communities often have overpumped them as another means of satisfying short-term water demands. The continuous pumping of groundwater to meet the needs of many coastal urban centers, such as Beverly Hills and Long Beach, has resulted in saltwater intrusions (Miller 1993, 235). By the 1940s the salt-fresh groundwater boundary had progressed up to 3 km (2 miles) inland in the vicinity of Los Angeles. The resulting saltwater intrusions have caused a loss of future options for the use of local groundwater supplies. While this does not per se represent land degradation, the decline of the groundwater resource has altered some inland areas where former freshwater springs have become saline.

One of the most interesting experiments currently underway in an attempt to counter some of the negative aspects associated with dam construction and the regulation of river flows is the planned large water releases from the Glen Canyon Dam on the Colorado River. The almost complete regulation of river flows in the middle and lower Colorado by a series of dams has resulted in a multitude of problems including huge losses of sand, invasion of exotic plants and fish, extinction of native species, appearance of an Asian tapeworm, and erosion of archeological sites (Blakeslee 2002). By releasing large quantities of water from dams, the goal is to mimic the unpredictable flow characteristics of natural rivers. The large flows scour the channel, help to restore predam conditions, and allow the native flora and fauna to flourish. The verdict is still out on how successful this new type of management strategy will be.

In summary, to meet the water demands of growing populations in urban areas as well as to meet the water demands of agriculture, distant river flows have been drastically reduced by the diversion of their waters into aqueduct systems. Whole ecosystems have been destroyed by the flooding of valley bottoms by the waters stored behind dams. In arid areas, reduction of river flows result in a host of problems in the mouth areas of rivers such as the Nile and Colorado Deltas.

Previously in this chapter groundwater overutilization for irrigation in the semiarid American Great Plains was discussed. In contrast, we shall now look at a similar problem in the context of humid areas in urban settings. Three examples are discussed: Long Island, New York; Houston, Texas; and Tokyo, Japan. Each illustrates some linkages between urbanization and unintentional land degradation under different environmental situations.

The bedrock of Long Island is overlain with unconsolidated deposits, up to a maximum of 650 m (2,000 ft) thick, of largely glacial and alluvial origin. Most of these deposits are permeable and permit a rapid infiltration of precipitation. With an average annual precipitation of more than 1,000 mm (40 in), and an evapotranspiration rate under 500 mm (20 in), a significant fresh groundwater reservoir between 37 and 75 trillion L (10–20 trillion gal) existed under the island prior to pumping (Heath, Fosworthy, and Cohen 1966). Under natural conditions an overall equilibrium existed between groundwater recharge and discharge into the ocean. This hydrologic balance resulted in a stable interface between fresh and salty water that was situated either at the coast or offshore depending on the nature and structure of the underlying surface deposits.

As the New York metropolitan area expanded onto Long Island, a number of changes in the groundwater system occurred. First, cesspool and industrial pollution infiltrated into the shallow portions of the water table. Second, increasing construction resulted in an increasingly impermeable surface. This, along with the development of sewer systems that steered the runoff directly to the ocean, lowered the recharge of the groundwater. Third, increased pumping of the underground water furthered contributed to the lowering of the water table. The result of these changes was that the saltwater-freshwater interface began to migrate landward in those zones where pumping exceeded recharge. Moreover, pollution in specific areas, especially along the western portion of the island that was part of New York City, reached sufficient levels as to prevent the groundwater from being used as a water source. In those areas where either polluted or saline waters now discharged into streams and ponds, the affected areas have clearly resulted in a degraded environment. The degradation of the groundwater in western Long Island resulted in the need to supply these areas from the New York municipal water system, which imports waters originating north of the city.

To reverse or at least stabilize the depletion of the groundwater in both the suburban and the remaining rural areas of the island, a number of policies were implemented. First, pumping of water is monitored to keep it in balance with recharge. Second, treated wastewaters are returned to the shallow glacial deposits to recharge the aquifer. Third, few storm sewers now directly empty into the coastal waters. Instead, a series of pits throughout the island have been constructed as catchments for storm runoff. These pits (ponds when full of water) are groundwater recharge zones. Because the function of the pits is solely to recharge the aquifers, these are areas that have degraded in terms of actual biomass production. Grasses, shrubs, and saplings now cover the land. With moisture conditions in these pits ranging from complete saturation (ponds) when filled with runoff to scorched during dry, hot summer periods, a degraded land condition exists in response to attempts to meet the water demands of this very urbanized area. This degradation is reflected by lower real estate values for the house lots that abut these "sinks." These pits clearly are a sacrifice zone established to save a degrading water supply. All of these policies have acted to arrest the decline of the groundwater reservoir, but they do not have the ability to restore the island's groundwater system to its original state.

Houston, Texas, is built over the poorly consolidated sediments of the Gulf Coast Aquifer. Furthermore, it is situated in a flat, low-lying coastal area. The pumping of water from an aquifer that is unconsolidated decreases the buoyant pressure that the liquid had contributed to the sediments. As a result, the sediments no longer support the pressure resulting from the overburden and they begin to compact. Once compaction occurs, it is almost impossible to restore the land to its former condition even if liquids are injected back into the sediments. Due to the withdrawal of large amounts of water from the aquifer, 1-2 m (6 ft) of subsidence has occurred in the Houston-Galveston area (Flawn 1970). As a result, some portions along the coastal zone are submerged below sea level. Perhaps the most serious problems of this nature occur in Japan.

Groundwater removal has resulted in urban sections of Tokyo and Osaka, where over 3 million people live, subsiding below the high tide level. In Tokyo parts of the city are subsiding at rates up to 500 mm/yr (20 in/yr), while parts of Osaka have rates of 76 mm/yr (3 in/yr) (Goudie 1981). With subsidence up to 4 m (13 ft) in Tokyo and 2.6 m (9 ft) in Osaka, without offsetting engineering works, these areas would now be susceptible to daily flooding (Forrester 1978).

As is evident from the examples in this section, satisfying an urban area's water demands can result in unintentional degradation of the environment. The submergence of coastal lands due to overpumping creates sacrifice zones. The need to set aside land in reserve for storm water catchment pits on Long Island, for groundwater recharge due to alteration in surface permeability associated with the change from a rural to suburban/urban area, results in lands within the pits being susceptible to frequent flooding. This limits the biomass productivity and utility of these areas. Land forever flooded behind dams to supply areas with ample water results in the reservoir areas becoming a sacrifice area, often in locales distant from the water-demand regions. These illustrations outline one by-product of urban growth: to insure the viability of an urban or agricultural area often necessitates the sacrificing of a component of the land resource for the benefit of the overall urban system. The resulting land degradation occurs because the overriding consideration in all the cases is to satisfy the water demands of urban communities. The protection and utilization of one resource, water, was achieved at the expense of another resource, land.

Rain-fed Agriculture

The world's overwhelming type of farming, rain-fed agriculture, depends on precipitation. In contrast to irrigated farming in which management can control the timing, intensity, and volume of water reaching any field, rain-fed agricultural systems must be able to cope successfully with the inconstant nature of precipitation that exists in most of the climates in which this form of farming occurs. Rain-fed agriculture exists through a spectrum of diverse geographic settings. Topographically, it is practiced on slopes ranging from extremely flat lands, such as in large areas of the Mississippi floodplain, to lands steeper than 30 degrees, such as in parts of the highlands of East Africa, the Andes, and the Himalayan zone. Climatologically, it is practiced in areas ranging from locales that experience almost daily rainfall and/or large amounts of precipitation to zones that receive relatively infrequent and small amounts of annual precipitation. Farming in these latter zones, under conditions in which annual precipitation is 500 mm (20 in) or less, is classified as a particularly risky type of rain-fed cultivation, namely dryland agriculture (OIA 1978). In the remaining portions of this section, rain-fed agriculture and land degradation are examined in one dryland and one humid case study.

In practice, dryland agriculture exists in areas experiencing precipitation in excess of the previously mentioned 500 mm (20 in) criterion. This is especially true in warm tropical areas that have high potential evapotranspiration rates. In India dry agricultural areas have precipitation values ranging from 350 mm (14 in) to 1,500 mm (59 in). Because of the seasonal distribution of rainfall in northwestern India, even the areas of higher precipitation (1,000–1,500 mm or 40–60 in) experience eight to nine months of potential evapotranspiration rates exceeding precipitation (Jodha 1988). Thus, even in these zones, moisture conservation measures, a critical component in dryland agricultural systems, are crucial for successful agriculture. Because moisture is the critical limiting factor in areas where dryland farming is practiced, when farmers are unable to manage their crops in synchronization with an area's precipitation variability, often crop failures and/or land degradation occur.

The environmental settings of dryland farming are usually risk-prone in the agricultural context. Because of limited moisture, without supplemental irrigation the likelihood of crop failure during periods of drought is high in these zones. Large variability of precipitation in dry climates is the norm. Thus crop failures or poor harvests during the drier years likewise are the norm. Successful farming strategies in these areas require that during drought the soil resource must remain protected for use during the wetter years. When crops fail or their growth is retarded during dry periods, plant cover is diminished. One result of this decrease in cover is that the soil is less protected from both wind and water erosion during drought conditions. Plowing and the removal of grassland cover for crops makes dryland agricultural lands susceptible to erosion during dry periods. The result is that accelerated soil erosion is often associated with dryland agriculture during dry episodes unless conservation practices remain implemented during these periods of moisture stress. When conservation is sacrificed for short-term demands, many lands degrade. Accelerated soil erosion reduces the resilience of vegetation to recover after each drought period.

To prevent land degradation from occurring under dryland farming, fallow periods were often an accepted strategy that evolved in these dry but agriculturally feasible settings. In the contemporary world, fallow periods are often reduced or totally eliminated. This puts traditional agricultural areas, where dryland farming was successfully practiced, at risk. Fallow and other traditional coping strategies in the farming system of the Luni district of India are examined in the next section. Prior to the twentieth century, these adaptive strategies had resulted in sustainable dryland agriculture. Alterations in the traditional system in response to contemporary conditions have upset a stable dryland farming system, and severe environmental problems resulted, making many parts of the area less viable for agriculture.

The Luni Block is a 1,989 sq km (770 sq mile) tract of land situated in a semiarid area of western Rajasthan. Agriculture was the dominant activity. Prior to the 1930s successful rain-fed agriculture was practiced throughout this area without significant deterioration in the resource base. Today it is an area that is environmentally very degraded (figure 5.11). Former productive fields and pasture are wastelands covered by dunes or rocky soils, or have experienced salinization.

The Luni Block is clearly situated in a marginal climatic area with regard to rain-fed agriculture, but prior to the European incursion and up to the 1930s, successful agriculture

Figure 5.11. Land degradation around a waterhole in the Luni Development Block in Rajasthan, India. The importance of cattle in the domestic economy guarantees that extreme pressure on particular landscape facets will cause local degradation. (Photograph by D. L. Johnson, fall 1976)

was in existence. Records show that a stable agricultural system existed from as early as the nineteenth century. The concentration of rainfall in a three-month period was crucial for agriculture. Even though precipitation is only 310–390 mm (12–15 in); (CAZRI 1976), its concentration in three months allowed for a short growing season. Farmers developed strategies to cope with both the low precipitation as well as the highly variable nature of moisture. As a consequence "the highly adaptive and evolved form of present day agricultural practices speak for the many centuries of human experience that must have gone into its making" (CAZRI 1976, 13).

Climatic conditions have remained constant since the 1800s through the modern period. Especially pertinent to the land degradation/desertification problem that this area is now experiencing is the fact that there is no indication of any progressive aridity; that is, there is no evidence of climatic degeneration occurring throughout the area (Walls 1980, 181). Climatic change is not the cause of this area's deteriorating land resource.

Prior to the 1930s, the population was relatively stable. Increases in population due to high birth rates were countered by frequent disease epidemics. For example, between 1911 and 1921, the area's population decreased due to outbreaks of disease (CAZRI 1976, 10). Population in Luni Block from the 1800s to the 1930s hovered at around thirty thousand. The introduction of immunization, and more recently antibiotics, illustrates the effectiveness of modern medicine in controlling diseases and premature deaths (Walls 1980, 181). But no concurrent innovations were introduced to curtail the high birth rate, which was needed to insure a stable population prior to these medical interventions. The result was that by the 1970s, the area's population had more than tripled to ninety-six thousand. One repercussion of the introduction of modern medicine into this agriculturally marginal area was that the local resources had to support an ever-increasing population.

The rapid increase in population during the twentieth century is likely the catalyst for the introduction of numerous modifications in the traditional farming system that, prior to these changes, had resulted in a long-term stable system. First, migration took place onto ever more marginal drier lands in the Block. Because these lands were more arid, farmers had avoided them previously. Their usage had been limited to either grazing or shrub forest, which provided complementary and needed goods to the farmers. Second, the period of fallow, crucial for allowing the soil to repair itself, was shortened or eliminated. Third, a change in the livestock mix occurred. Buffalo and sheep declined while goats increased. Goats are very efficient browsers. Not only can they thrive where other livestock cannot, but if they are not well managed, their efficiency can strip the land of much of its plant cover. Fourth, the increased gathering of firewood to meet the needs of the growing population furthered the decline of the area's vegetation cover (Walls 1980, 187; table 5.2).

Besides population pressure, other factors contributed to the expansion of agriculture into ever more marginal areas. Among these were the government's need for cash crops, short-term pressures of market economies, and the lure of quick money to be gained by bringing new areas under cultivation (Walls 1980, 183). These economic incentives remove traditional constraints that encourage rural residents to limit exploitation of resources to what is needed for basic food and income needs. Thus, even on land that remained in long-

Table 5.2.   Exploitation of Woody Biomass (metric tons)

	1963	*1973*
Fuel wood requirements	32,703	42,732
Other wood demands (building, fences)	1,662	2,338
Remaining trees, shrubs, or forest, fallow, on degraded land	153,695	81,214

Source: CAZRI (1976, 22).

term fallow or common rangeland, farmers, with resources to invest, were encouraged to accumulate animals that produced milk and meat products in demand in urban areas. This resulted in steadily increasing pressure through overstocking (Jodha 1985, 261–2).

Also important were the changes in the social conditions of arid India. Jodha (1985) implicates land reform as a prime factor in promoting land degradation in Rajasthan. Before reform in the 1950s, feudal landlords controlled access to common property resources by imposing heavy taxes and rents on peasants who wanted to convert rangeland into cropland. The high cost of the conversion made it economically undesirable to open new cropland in marginal dry areas. Land reform removed this constraint on use, and village councils that were supposed to function as substitutes for the old landlords proved unsuited to the task. Changing caste relationships also played a role (Jodha 1985, 258–59). Many Untouchables who had been restricted to lower-status craft occupations were liberated when the social control of the landlord class declined after 1950. Their ambition, frequently realized, was to become farmers—even if it meant accepting suboptimal farmland redistributed from former communal property. Members of the military caste also lost work in the 1950s, as the armies of princely states were eliminated or consolidated into the national armed forces. These higher-status individuals affiliated to the landlord class were frequently compensated with land resources drawn from the drier rangeland. Technological innovations contributed to conversion into cropland, since tractors were introduced into the region in increasingly large numbers after 1950. No longer dependent on animal power to plow land during the short planting season, farmers were able to bring into production increasingly large tracts of land, forcing animal herders to concentrate on reduced amounts of the poorest quality land. Thus social and technological change along with population growth played a role in both Luni Block's and Rajasthan's pattern of destructive creation.

The lands of Luni Block, while harsh and fragile, had been successfully farmed in a sustainable manner through the development of a complex traditional agricultural system. A set of crucial strategies evolved within the traditional milieu that met the short-term food and material needs of the inhabitants, while preserving the land resource over the long-term. Four major components of the traditional mixed cropping-livestock agricultural system were:

1. cultivated fields
2. pasture and fallow lands
3. xeromorphic forest
4. water resources.

The rational management of the interactions between these four components was crucial to the maintenance of a sustainable agricultural system. The primary function of cultivated fields was to produce foodstuffs. Wheat, pearl millet, and sorghum, all crops that can survive dry intervals during the growing season, were the main crops. The residues of the crops were utilized for livestock fodder. Fields in fallow and pasture provided grazing areas while soil fertility was restored under these conditions for future cropping. The groundcover of the fallow and pasturelands also protected them from wind erosion. The xeromorphic thorn forest provided both firewood and building materials for the inhabitants of the Luni Block. The trees also curtailed wind erosion on these lands. Both surface and groundwater were utilized as water supplies for the inhabitants in the Block. Surface water was harvested and stored in ponds. Each village had either one or two ponds; in an average year, enough water was stored in the ponds to last from three to six months (CAZRI 1976, 33). After the ponds dried up, the villagers dug shallow wells near the pond sites or relied on deep brackish wells until the advent of the next rain season.

To meet the increasing demand for foodstuffs with the increasing population due to medical improvements and intramigration, cultivation was altered. Cultivated land increased greatly in the arid western districts, former areas of pasture (CAZRI 1976, 13). Fertilizers were too expensive with the result that productivity declined rapidly due to decreases in fertility even when precipitation was adequate. With the conversion of these lands into farming, bare ground was now exposed and during the dry season wind erosion began to be significant. Also changes in the cropping system resulted in increasing salinization in other parts of the area (Walls 1980, 189). Pasturelands experienced direct consequences from agricultural expansion onto former pasture and intensification on traditional farmlands. The amount of land available for grazing was reduced. With no reduction in livestock, overgrazing occurred and wind erosion became effective on these lands. Woody biomass also decreased with the growing population in response to the needs for fuel, feed for livestock, fencing, and hut construction. Not only was the aboveground component of the vegetation removed by the inhabitants, but roots were dug up and extracted too! Thus these lands also became susceptible to wind erosion.

Land degradation/desertification in the Luni Block became a problem and made the area a case study for the United Nations Conference on Desertification in 1976. From an agricultural perspective, the Luni Block is situated in a hazardous setting. Climatically, it is at the extreme minimal precipitation limit of rain-fed agriculture. Its soils are light to medium in texture and when exposed to wind are highly susceptible to erosion with dune formation and a hummocky landscape resulting. In addition a thick, hard calcium carbonate layer underlies a large part of the area. Despite these risk conditions, a successful rain-fed agricultural and livestock system evolved over the millennia.

By the mid-1950s, in response to rapidly changing conditions, the traditional farming system could not meet the demands of the inhabitants or needs of the nation. Agricultural expansion and intensification, the human responses to these demands, placed stresses on the environment that were beyond its capacity. Decreases in soil fertility and overgrazing resulted in a dramatic decrease in the area's groundcover. Wind erosion became widespread. The result was that the "new" farmlands were largely abandoned by the mid-1970s (CAZRI 1976). With the resulting severe degradation (desertification), manifested by dune formation, a hummocky landscape, widespread salinization, and exposure of the underlying calcium hardpan layer at the surface, the area did not revert back to serving as pastureland. A crucial component of the sustainable traditional dry farming and livestock system had been lost for an extremely brief period of agricultural expansion, when a minimal short-term gain of additional crops occurred. What was an extremely rational and successful sustainable agricultural system, even though yields were not high, evolved into an unstable and degrading system. Stressed beyond its capacities to produce sufficient food and other products, the Luni system is no longer able to meet the present and future needs of the area.

The last thirty years have added more nuanced perspectives on the past, present, and future of environmental and land-use change in Rajasthan. Often the postcolonial scientific elite in India, including foresters, evaluated the condition of rural resources in Rajasthan in quite a different way than did local resource users (Robbins 1998, 2000, 2003). Moreover, the debate over who can best control and manage local resources is not simply one dividing state agents from local residents in pursuit of their daily livelihood needs, but also pits local groups divided by caste, ethnicity, gender, and class against each other. Foresters, for example, regard the spread of mesquite (*Prosopis juliflora*), an introduced North American dryland woody plant, as an effective management tool that provides important land cover and fuelwood resources, whereas local resource users often find it undesirable in comparison to native tree species (Robbins 2001a). When institutions that formerly controlled resource access and prevented degradation successfully begin to loose their effectiveness, ". . . it is typical for farmers to blame herders, landlords to blame outcastes, and women to blame men" (Robbins 1998, 415). Yet many local institutions remain viable in protecting communal resources, and some of the traditional interaction patterns continue to promote environmental health. Thus Agrawal (1999, 109–10) demonstrates that the dung-stubble loop between farmers and herders remains viable and important for both groups. Manure sales are an important income source for migratory Raika sheepherders, while organic fertilizer input is an important nutrient supplement for the fields of both irrigation farmers and low caste agriculturalists. The result of environmental change in semiarid Rajasthan has produced a mosaic of mixed land uses, some improving, some declining in condition in response to a subtle combination of local, site-specific circumstances. Such "impure" landscapes (Robbins 2001b) make it exceedingly difficult to detect a uniform, directed trend in environmental status, but rather suggest that the contemporary picture of land degradation in Rajasthan is more complicated than that painted for the region in the UNCOD Luni Block data.

Rain-fed agriculture in the humid midlatitude area of the American Southeast is a cautionary tale from a sharply contrasting habitat. Some scholars rank the destruction of farmland due to wind erosion, and the resulting Dust Bowl in the semiarid American southern plains during the 1930s, as one of the three worst ecological blunders to agriculture (Borgstrom 1973, 23). Yet, throughout the 1930s, the humid American Southeast, and not the southern plains, experienced the most extensive and severe soil erosion (Healy and Sojka 1985). Farming in the region resulted in millions of hectares of cropland becoming worthless for agricultural activities. By 1938, of the nation's 20 million ha (50 million acres) of abandoned farmland due to degradation, a significant proportion was in the Piedmont area of the southeast.

Land degradation in the Piedmont and hilly uplands of the Southeast differs markedly compared to Luni Block. In contrast to the seasonally desolate landscape with few options in the semiarid area, land degradation in this humid area of the United States resulted not in complete devastation, but rather the elimination of many land-use options. In the Southeast, the region suffered serious but relatively moderate land degradation in spite of high soil erosion. This different outcome is a result of both physical and human contrasts between a portion of semiarid India and a humid area within the United States. As the Piedmont region suffered the greatest land degradation in the Southeast, this is the focus of our next inquiry into unintentional destructive changes of the environment resulting from human misuse.

The southern Piedmont begins in southern Virginia and stretches some 1,200 km (750 miles) until it ends in eastern Alabama. It is a rolling hilly region primarily located between the higher and steeper Blue Ridge Mountains in the west and the lower and flatter Atlantic Coastal plain in the east. The local relief in the region ranges from 15 m (50 ft) up to 60 m (200 ft). Its climate is both humid and mild with a 200–240 day frost-free period and annual precipitation up to 1,500 mm (59 in).

Prior to European settlement the Piedmont was almost 100 percent forested. "Erosion appears to have been negligible on the Southern Piedmont of aboriginal times as indicated by the clear streams, the presence of dark, mature bottomland soils, and the fact that present erosion rates of undisturbed forested areas are minimal" (Trimble 1974, 34). Beginning around the 1700s and continuing up to the 1930s, exploitation of Piedmont soils for short-term profit became the modus operandi throughout most of this region. From its inception, Piedmont agriculture developed into an extensive versus an intensive type of farming system. Unlike Luni Block, an abundance of inexpensive land was available for the local inhabitants. With relatively high labor costs, farmers developed cropping practices that concentrated on yields per farmer and not production per area. Land was cleared, farmed, and abandoned once yields did not satisfy needs. Farms, when deserted, were left without vegetation cover to protect the remaining soil. Thus abandonment, in itself, was a highly deleterious farming practice that exacerbated soil loss. The exposed soils on the abandoned lands quickly eroded due to the abundant rainfall and the soils' relatively high erodibility properties. The continuous misuse of land over a 150-year period was made possible by readily available fertile land immediately to the west. Once "exhausted" for agricultural needs, land was sold and new territory was cleared of forest to the west (Craven 1926). This was the general cycle of tobacco, cotton, and mixed crop farming throughout the Piedmont. Immediate profit with minimal regard to the land resource and its sustainability is one legacy of the colonial and early American periods in this region. Soil erosion was a widespread problem during the early history of the United States (Hambidge 1938). Of all the regions, the Piedmont may have experienced the nation's most serious soil losses. Widespread clay soils, the deeply weathered subsoil, the small amount of level land, and ubiquitous poor farming practices resulted in widespread degradation, at times culminating in extensive gully formation (Fenneman 1938). The lack of a national conservation ethos resulted in 95 percent of the uplands experiencing some degree of soil loss, 65 percent of the Piedmont losing its topsoil, and more than 10 percent of the area also losing its subsoil. About 1.1 million ha (2.7 million acres) of land had active gullies before 1930 (Bennett 1929). Degraded landscapes with gully scars, poorly vegetated hillslopes, and muddy rivers due to the accelerated erosion occurring throughout this region were the rule throughout the Piedmont.

One result of the exploitive nature of the farming system and the resulting land degradation was that between 1925 and 1960 harvested cropland decreased in every county, except one, in the southern Piedmont. In parts of the region, acreage in row crops declined by over 90 percent (Trimble 1974, 97–98). Yet while erosion and salinization resulted in a degraded Luni landscape leaving only reduced economic opportunities for future generations, in the Piedmont today only rare and isolated patches of bare soil can be found. With complete abandonment of land in the Piedmont, even though severely eroded, climatic conditions and the remaining weathered subsoil permitted natural processes to begin a healing process. The area, through a slow transition process from weeds to brush and finally back to forest, gradually increased its biomass production potential. This natural land healing was additionally reinforced, beginning in the late 1930s, through the intervention of major national conservation programs and by the introduction of a changing crop mix on the remaining cultivated lands (Bennett 1943). Gully and sheet erosion were curtailed through both engineering works and the planting of soil-conserving vegetation. In particular, kudzu—which has since become a major problem (Alderman 2004)—pasture, pine trees, soybeans, and wheat replaced cotton and corn. These conservation interventions have resulted in a region where once again streams are generally clear, forest and pasture occupy a majority of the former highly eroded croplands, and tilled cropland reflects a "new" farming system that emphasizes a variety of conservation practices and new crops. No longer is a short-term exploitation of the region's remaining agricultural lands the norm. Interestingly, a study undertaken in the 1980s found that a higher economic return occurs when pasture and cropland are converted to pine plantations in some parts of the Piedmont (USDA 1983). Thus the changes in land-use practices are rational both from economic and environmental perspectives.

However, today's land use, albeit productive, still strongly reflects its colonial and American heritage of land misuse and land degradation. First, existing soils on the uplands and hillslopes have a lower potential than they possessed prior to the clearing of the area. Thus potential options only viable with soils having a higher fertility are lost. Second, the sediment transport today in streams, while comparable to rates under the virgin forest, has initiated a new cycle of stream erosion in parts of the region. The excessive erosion prior to the 1940s resulted in thick alluvial deposits on the valley bottoms. Because of effective conservation practices that rehabilitated the degraded uplands and slope areas, the resulting clearer stream flows have resulted in renewed stream competency. "Streams (have) incised themselves into the modern alluvium lowering their beds as much as 3.7 m (12 ft). There has been intensive erosion of the friable stream banks (composed of post-1700 sediments) taking trees, fences, and good pasture or cropland" (Trimble 1974, 118). This contemporary instability along some of the bottomlands, which is the result of the interaction between the prior land degradation in the uplands/slopes area and the concomitant deposition of sediment on the valley bottoms (deposition in the form of unintentional creative destruction as new productive farmland was created), and the modern rehabilitation practices throughout the region, reduces the productivity of these floodplains. In particular, wetlands are created. While less productive from the human perspective, these new wetlands open up niches for other plant and animal communities. This continuing process of wetland expansion illustrates some of the complex interactions that exist in nature. The contemporary conservation strategies directed to rehabilitate the eroded upland/slope zones have inadvertently initiated accelerated fluvial erosion on the productive bottomlands. From the human perspective, the resulting expansion of wetlands results in a degrading land resource that had originally been "improved" at the expense of land degradation in the region's upland and slope zones.

The Piedmont, unlike Luni Block, represents relative versus absolute land degradation. The Southeast's humid climate results in a more resilient environmental condition than in semiarid northwestern India. It is more forgiving of human transgressions. Even though upland and hillslope soils were severely degraded, because moisture is not in short supply due to the humid climate, vegetation has been able to reestablish itself in a relatively short time. Furthermore, because the southern Piedmont exists within the context of a strong national economy, resources derived from outside the area were made available to facilitate the rehabilitation process. Finally, while the Piedmont is again a highly productive area, it is less productive than it would have been if the degradation had not occurred. In this sense, the full range of natural resource based options was sacrificed for the enjoyment of present advantage. Instead of exporting environmental costs to a different spatial unit, these costs were transferred across time and inflicted upon future generations. This is a variation on the sacrifice zone pattern that is more characteristic of destructive creation. In this sense, resource-use practices of the Southern Piedmont reflect the intergenerational impacts of destruction creation that are more characteristic of the contemporary era.

Terrace Systems

Cultivation on steep lands encourages rapid surface water runoff. Two ramifications of this are excessive soil losses due to both the quantity and velocity of the surface flows draining the steep slopes and relatively low infiltration of precipitation due to the brief time period that the water remains on the slope. Unless this runoff is controlled, eventually land degradation occurs from the loss of the upper fertile soil horizons and/or gully formation. Additionally soil textures become increasingly coarse as the fine soil components are eroded to a greater extent than the less fine ones (Lewis 1981). A widely recommended agricultural intervention to prevent high soil losses on steep slopes and encouraging infiltration of moisture is terracing, a practice that topographically alters the landscape and increases its potential instability.

Terracing dates back to antiquity and has undergone an imperfectly understood pattern of evolutionary development and diffusion (Spencer and Hale 1961). The Nabataeans utilized terraces to increase infiltration in an extremely dry area. In China and Ethiopia, stone terraces have a long history as a conservation practice to control soil erosion. In the Andes, Inca development of terracing intended to trap and control runoff as well as to control soil loss (Pawluk, Sander, and Tabor 1992). Terraces are generally constructed by cutting into the soil on a slope segment and then using the material removed in the cut to add fill along the downslope portion of the terrace. When water conservation is a purpose of the terrace, the bench portion is either constructed level or slopes inward toward the hill (figure 5.12). Where the major purpose of the terrace is the reduction of the erosion poten-

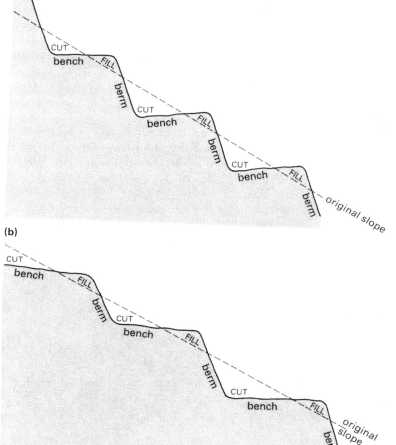

Figure 5.12. The cutting and filling of hillslopes required for terrace construction make these areas less stable if not constructed properly and regularly maintained. The benefits of terraces include erosion control, the development of flat surfaces on which crops can be cultivated, and moisture retention.

tial, the bench portion is either level or outward-sloping. In all instances the bench is constructed at an angle that is less than the original hillslope. Construction of erosion control terraces usually requires runoff ditches and/or conservation banks to disperse the runoff into thin, nonerosive flows over the terraced fields. The goal is to prevent runoff from concentrating its flow, which would create either rills or gullies.

By altering the configuration of hillslopes, terrace construction creates slope angles that are both less than (bench) and greater than (terrace berm/wall) the original hillslope. Since the natural terrain likely represents the most stable morphology (Hack 1960) for the region, terracing inherently creates unstable conditions. This instability is found in two areas: (1) the bench areas, with their lower slopes, become zones that have a higher potential for deposition;

and (2) the berm zones, with their steeper slopes, are prone to erosion. Only through continuous maintenance will terraced lands remain stable. As but one example, terraces that for centuries were stable are now being destroyed in many parts of the world from Spain to Yemen (Vogel 1987). Due to rural out-migration, farmers no longer have the available labor to maintain all of the terrace walls and drainage channels in these areas. Runoff that previously was controlled by continuous terrace maintenance now results in rill and gully formations. These phenomena, the result of uncontrolled sheet wash, are slowly destroying the terraces. The bench areas of the terraces are becoming steeper while the steeper terrace walls are gradually being destroyed. Both of these processes remove the area's topsoil, which is deposited along the valley bottoms. Thus once maintenance becomes slack, terraced land often leads to accelerated degradation along the affected hillslopes (figure 5.13). Ultimately, the destruction of the terraces will restore the hillslopes to an equilibrium condition that no longer requires the human input of maintenance. Albeit the new equilibrium condition will have a drastically altered soil, hydrological, and vegetational status compared to the conditions existing prior to the human intervention of terracing.

Terracing requires major labor inputs or the use of heavy machinery to reconfigure the hillslope and construct the requisite drainage ditches. This is even true for the *fanya juu* terraces in Kenya (Lewis and Berry 1988) and the *fosses aveugles* in Rwanda (which were largely destroyed during the political instability of the 1990s), where the majority of the cut and fill required for terrace formation was done by erosion and deposition processes (figure 5.14).

Figure 5.13.   Terraces on steep slopes in Rwanda begin to erode when either maintenance declined or excessive rainfall occurred. (Photograph by L. Lewis, 1990)

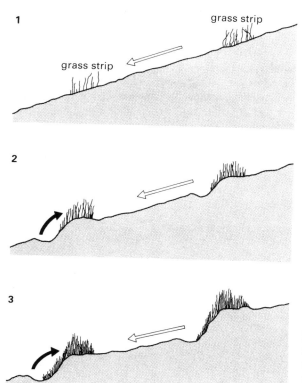

Figure 5.14. Grass strips are a low-cost method employed to reduce erosion. Over time, the gradual accumulation of soil behind the grass strip transforms a sloping surface into a terrace. (From Lewis and Berry 1988, by permission of L. Lewis.)

Similar methods of terrace construction are employed at these two sites. First, a drainage ditch is dug orthogonal to the direction of the downslope. The excavated soil from the ditch is placed immediately above and parallel to the ditch. Grass and shrub strips are planted in numerous rows along the low ridge formed by these excavated materials. Runoff from the field above the grass strips transports soil from the field to the grass strips. Contact with the plants decreases the surface water velocity, which encourages deposition in the grass strip zone. Over a number of years, the erosion on the field results in a proportion of this soil being deposited in the grass zone. This material, along with the continuous maintenance of the drainage ditch from which the farmer throws the soil deposited in it up into the grass zone, results in the formation of a berm along the grass strips.

If the fanya juu/fosses aveugles terraces are to be successful, care must be taken to insure that the spacing of the grass strips is close enough in order that the soils eroded from the upper portion of the bench will not bring infertile soil into the root zone of the crops. In many areas of central Kenya, thick volcanic soils exist, and the erosion and deposition of these fertile soils to form terraces has made the fanya juu technique successful (Lewis and Berry 1988). But in parts of western Rwanda, where soils are both less fertile and thinner, this method of terrace formation contributed to the land degradation problem (Lewis 1992).

Often distances between grass strips were too great given the thickness of the soils (figure 5.15). This has resulted in the highly acidic and infertile B-horizons being brought close to the surface along the back portions of many of the terrace benches. One outcome of this has been poor crop yields on the back portions of these terraces. To increase food production on the benches' back portions, during each growing season many of the farmers remove some of the better soils from the berm immediately above the zone of poor soils. This material is then spread along the backbench and acts as a fertilizer. While improving crop yields in the short-term, this counters the long-term purpose of the terrace, namely reducing the downslope movement of soil. The human erosion (removal of soil from the berm) and deposition of soil spread over the back portion of the bench results in the movement of large quantities of soil each year downslope. Thus terrace construction in some Rwandan areas represented another contributing factor to the widespread degrading condition of many highland areas in western Rwanda (Lewis and Nyamulinda 1989). This is one example of how topographic changes on stable steep lands can promote inappropriate interventions that contribute to land degradation by encouraging the very unstable conditions that the terrace technology was intended to prevent.

When farmers terrace the land it is done with the intent of growing specific crops that likely would not be sustainable or possible under natural conditions. Therefore, as in any agricultural endeavor, terrace lands are covered by vegetation that has replaced the natural

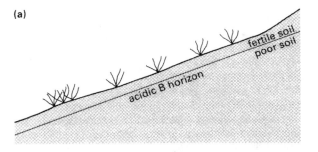

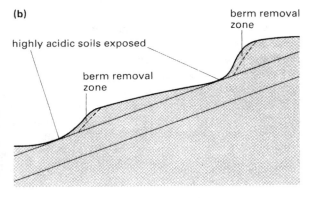

Figure 5.15. In Rwanda, efforts to expand the width of terraces and improve their fertility by spreading topsoil on them often exposes infertile subsoil layers whose low fertility decreases agricultural yields. (From Lewis 1992, copyright John Wiley & Sons. Reproduced with permission.)

groundcover. Through this alteration, the previous ecosystem is greatly altered or destroyed and a new one is created. In northwestern Rwanda, as the foothills of the Virunga Mountains have been cleared and terraced for agriculture, the forest ecosystem has been destroyed. This created landscape permits the people to increase agricultural production and meet their food needs as their population grows. However, the destruction of the tree cover has also had major deleterious effects. It has destroyed a significant proportion of the habitat of the mountain gorilla, now protected but highly endangered. Even though the remaining forested lands are within a national park, it remains to be seen if enough of the mountain gorilla habitat remains to insure their long-term survival. The hillslope terracing and forest removal process in northern Rwanda along the Ugandan-Congo border illustrates an essential feature of creative destruction. It inherently results in both winners (those that benefit from the change—farmers) and losers (the mountain gorilla and other wildlife and plants).

Intensive Animal Husbandry

Traditional nomadic animal husbandry is an extensive operation. Animals are moved over considerable distances in an ecologically astute fashion to find grass and water. Most often located in drier environments, this form of animal production is a gigantic rotational pastoral system carried out over great distances and, while the animal populations involved are frequently large, the human population engaged in the traditional livelihood system is often quite small. Traditional mobile animal production systems are extremely well adapted to the highly variable, random, disequilibrium ecology of their primary habitats (Behnke, Scoones, and Kerven 1993; Roe, Huntsinger, and Labnow 1998). The key to low-intensity traditional mobile pastoral systems is that the animals travel to find their food rather than have their food brought to them. In this way herds and herders are generally able to escape the impact of local drought and temporary land degradation, returning to these areas at different times of the year when natural resources have had time to recover. Required to work hard to reach sustenance, animals in traditional pastoral systems graze extensively on natural forage found in relatively open landscapes, produce relative low milk yields, yield less marbled (fat-filled) meat, and usually take a relatively long time (about three years for cattle) to reach marketable weight and size.

Originally many of the animal husbandry systems that emerged in North America after the Europeans began to colonize the region were extensive in nature. Particularly in the interior, cattle and hogs were turned loose into nearby common land to forage for themselves. Periodically these animals were rounded up and driven, often for considerable distances, to the markets of more densely settled coastal communities (Carlson 2001). These seasonal drives were the primary way to get cattle to market and they represented the major source of cash income for the dispersed population of the interior. Impacts on the environment by these cattle were temporary and minor, since nature set the limit on livestock numbers and artificial efforts to enhance herd size were negligible. Only limited capital and labor were invested in herd development and management; animal numbers were regulated by natural fluctuations in locally available forage, and the profits gleaned from the enterprise represented a much appreciated bonus supplementing a mixed economy with few other commercial possibilities beyond home-brewed alcohol (moonshine). The cowboy culture of the open range and long-distance cattle drive, which emerged in the Great Plains after the American Civil War, was a variant on this time-honored practice. While more sedentary forms of animal keeping quickly replaced the era of open-range herding, the legacy of that era continues to color local identity as well as outsider perception of the American West (Knight, Gilgert, and Marston 2002; Shoumatoff 1997; Starrs 1998).

Intensive modern animal husbandry systems work in exactly the opposite fashion to extensive systems such as pastoral nomadism or traditional ranches. Zero grazing, a system in which animals never exit a confined feedlot environment to access naturally growing pasture, is the ultimate goal. With livestock kept in close, controlled confinement, intensive systems bring fodder to the animals. Restricted movement enables the animal to concentrate on growing as fast as possible. With few calories burned off by physical activity, modern meat is marbled with a high fat content, which results in a more tender meat than its range-fed cousin. Because large numbers of animals are kept in close confinement, the danger of disease is increased and the risks of epidemics sweeping through a herd are considerable. Sick animals grow less quickly to slaughter weight and consume more fodder over their longer trajectory to the abattoir than do healthy ones. This longer fattening process erodes the profitability of the intensive operation. To maintain a healthy animal population, managers resort to the prophylactic use of pharmaceutical drugs and, whenever they are legally allowed, insert growth-inducing additives into the feed. The long-term consequences of these practices for human health have received considerable attention for some time (Schell 1978), although their ultimate impact remains incompletely understood. In some quarters, the higher consumption of animal fats in marbled meat in the United States is blamed for a deterioration of consumers' health and an increase in the incidence of heart attack, cancer, and strokes (Rifkin 1992, 171)—although a number of equally important general life style variables, such as lack of exercise, smoking, and supersized meals, are undoubtedly also implicated.

Every effort is made to organize modern livestock production as if it were a factory (Mason and Singer 1990). At the scale of the farm, the animal-rearing process is divided into a series of segments linked to the animal's stage in the life cycle. Inputs are fed into animals that are moved by stages from one holding area to another until the ultimate goal is attained and the animals are sent to market or designated for a specialized use such as incorporation into a dairy herd. The feed funneled to the animals is scientifically selected to match the nutritional needs of the beasts at each stage in their development. Clearly whenever waste products from some other components of the larger agricultural production system can be added, this improves economic and energetic efficiency. Central to the organization of intensive factory farming is severe restriction in the movement of the animals, which invariably are kept in close confinement and are prevented from wasting energy on superfluous movement.

Pork production is a good example of the factory farm system (figure 5.16). As an omnivorous (but not indiscriminate) scavenger with a biological structure that efficiently translates feed into weight gain, the pig is an ideal candidate for this type of system. On average, it takes the relatively efficient U.S. livestock industry about 6.9 kg (14 lb) of feed to produce each half kilo (1 lb) of pork that reaches the tables of American consumers (Durning and Brough 1992). Since up to 85 percent of the cost of commercial pork production can be expended on feed (Pond 1983), by reducing the cost of feed inputs, produc-

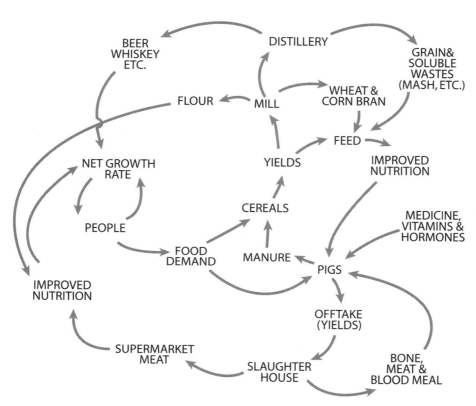

Figure 5.16. Modern pork production involves keeping a large number of sows and their offspring in close confinement. (Diagram by Anne Gibson, Clark Labs)

ers can contribute most directly to improvements in profitability. Thus waste products such as bran from milling grain can be added to the corn, soy, sorghums, and other grain products that constitute the bulk of hog inputs. This indirectly captures for humans a nutritionally important product via an intermediate processor, the pig, rather than people having to consume them directly as a less-tasty dietary supplement sprinkled on one's morning cereal. Similarly, the waste products of distilleries and other agricultural processing plants can be recycled to hogs as a component of their diet. In addition to disease-related concerns, the protein deficiencies of a diet containing a great deal of corn mean that significant vitamin, protein, and hormone supplements are considered necessary to keep the modern hog happy, healthy, and rapidly growing.

The factory farming system can pose serious land degradation problems at the local scale of the farm. It also can threaten the health of regions as entire landscapes are altered from biologically diverse assemblages of wild and domesticated plants to fodder-oriented monocultures to sustain and feed a vertically integrated meat-production industry. Moreover, these environmental problems are not new, but emerged early in the history of industrial-scale

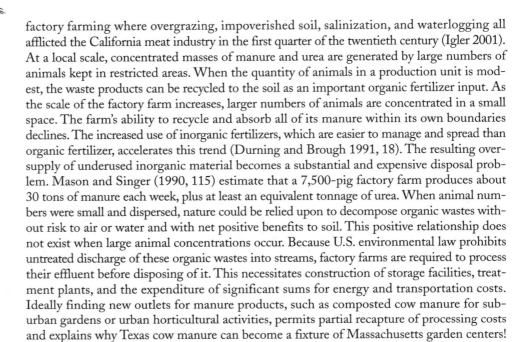

factory farming where overgrazing, impoverished soil, salinization, and waterlogging all afflicted the California meat industry in the first quarter of the twentieth century (Igler 2001). At a local scale, concentrated masses of manure and urea are generated by large numbers of animals kept in restricted areas. When the quantity of animals in a production unit is modest, the waste products can be recycled to the soil as an important organic fertilizer input. As the scale of the factory farm increases, larger numbers of animals are concentrated in a small space. The farm's ability to recycle and absorb all of its manure within its own boundaries declines. The increased use of inorganic fertilizers, which are easier to manage and spread than organic fertilizer, accelerates this trend (Durning and Brough 1991, 18). The resulting oversupply of underused inorganic material becomes a substantial and expensive disposal problem. Mason and Singer (1990, 115) estimate that a 7,500-pig factory farm produces about 30 tons of manure each week, plus at least an equivalent tonnage of urea. When animal numbers were small and dispersed, nature could be relied upon to decompose organic wastes without risk to air or water and with net positive benefits to soil. This positive relationship does not exist when large animal concentrations occur. Because U.S. environmental law prohibits untreated discharge of these organic wastes into streams, factory farms are required to process their effluent before disposing of it. This necessitates construction of storage facilities, treatment plants, and the expenditure of significant sums for energy and transportation costs. Ideally finding new outlets for manure products, such as composted cow manure for suburban gardens or urban horticultural activities, permits partial recapture of processing costs and explains why Texas cow manure can become a fixture of Massachusetts garden centers!

But large concentrations of manure, despite regulations mandating treatment, frequently seep into and negatively impact the local environment. The litany of these impacts is considerable. At one level are the odors, whose noxious scents frequently offend the nostrils of neighbors. The combination of smells and sounds generated by early morning farming activities are often anathema to nonfarming suburbanites, and frequently produces local antipollution legislation regarded as impractical and punitive by farmers and agrobusiness managers. At another level are the pests, birds, rodents, and insects attracted to the farming operations (Mason and Singer 1990, 124). These unwanted guests are a threat to both farms and neighbors through their potential to facilitate the transmission of disease. Run-off from farming operations carries unwanted and undesirable contaminating organic material from farms to adjacent land, streams, and ponds. Repeated heavy use of organic manure slurry to fertilize fields, combined with a high groundwater table, such as occurs in the southern Netherlands (Durning and Brough 1991), can produce excessive concentrations of nitrates and phosphates in the soil and groundwater. Too much nitrogen and phosphorus in surface aquatic ecosystems can promote the growth of aquatic plants and algae, resulting in oxygen-deficit conditions inimical to the survival of fish and other aquatic creatures. Ammonia released to the atmosphere from the large concentration of animal factory farms in the southern Netherlands is implicated as a major source of acid rain and acid deposition in the soils of the region (Durning and Brough 1992, 71), and is likely to be an important regional factor in more dispersed, less readily measurable context wherever large concentrations of animals are found.

To date managing the sewage stream flowing from animal factory farms is at best an imperfect process. Because inorganic fertilizers are cheaper to use and easier to manage than their organic alternatives, manure wastes are regarded by most factory farm managers as a problem rather than a potential resource. Until this attitude changes and animal wastes are perceived as an important source of by-product income from the main production enterprise, wastes will at best be captured grudgingly and incompletely. Serious local impacts on soil, groundwater, and air will be the consequence.

Managing Nature: Parks, Wildlife, and Nature Protection

Nature and culture interact to produce distinctive landscapes. What we see of nature and culture in those landscapes is a product of our values and fundamental beliefs about what should rightly be there. These perceptions, in turn, are the product of myths and memories generated over centuries of cultural accretion and are found reflected in artistic, literary, historical, naturalist, and mythological representations of the environment (Schama 1996). Cultural differences in what people perceive to be nature and natural also affect whether they believe that particular uses of resources produce land degradation or whether the technologies that permit that exploitation and development of nature are inherently good or bad.

One situation in which nature and culture struggle to coexist is the nature park. In North American culture, nature is often viewed as a distinct, sacred entity, whose survival depends on isolation from the profaning influence of humankind. The goal is to keep observable human impacts as minimal as possible, although even that decision represents an intervention with significant consequences. From this strict constructionist perspective, all but the most insignificant human presence constitutes a form of degradation. The archetype is the wilderness area, a concept codified as early as 1929 during the administration of U.S. Secretary of the Interior Ray Lyman Wilbur, who was determined that parts of Glacier National Park, as well as designated wilderness areas in older parks such as Yosemite and Yellowstone, should be preserved in perpetuity in primitive condition (McClelland 1998). Such areas were to be cut off from all but the most minimal, recreational uses by extremely limited numbers of people. Here only a privileged, priestly few were to be allowed to worship and commune with the pure, preserved wild landscape. This is particularly true in the western United States, where parks and wilderness areas are large and the distance to substantial population concentrations is great. Export to Africa of this ideal of nature as wilderness has resulted in the exclusion of indigenous populations from the use of land that had long been within their sphere of exploitation. The concentration of resource use outside the park boundary inevitably produces increased degradation, and is frequently met by many acts of resistance to the park in the form of poaching, theft, and invasion (Neumann 1998). This type of perpetual struggle at the margin of the nature park has led Brandon, Redford, and Sanderson (1998) to question whether rigidly bounded parks in the neotropics can hope to survive if they fail to incorporate socioeconomic issues into park planning and allow for reasonable exploitation of park resources by adjacent indigenous populations. Employment of this strategy by developers of a national park at Mt. Sinai in Egypt, whereby local Bedouin not only serve as park rangers but also can continue to graze and cultivate within park boundaries, promises to provide protection for a unique and endangered local flora and fauna (Hobbs 1995). In eastern North America, wilderness areas, such as the Bristol Cliffs Wilderness Area in Vermont, are often much smaller in scale and are more likely to be an artifact of human perception and re-creation and recognized as such rather than pristine habitats (Elder 1998).

In other types of national parks and public lands, the conflict between users and protectors, loggers and spotted owls, ranchers and the anticattle apostles of the restoration of virgin, pre-European settlement ecologies, tourists and rangers rages unabated. On the whole, this struggle has tended to favor those who wish to develop park landscapes and make these spaces accessible to a wider public (Sellars 1997). Certainly communities located close to the park boundary, and particularly to the major points of entry into the park, can benefit economically from their location (Machlis and Field 2000). In instances where economic growth has occurred without sacrificing aesthetic considerations, local character, and environmental quality, a strong planning function that uses land-use restrictions and habitat conservation initiatives is essential to success (Howe, McMahon, and Propst 1997) and difficult to maintain. Addressing the often uneven and unexpected consequences of "green" tourism in the Third World, Honey (1999) points out the frequently tragic consequences for both people and environment of trying to balance ecotourism and sustainable development. But demands for wider, straighter, and faster access roads, with all of the attached infrastructure of contemporary suburban civilization, continue to encounter fierce intellectual and philosophical opposition (Abbey 1968; Frome 1992) from those who prefer a less cultured, rougher-edged nature and fear for the ultimate survival of the environment that made these spaces attractive in the first place.

In Europe cultural landscapes are viewed as being as much a part of nature as are rocks, trees, and wildlife. Only in the last dozen years has an analog, the national heritage corridor, emerged in the United States (Conzen and Carr 1988). Perhaps because space is less abundant, fewer untouched habitats are found in Europe. Indeed, the humanized landscapes that developed over centuries are regarded as having inherent "natural" value and considerable effort is placed upon retaining both the remnants of nature and the cultural traditions and livelihoods that have crafted contemporary landscapes. So blurred is the distinction between nature and society that many areas, such as the Broads of England's East Anglia, now revered as natural habitats meriting protection, had their origin in humankind's "destructive" alteration of nature (see chapter 7). Such attractive landscapes, composed of complex, intertwined threads of natural and human origin, are often most threatened by the pressures of tourists drawn to the very aesthetic features that make the landscape distinctive. How to prevent large-scale economic change and industrial development and pollution from adversely impacting nature was significantly assisted by the growth of the European Union and its environmental legislation and environment-based educational programs (Bromley 1997). Efforts to use local natural/cultural landscapes as a basis to promote sustainable ecotourism and rural economic growth occurred in many countries. Often these efforts required compromises, such as the hardening of popular paths with all-weather pavements in order to prevent erosion (Shipp 1993, 27). Frequently planning procedures mandated the establishment of zones within and adjacent to parks that regulated the intensity of use. To some areas access was virtually impossible for all except scientists conducting experimental and monitoring projects. Other areas were virtually sacrificed by permitting extremely intensive use in order to control impacts elsewhere. Quiet zones and pedestrian and cycling zones often represented intermediate areas of more modest impact. The result is a natural landscape of blurred boundaries, where people live and work at economically productive activities within the park, in which both local residents and visitors have a vested interest in the myriad of ways in which nature and society can coexist without causing destructive creation.

Thus different cultures conceive of nature and human livelihood in different ways. These contrasting and competing perceptions result in culturally prescribed policies for managing and interacting with nature. These variable perspectives on the relationship between nature and culture are multifaceted and are susceptible to variable interpretations. Cross-cultural comparison of different ways of balancing nature's needs and society's demands suggest that nature parks can play a significant role in rural landscape management and nature park policy development.

Summary

Efforts to intensify production in agricultural systems and to alter the nature of common property resource systems have produced serious land management problems at the local scale, and these are featured in this chapter. While most of these land-degrading impacts are local in nature, they are the product of more than local decisions. Often land degradation reflects the impact of much broader trends operating in economy, society, and environment. Taken in aggregate, the local disturbances generated in response to broader processes can build into problems that are regional in scale and transcend the peculiarities of local habitats. It is to these examples of regional and global land degradation trends that we turn to in the following chapter.

Land Degradation
Regional and Global Examples

IN 1800 BEIJING was the only city with a population exceeding 1 million. Worldwide, only seventy-four cities had populations greater than one hundred thousand and, with the exceptions of México City and Cairo, all were located in Europe or Asia. Today over two hundred metropolitan areas have populations in excess of 1 million with every inhabited continent having a share in this increase. This urban growth is associated with the surge in overall population growth made possible in part by increasing food production as well as increasing industrial and commercial activities, by-products of industrialization and technology advances. Once, urban growth was primarily concentrated in the industrial countries. Today it is a worldwide phenomenon, with some of the highest rates of urbanization occurring in the developing world. Land scarcity in rural areas and the search for economic opportunities are two important factors contributing to rapid population growth in the less developed nations. No matter what the driving forces causing urbanization may be, as populations increase urban areas expand areally.

Urban Growth

Urban areas in the developed world, with their greater transport systems, usually cover larger areas than those situated in areas of lower wealth. The Los Angeles-Riverside urbanized area with a population of 14 million covers 5,959 sq km (2,299 sq miles); Tokyo-Yokohama population of 2.995 million covers 4,432 sq km (1,710 sq miles); and the São Paulo area with 1.68 million population includes 2,462 sq km (950 sq miles). México City's metropolitan area is a classic example of urban sprawl and illustrates what is occurring elsewhere whether it is Lagos in Nigeria, Paris in France, or Dallas-Fort Worth in the United States.

In 1960 with a population slightly over 6 million, 75 percent of México City's inhabitants were living within the Federal District. In 2000 with a population in excess of 17 million only about 45 percent of México City's population lives in the Federal District, with most growth occurring in the suburbs as the area spreads into the surrounding states (Demographia 2001). Urban sprawl, associated with modern urbanization, consumes significant areas of once highly productive agricultural lands to house increasing populations and to provide land for other urban activities. As land use is metamorphosed from rural to urban, a high proportion of vegetated land cover is replaced by sterile concrete and asphalt, productive wetlands are filled and replaced by urban structures, and pollution associated with many urban activities results in a spectrum of chemical and biological changes that often degrade the affected areas (Flawn 1970, 119).

During the establishment of older urban centers, a critical criterion that determined their location was proximity to fertile land (Mumford 1961). With the ongoing urban explosion, since today for the first time in the history of humankind the urban populations worldwide are greater than the rural, this juxtaposed highly productive farmland is being destroyed even in countries that have a shortage of fertile lands. Urbanization, with all of its positive and negative feedbacks, is truly a worldwide phenomenon. It is one of the most critical events—including both anthropocentric and nature-driven processes—shaping the Earth's surface. Running the whole gamut from the ancient agricultural lands in the lower Nile Valley, the pre-Columbian *chinampa* agriculture in the Basin of México to the modern citrus groves near Orlando, Florida, productive agricultural lands are being consumed by urban sprawl. Today the highly productive chinampa agriculture (Xochimilco's floating gardens) occupies less than one-hundredth of the area that it did during the Aztec Empire (Oterbridge 1987). The continuing expansion of México City would have led to the complete disappearance of these agricultural lands if it were not for their tourist attraction. Similarly, with the growth of Disney World and other activities attracted to the Orlando area directly and indirectly connected to it, orange groves are rare in the immediate area as they have been cut down for more economically profitable nonagricultural uses.

From the strictly economic perspective, the conversion of lands with high and moderate biomass productivity to urban uses with far less biomass production potential is extremely rational. Yet the reduction in biomass production can be viewed as land degradation even though from the human perspective the area's utility could have increased. The consumed lands required for urban functions become a *sacrifice zone* due to urban sprawl and functions required for urban areas such water supply. As urban areas grow, a consequence of the majority of humans worldwide now living in urban areas, they require the development of ever longer supply lines to satisfy cities' energy, food, water, and material needs. In the American Southwest, a largely arid or semiarid area, rapid urban growth is placing increasing demands on the area's limited water supply. Some cities are purchasing water rights from farmers to meet their water needs. This reallocation of water from irrigation to urban use results in the contraction of productive rural lands in order that urban growth can continue unabated. Not only are the lands adjacent to urban areas affected, but also to satisfy the energy, water, and material requirements of cities, additional sacrifice zones often result. These areas frequently are not directly connected to, and sometimes are quite distant from, the urban area that caused their occurrence (figure 6.1).

Figure 6.1. Open pit mine in Arizona. The pursuit of minerals produces a permanent scar on an arid landscape and poses a serious threat to groundwater. (Photograph by L. Lewis, June 2002)

Satisfying the Needs of Urban Areas

To exist cities require construction. Building materials are shipped in large quantities into any active city. For example, in Los Angeles during a period of rapid growth in the 1960s, per capita consumption of sand and gravel was in excess of four tons annually (Reining 1967). The sand and gravel pits supplying these materials usually experience various degrees of environmental deterioration as the sands are removed. These pits result in a pockmarked landscape usually characterized with degraded vegetation and bare ground. The pitted areas are examples of sacrifice zones that, although spatially detached from the urban area, represent land degradation resulting directly from urban activities (figure 6.2).

Two ubiquitous by-products of urban construction in general are that topography is greatly altered, and the permeability of the area, especially in nonresidential areas, is greatly reduced. To facilitate urban activities, a common practice is to alter an area's morphology either to improve drainage or to lower the relief through cut-and-fill operations. Exemplary of the topographic alterations needed to meet the needs of an urbanized society, between 1803 and 1995, 2,125 ha (5,250 acres) of land were reclaimed from the immediate waters surrounding Boston (Seasholes 2003, 2) as a way of meeting both the space needs of the growing population and

Figure 6.2. A gravel pit in Northboro, Massachusetts. In order to exploit subsurface gravels for construction material, a degraded surface is left behind. (Photograph by L. Lewis, July 1994)

requirements of the infrastructure, especially airport expansion (figure 6.3). Cut operations were also widespread throughout this period. Originally, the central core of downtown Boston had a hilly glacial till topography. Most of hills tops were lowered to use the material for filling up the Back Bay ponds and wetlands to increase land for urban expansion as well as to facilitate construction (Wilkie and Tager 1990). The original size of the peninsula in which Boston was first sited was more than doubled in the process. The Charles River and its adjoining landscapes of marshes and mudflats were converted from a tidal habitat encompassing hundreds of acres to the current completely managed park landscape that delights the visitor today (Haglund 2003). A remnant of Beacon Hill, the site of the seat of Massachusetts state government, today rises only slightly above the broadly leveled downtown landscape of modern Boston, which bears slight resemblance to the original wetland and hilly terrain of the area. In general, extractions of earth materials in cut operations remove an area's topsoils and brings less fertile materials onto the surface that normally are covered. This represents a permanent degradation in the potential of the land for biomass production. The degradation is reflected in most landscaping operations for modern office buildings built in any city. To establish plants for aesthetic reasons, topsoil is usually transported to the building sites and placed on the grounds before any significant vegetation growth is possible (figure 6.4).

Not only are topographic changes and decreases in fertility of surface materials features of urbanization, but also a significant percentage of urban surface materials bear little resemblance to their natural properties. Almost any modern urban center has had a large proportion of both its vegetation and soil replaced by concrete, asphalt, bricks, and other building materials. With the exception of parks, close to 100 percent of the surface of a modern city's commercial core is completely occupied by buildings, sidewalks, and streets. Land values are

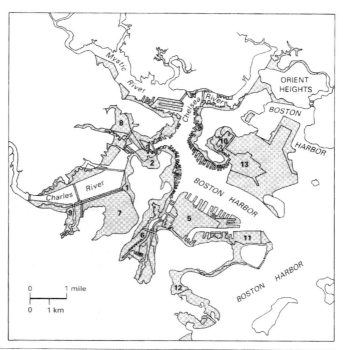

	Location	Approximate acreage	Dates		Location	Approximate acreage	Dates
1	West Cove	80	1803–1863	8	Charlestown	416	1860–1896
2	Mill Pond	70	1804–1835	9	Fenway	322	1878–1890
3	South Cove	86	1806–1843	10	East Boston	370	1880–1988
4	East Cove	112	1823–1874	11	Marine Park	57	1883–1900
5	South Boston	714	1836–1988	12	Columbus Park	265	1890–1901
6	South Bay	138	1850–1988	13	Logan Airport	750	1922–1988
7	Back Bay	580	1857–1894				

Figure 6.3. Areas adjacent to Boston's valuable commercial and residential central city districts have steadily been converted into developed land by a combination of enclosure, drainage, and landfill. (Copyright © 1991 by the University of Massachusetts Press. Reproduced by permission.)

too high for most other activities. Situated in the heart of downtown Boston, Boston Common, where once it was permissible for livestock and horses to graze, is now surrounded by a densely populated urban conglomeration that is largely inhospitable to most living creatures except urbanized humans, their pets, and rodents. Residential areas undergo similar changes, but at a lower intensity. Suburban areas experience marked changes but at a far lower intensity with more open space. Yet even in these areas biomass production is curtailed albeit far less drastically than in more urbanized areas. Thus, without even considering *brownfields* (derelict urban land with toxic residuals from former activities), urbanization clearly results in lower biomass production and hence land degradation. Drastic alterations in their vegetation and topographic situations similar to those in Boston are an intrinsic

Figure 6.4. The only vegetation visible in the main square in front of Leuven, Belgium's Town Hall is brought in from the countryside by truck and artificially maintained. (Photograph by L. Lewis, March 2003)

property of twentieth and twenty-first century urbanization. These changes occur not only in wealthy developed nations but also are found in urban zones in the less developed nations too (Lewis and Berry 1988).

As cities increase in population, a number of changes take place in the relationship between the urban area and the environment. As urban populations grow, a lower percentage of urban demands with regard to energy and water supply, building materials, food, transportation, and waste disposal can be provided from local resources. Not only do the impacts of each of these activities on the environment increase in magnitude as population increases, but also they often occur in distant places and in areas with relatively low populations. When the impacts result in areas outside of the urban area's immediate region, because of the spatial discordance the urban population driving the negative environmental impacts is often shielded from them. This acts to dampen immediate corrective responses to the environmental impacts driven by urbanization.

Energy

Electrical energy, a necessity for any urban area, primarily is generated by hydro, coal, gas, oil, or nuclear fission. Each of these power sources impacts the land resource. Reservoirs required for hydroelectric generation flood lowlands. In the process hectares of fertile lands and areas of high biodiversity are often destroyed (Barrow 1991). These flooded areas become sacrifice zones for the well-being of a distant urban area. Associated with the mining of coal for thermal power plants are a variety of environmental consequences that have direct bearing on the land degradation problem. Spoil, a by-product of mining operations, covers large tracts of land and renders them largely useless for any beneficial land use (figure 6.5). In the past and in many contemporary areas where these spoils are derived from coal mining operations, these spoil heaps are areas that generate acidic waters. The acidic runoff from the overburden degrades surface and ground waters, often killing vegetation and rendering the surface water unfit for even minimal uses. Such was the case in numerous coal-mining areas in western Pennsylvania, West Virginia, southeastern Ohio, and southern Indiana and Illinois. Massive land degradation resulting from both the mining of coal and its utilization in power generation to meet urban and industrial needs has occurred in Saxony (Germany), Silesia (Poland), and Bohemia (Czech Republic). The land degradation resulting from these activities following World War II, when these areas were part of the Eastern European Bloc, was of such a magnitude that this part of central Europe has been referred to as a "Bermuda triangle of pollution" (Ministry for Economic Policy and Development of the Czech Republic 1991) and has required enormous efforts over more than a decade to ameliorate. The other mineral energy sources—oil, gas, and nuclear fuels—

Figure 6.5. Coal tailings resulting from mining operations are a frequent occurrence in the landscape of northeastern Flanders. (Photograph by L. Lewis, April 2003)

likewise can have deleterious effects on the land resource. Ramifications of the extraction of these mineral fuels are discussed in following sections.

The major source of energy is wood, often in the form of charcoal, in many Third World cities. With the explosive growth of many urban areas in these parts of the world, demand for wood has increased. In fact, demand is growing much faster than forests can reproduce. Around a significant number of towns and cities, the environmental impact of this phenomenon is intense. For example, in a twenty-year period between 1960 and 1980, the zone of wood exploitation for charcoal moved over 300 km (185 miles) south of the Khartoum market (figure 6.6). Its southward movement in this semiarid setting has been a factor in the desertification that occurred in parts of the region (Berry 1983).

Many nuclear plants are now over thirty years old. Besides the concerns of radiation leaks, especially since Chernobyl, disposing of nuclear waste from nuclear plants is a global dilemma. In the United States, more than 20,000 tons of the nation's nuclear waste is to be disposed in Yucca Mountain, Nevada (USGS 2000). Some of this radioactive waste is derived from energy generation. Wherever it is deposited, land degradation clearly occurs as the area's utility for future options is minimal due to the long duration of the radioactivity. In addition, there are other concerns associated with nuclear energy—which seem relatively benign compared to radioactivity—but their environmental impacts are major. Nuclear plants require huge quantities of water for cooling. The Salem, New Jersey, plant, situated on an artificial island, requires 3 billion gallons of Delaware Bay water every day

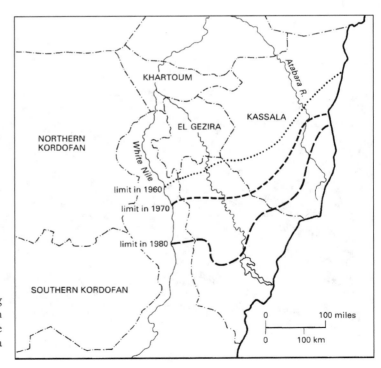

Figure 6.6. The continuing demand for charcoal in Khartoum results in an ever-expanding zone of tree cutting in Sudan's savanna landscapes.

(Vargo and Gallagher 2002). The release of this heated water, while clean, raises the water temperature and affects local ecosystems. A nuclear plant in the Philippines produced sufficient temperature change due to this type of heat pollution to kill a portion of a coral reef.

Building Materials

Construction materials—that is, sand, stone, and clay—are needed in large quantities to meet the various needs of urban areas. Besides the buildings themselves, all forms of transport infrastructure need construction materials if supplies are to flow both toward and out of cities. Modern interstate highways sometimes cover over 10 ha of land for every kilometer segment (40 acres/mile). Every bit of land covered by roads, vital to the viability of the urban areas, represents a zone sacrificed for urban success. Furthermore, as with other forms of mineral extraction, construction materials needed to build and maintain the highways scar the land and add dust to the air. Unlike some forms of mineral extraction and their tailings (for instance coal), which often kill vegetation, the production of construction aggregates is relatively benign. But a degraded, pockmarked landscape devoid of soil cover is often left behind after the mining operation ends. In New England, many former gravel pits and quarries became waste disposal sites for the urban area that had been the demand catalyst for the extraction of construction materials in the first place. Today with government controls that require the protection of groundwater supplies, this practice has ceased. Some of the former landfill areas have been restored to limited types of urban development, such as playing fields. In other areas, because it is not know what is exactly in the landfills, reforestation of the sites is the only practical option. Because land-use options are more restricted than prior to the mining operations, even if there has been a restoration project, they represent a clearly degraded condition.

Tourism and Environmental Quality

One of the world's largest industries, tourism directly employs 3 percent of the world's total workforce or 8 percent if indirect jobs are included, that is, one in every twelve workers. Tourism almost tripled between 1970 and 1992 and continued to grow unabated through the year 2000 at about 5 percent annually (Rekacewicz 2001); and it is estimated that growth will again reach 5 percent by 2004 (OTTI 2002). At first glance, tourism might seem environmentally benign as tourists seek out sunshine, historic sites, popular urban venues, interesting landscapes, or recreational areas, all of which appear not to be consumptive in terms of resources. While it is true that most prime tourist activities do not directly consume resources (sunbathing, sightseeing) and that they do not have the "smokestack" image of manufacturing or mining operations, tourism generates significant levels of pollution, consumes large amounts of natural resources, and clearly has caused land degradation such as blighting countrysides and polluting beaches as well as offshore waters. Negative environmental responses to tourism flip-flop destinations from attractive to out-of-favor tourist areas. Thus one of the biggest beneficiaries of curtailing land degradation associated with this economic activity is the tourist industry itself.

Tourism impacts the environment on several scales. At the local level, competition for scarce resources such as freshwater and land, air and water pollution, and landslides and avalanches may occur. Regional level impacts include loss of habitats and biodiversity. At the global level, emissions from road traffic and deforestation are likely to contribute to climate change (MacGillivray 1995). In Europe nationally protected areas including national parks cover over 8.5 million sq km (3.8 million sq miles). In the United States, national parks alone include 47 million acres (190,200 sq km) (National Park Service 2001). The very attractiveness of some of these parks and the large numbers of people that visit them threatens the very reason why both the lands were first protected and why tourists come to the areas. Learning to value these areas for their intangible values as well as their habitat features is a conflict-laden and controversial process (Harmon and Putney 2003). National parks and protected areas (for example, national forests and monuments in the United States, the Alps in Europe) because of their physical properties likely will suffer environmental damage when tourists come in large numbers. Box 6.1 lists some examples of tourist activities having negative environmental impacts. Problems are usually greatest in areas, such as the Great Smoky Mountains National Park, that are easily accessible to urban centers. But in contrast increased usage with resulting higher fee collection and increased incomes for the local population can result in increased safeguards to the environment. Such is the case in Majorca (Spain) where 35 percent of the island has been classified as protected since 1991 due to both international and local desires to protect the area (Balearics Tourism Authority 1992).

Coastal beach areas in warm climates, and some not so warm, are popular tourist destinations. Cape Cod, the whole Florida coastline, the Caribbean, and the Mediterranean are just a few examples of coastal attractions. The Mediterranean in particular exemplifies some of the environmental problems associated with tourism. The Mediterranean basin attracts 35 percent of all international tourists worldwide (MacGillivray 1995). Beaches are a major attraction for people vacationing in this area. While problems are found everywhere along the Mediterranean, including France, Spain, and Italy, a brief discussion of the North African coastal area follows as the countries in this area with their weaker economic position have had great problems in adjusting their infrastructure to the influx of tourists with concomitant negative environmental impacts.

During the tourist season, the coastal regions along the whole Mediterranean Sea have a population of 200 million. Throughout the year approximately 120 million tourists, or about 35 percent of all international tourists visit this area (UNEP 1991). The hotels, guest houses, second homes, roads, parking areas, and shops directly oriented to tourism cover over 2,200 sq km. Current growth at the end of the twentieth century was particularly strong along the North African coast. Most tourism in this area is concentrated from June through August. The Mediterranean Action Plan (MAP) coordinating unit estimates that in twenty-five years, unless current trends change greatly, between 35 million to 52 million tourists over today's numbers will come to the Mediterranean coastal and littoral zones in the five-week peak season from July to August (UNEP 1991). This is roughly equal to the combined populations of Algeria and Morocco and is greater than all of the North African countries except Egypt. The infrastructure provided for the permanent population in the North African countries cannot cope with this inflow of population. Local authorities are unable to afford sewage facilities to handle the short concentrated period of tourism. To build adequate facilities means that they would be greatly underutilized for ten months out of each year. As a result, during the tourist periods, much of the sewage goes directly to the sea. This

BOX 6.1. Loving them to death: tourism and recreation in protected areas in Europe and the United States

New developments: Nature and national parks are already under extreme pressure from the number of visitors, demand for outdoor activities, and development of tourism facilities (for example, large hotels on the Mazurian Lakes, Poland, hotels in the Grand Canyon—a freeze on construction now exists at the Grand Canyon).

Overcrowding occurs at peak periods. In Hohe Tauern, Austria, visitors are concentrated in six months, while in Ojcow, Poland, 60 percent are there during the three summer months. Shenandoah National Park, Blue Ridge Parkway, and Great Smoky Mountains National Park are extremely heavily used and car traffic can reach road capacity during peak periods. Most visitors are daytrippers in many parks arriving by car (over 90 percent to the Hohe Tauern), leading to traffic congestion, congestion of car parking space, and litter problems.

Changes in habitats of native animals: Wildlife attracts visitors, who disturb breeding patterns, leading to a fall in animal numbers; as species become rarer, more people come to "See it while you still can."

Path erosion and wear and tear arise from walking or mountain biking. Popular routes such as the Pennine Way in the UK, and the Appalachian Trail in the eastern United States require extensive maintenance and repair.

Introduction of exotic species: For example, in Tenerife, Canary Islands, more than a dozen species have invaded the Teide National Park, having been carried as seeds on vehicles, clothes, or tents of visitors. Kudzu, mimosa, and the European wild boar in Great Smoky Mountains are exotics creating problems in this park.

Conflicts exist between tourism and nature conservation, traditional hunting and agriculture. For example, the Coto Donaña National Park in southwest Spain is an important breeding site for many of Europe's birds, and home to endangered species such as the imperial eagle, (*Aquila heliaca*) and Spanish lynx, (*Lynx pardina*), but is now threatened by water extraction for tourism and local agriculture. In Triglav National Park, Slovenia, 2 million tourists outnumber residents during the summer by almost 1,000 to 1. Excessive camping in Zion National Park has necessitated closing down many of the campsites to minimize damage to the park.

Sources: MacGillivray 1995, 493; The National Park Service (USA) 2001.

alone results in major beach pollution (World Resources Institute 1990). The closure of some beaches during the season due to pollution puts pressure on the remaining open ones.

Another environmental consequence of tourism is the huge amount of water tourists consume. July through August, the peak tourist season, is also the period of maximum high temperatures and minimal precipitation. A tourist in a luxury hotel consumes between 300 to 600 liters of freshwater daily. This is in excess of what the same individuals would consume in their home areas. "Such excessive water draw-off has already seen a lowering of ground water levels in the Hammarnet region of Tunisia, leading to a social water waste-land and the abandonment of some cultivated land in other regions" (UNEP 1991). Connected with tourism in this area and elsewhere has been an explosion of golf courses.

Almost all golf developments cause environmental damage. To meet the demands of "championship" caliber golf courses, their construction requires earth-moving operations to create the rolling terrain with water hazards that golf aficionados crave. Farmland and forest land is converted, wetlands are drained, and desert areas are irrigated. Vegetation is altered as grasses replace existing flora, while heavy use of fertilizers, employment of pesticides and herbicides, and continuous mowing of the fairways, roughs, and greens result in a loss of floral and faunal diversity. In humid areas, pesticides may infiltrate into groundwater or runoff and enter streams or sewer systems, thus degrading local water supplies. In arid areas such as the Mediterranean, Southern California, and Arizona, water requirements reduce alternative uses of the water. In response to the water shortfalls in these areas, recycled water is being used more and more to sustain these recreation areas. Of course, the result is that unless care is taken in the management of this water, the increased salinity of the water will prevent further recycling of the water for these purposes.

Along with coastal areas, mountains have a strong attraction to tourists. Clearly road construction to make these areas more accessible results in numerous environmental impacts and land degradation. Hiking in the summer months and skiing in the colder periods bring large numbers of visitors to mountain areas, all of which activities have a very high potential for accelerated erosion, due to the local high relief and steep slopes, whenever the natural vegetation cover is disturbed. The extreme sensitivity of mountain areas requires special managing strategies. Yet even in countries such as Switzerland—which are wealthy, have a strong tradition of following rules, and have instituted strong environmental interventions—the intense recreational activities occurring in the Alps result in ongoing accelerated erosion, in some places contributing to extreme events such as landslides or avalanches. For hiking activities, the greatest problems occur along heavily used trails and camping sites. The trails themselves become eroded and generators of concentrated surface flows if not well designed. But the cumulative impacts of skiing have particular importance in terms of land degradation. Construction of ski trails, parking lots, and access roads to the ski areas, along with housing construction result in the removal of trees. The outcome of tree clearing is habitat destruction, alteration of surface water runoff, and increased landslides and snow avalanches (box 6.2). While all skiing impacts the environment, downhill as opposed to cross-country skiing alters the environment to a greater degree in terms of land needed, energy required for infrastructure (ski lifts, lodges, snow making), and vegetation changes. Tree clearing for downhill trails in particular can create problems with regard to erosion and water runoff if trails are not completely covered with grasses during the warmer snow-free periods. Traffic congestion and air pollution problems in some Alpine areas have resulted in establishing car-free zones and other restrictions placed on parking lots and hotel expansion, which restricts resort expansion. The net result of these actions is to curtail land degradation in those areas, but to shift the problems elsewhere such as the Pyrenees (Andorra, Spain) and Central and Eastern Europe.

Tourism is clearly not as benign to the environment as might appear at first glance. Many activities require major infrastructure investments to transport people to the areas as well as

Box 6.2. Impacts of Skiing in Europe

Mountains are the last remaining areas in Central and Eastern Europe that are relatively untouched by human activity. They support a rich biodiversity, although relatively few species can survive in the harsh environment. Recently introduced human activities related to skiing threaten the balance of delicate ecosystems by introducing pollution, deforestation, soil erosion, and human disturbance.

Forest clearance and increased incidence of avalanches: Some 100 km^2 of forest have been removed throughout the Alps, which has led to higher incidence of avalanches. In Austria, the creation of 0.7 km^2 of ski runs in one year (1980) for the Winter Olympics contributed to a major mudslide in the cleared area in 1983. New resort construction may involve bulldozing, blasting, and reshaping of slopes, leading to a higher incidence of avalanches.

Visual degradation of landscapes: Natural forest barriers may be replaced with unsightly concrete, plastic, and wooden barriers associated with developments.

Inappropriate development: The uncontrolled development of ski centers in some Central and Eastern European countries (for example, in the High Tatra mountains, Poland) is threatening the environment.

Loss of habitats and disturbance of endangered species: Operation of lifts, off-trail skiing, use of all-terrain vehicles, and compaction of snow on trails disturb rare species such as black grouse (*Tetrao tetrix*) in the French Alps, which can be decapitated by overhead wires, displaced from their breeding grounds, and can face competition for space with cable car installations.

Sewage disposal and water pollution: In the French Pyrenees, the sewage from summer tourist resorts discharges directly to streams and leads to water pollution. In the Alps chemicals used in preparing thirty-six glaciers for skiing have been reported to lead to increases in nitrogen and phosphorus levels in drinking water. Exhaust from private cars and coaches have led to death of trees and to wildlife damage. In Switzerland 70 percent of domestic and foreign tourists arrive at their destination by car.

Unsustainable use of water: By 1992, four thousand snow cannons were producing artificial snow to lengthen the ski season in the Alps, using (and competing with other uses for) 28 million L of water per kilometer of trail. In Les Meunières, France, 185 cannons installed for the 1992 Olympics were supplied by drinking water sources. Artificial snow melts slowly, reducing the short recuperation time for Alpine grasses and flowers. Furthermore, skiing in sparse snow conditions contributes to erosion and damages sensitive vegetation. The result is a severe reduction in water absorption and holding capacity of mountain slopes, and increased risk of runoff and avalanches.

Overdevelopment of tourism: By 1991 employment in the Swiss tourist industry in Alpine areas accounted for probably one in every three jobs. Over 75 percent of accommodation is in the so-called para-hotel industry, outside the regulated hotel sector (holiday apartments, holiday houses, and campsites).

Source: MacGillivray 1995, 494.

house and support the tourists' life styles, once they arrive. Also, numerous attractions themselves result in environmental changes. Golf and skiing are but two such examples. For example, theme and leisure parks, ecotourism, camping activities, as well as urban tourism all have both direct and indirect environmental effects which result in land degradation both locally and spatially not connected. Thus tourism as other industries must be well managed to minimize its deleterious environmental impacts in the land, water, and air domains.

Energy Production and Other Raw Materials

Mining

The Industrial Revolution required natural resources both for the energy to power the machinery that replaced human labor as well as to acquire the materials needed for the manufactured products. Increased mineral utilization was a foundation of the Industrial Revolution. This need resulted in a rapid exploitation of the Earth's minerals in the 1700s. Between 1750 and 1900 mineral use increased tenfold. Since 1900 mineral use has increased an additional thirteenfold (Young 1992). As minerals are part of the environment, mining them without modification of the environment is in most cases unrealistic.

Modifications need not result in land degradation. New habitats can be created during mining operations by altering drainage and topography in harmony with the geomorphic and climatic situations (Barrow 1991, 229). Nevertheless, to prevent land degradation from occurring, mining operations need to be controlled by strong legislation. This is because higher costs are associated with the extraction of the mineral if environmental damage is to be prevented. Given that mineral resources are where you find them, not where you would like to find them, prevention of deleterious effects associated with their removal must be part of the initial plans of the mining or drilling operation. Today in countries with a strong "green" lobby, as a consequence of good planning and strong environmental values, the potential environmental consequences are sometimes deemed too costly to society, and mineral extraction may be prevented. Such has been the case in protected wildlife habitats along the north coast of Alaska, some coastal areas of California, and coastal New England where, to date, exploitation of potential petroleum has been prevented. But the overall reality from the worldwide perspective is that, in most cases, the short-term economic benefits of mineral extraction are all powerful and mining or drilling operations are initiated.

There are two reasons why land degradation generally results from mineral extraction. First, industrial development remains the highest priority for most nations. "The American Colossus was fiercely intent on appropriating and exploiting the riches of the richest of all continents, grasping with both hands, reaping where he had not sown, wasting what he thought would last forever" (Pinchot 1947). The viewpoint expressed by Pinchot is widely followed by many nations in their pursuit of short-term economic rewards. A second reason, somewhat related to the previous one, is that short-term economic benefits, such as reaching production goals and employment, delegate environmental considerations to the back burner (Ministry for Economic Policy and Development of the Czech Republic 1991). To insure that the deleterious aspects of mining operations are minimized requires, in almost all cases, conservation strategies that would increase the overall cost of the mineral extraction. Worldwide, prices of most mined minerals usually reflect only the direct immediate costs such as fuel, equipment, and labor as well as the acquisition of mining rights. Rarely are provisions made to include the costs of destroyed landscapes in the price determination of minerals. This is no longer true in many of the countries of North America and Western Europe where environmental legislation now requires provisions for reclamation during and after mining operations. In Germany, after the unification, the government incurred high costs as they attempted to rectify the environmental problems left behind by the former German Democratic Republic so that the former mined areas would meet the country's already established standards.

Mining's Direct Impacts on the Land Resource

Mineral extraction by mining takes place either as a surface or an underground operation. Both methods can alter the preexisting environment to a sufficient magnitude that land degradation results. With over 500,000 ha (1.24 million acres) of land directly disturbed by mining activities each year (Young 1992), many areas in close proximity to mining operations absorb the environmental costs of mineral extraction for the users. Some implications of these land alterations are discussed in the following sections.

Subsurface mining is an ancient activity. The salt mines near Hallein (Austria) and Bad Reichenhall (Germany) were active in prehistoric times. A coal mine in Glamorgan (Wales) began operation prior to 1400. By 1600, Wales was exporting coal to France (White 1991). With the Industrial Revolution, mining operations became more extensive to satisfy both the demands for materials (iron and lead) and energy (coal). Regardless of the mining operation, all subsurface mining moves materials from beneath the surface to the surface. Not only the mineral desired, but other earth materials cleared from the shafts and passageways in the mine have to be brought to the surface. In areas where groundwater is superimposed upon the mineral vein being mined, pumping to lower the water table also is required. All of this extracted material is deposited on the surface or transported away by the surface hydrologic system. Regardless of the mining method, the mineral being mined, surface subsidence and mine collapse are common environmental adjustments associated in many mining areas. Subsidence alters drainage patterns and damages surface infrastructure (e.g., roads, and utilities) and buildings in built-up areas. Collapse drastically alters the surface configuration of the affected areas and degrades the immediately affected areas. Mining operations may result in surface waters infiltrating into the subsurface voids, especially after the mine is abandoned and pumping ceases. In some areas of Appalachia, large underground impoundments of water have filtered into coal mines. These waters become very acidic and if they reach the surface via subsurface flows, their low pH values devastates the aquatic systems impacted by them.

Another aspect of underground mining is the solid waste materials deposited on the surface as tailings. These materials are often piled up in large mounds in close proximity to the mine. The composition of many tailings can contain toxic minerals such as mercury or iron sulfide. Water percolating through these tailings results in water quality problems downstream. Tailings dating from lead-mining operations in Roman times are still poorly revegetated due to their chemical attributes (Barrow 1991). The mountains of slag and shale deposited throughout Pennsylvania, Flanders (Belgium), and Wales (figure 6.7), if left to natural processes, would take a thousand years before the evidence of the past mining activities

Figure 6.7.   Coal mine tailings near Hirwaun, Wales, United Kingdom, where the last operating shaft mine in Wales has now been closed. (Photograph by L. Lewis, June 1994)

was covered by vegetation and forgotten (White 1991). Massive and expensive reclamation of these coal tailings throughout Wales and Flanders is well underway (figure 6.8). The coal tips have been leveled and the land is usually converted into some type of commercial use. While this allows some types of economic activity to now occur in an area which had been environmentally degraded previously, it is not a complete reversal of land degradation as its biological productivity potential remains constrained and land-use options remain limited.

The reversal of land degradation in these former mined areas is important because the highly acidic tailings not only result in rather sterile conditions on the tips themselves, but also runoff from precipitation becomes acidic and kills flora and fauna especially in the brooks receiving these waters. Many lands and streams within the Appalachian coalfield areas of western Pennsylvania, West Virginia, eastern Kentucky, and eastern Ohio have suffered from these acidic waters. Environmental legislation limits the damage of current mining, but Appalachia still is left with a filthy legacy of degraded landscapes dispersed throughout the area. Current surface mining in the area continues to negatively impact some of these areas.

With technological improvement in earth-moving equipment, surface mining for economic reasons is favored over subsurface mining, whenever possible. Compared to underground mining, the cost per unit of production is considerably less in surface mining operations. Additional advantages include: (1) a larger percentage of mineral recovery in surface mining, often close to 100 percent, and (2) greater safety of mining operations rel-

Figure 6.8. Land reclamation of coal wastes in Flanders is being funded by the European Union. (Photograph by L. Lewis, April 2003)

atively free of accidental explosions and mine collapse. Minerals with a low value per unit volume, such as sands, gravels, and phosphates, are only surface mined. In contrast to the short-term economic advantages, since surface operations always significantly disturb the surface, large-scale land degradation has resulted from these activities in almost every nation. Figure 6.1 (see page 165) illustrates surface alterations caused by open-pit mining, which in this case are visible from a height of over 9,100 m (30,000 ft).

The oceanic nation of Nauru best exemplifies the economic benefits of surface mining as well as the environmental devastation that all too often accompanies this type of mining. Through the mining of rich phosphate deposits found on the island, Nauru has achieved a gross domestic product (GDP) per capita of about $10,000. But over 85 percent of the phosphates already are mined, leaving the center of the island, or 80 percent of its total area of 21 sq km (8.1 sq miles), a wasteland of coral pinnacles, some of which are covered with secondary vegetation. The long-term economic viability of the island rests on the island's investments of its phosphate receipts. The mining operation represents a classic example of land degradation. By taking advantage of the land degradation in its center, a new form of short-term income for the island has come about—using it as a holding area for illegal immigrants caught going to Australia.

In the United States, surface mining has disturbed over 23,000 sq km (5.6 million acres). Coal mining accounts for 41 percent, while sand and gravel removal is responsible for 26 percent (Tank 1983). In 1999 only 3.3 percent of nonfuel mining (extraction of sand, gravel, copper) was subsurface and 96.7 percent was surface (Moore 2001). With the development of efficient earth-moving equipment, such as large power shovels that can hold over 95 cu m (125 cu yd) of earth and rock, deep mining is continuously declining relative to surface mining operations.

Over the last decade, the nature and extent of coal mining operations in Appalachia has changed significantly. Increasingly, individual surface mines larger than 1,200 ha (3,000 acres) have been proposed, and technology has enabled machines to remove the tops of mountains (mountaintop removal), with excess overburden material typically disposed into adjacent stream corridors (valley fills). In some instances, valleys of up to two miles long have been completely filled, covering perennial streams, wetlands, and tracts of prime upland wildlife habitat. Cumulatively, tens of thousands of acres are believed to have been affected by these operations. Further, the steep terrain in the Appalachian coalfields, where most of this coal-mining activity occurs, is believed to offer few economically feasible disposal alternatives for the excess overburden and mining spoils (USEPA 1999).

Because of the environmental damage and land degradation associated with this type of mining, a lawsuit was pursued to stop mountaintop removal mining, which was also creating much land degradation damage to rivers within West Virginia. In May 2002 a federal court ruled that the Army Corps of Engineers could no longer give out permits allowing mining companies to dump the excess rock and dirt from a mountaintop removal into a stream valley. "The judge said the corps issued permits that allowed companies to fill 87 miles (140 km) of streambeds with mine waste. Past . . . permit approvals were issued in express disregard of the corps' own regulations and the (Clean Water Act), as such, they were illegal" (Watson 2002, 2). The result is that no new mountaintop mining is, for at least now, possible. Environmental considerations outweighed economic considerations in this court case.

It is difficult to prevent land degradation whenever surface mining occurs. First, huge depressions result in some operations (see p. 165). The Kennecott Copper Mine in Bingham, Utah, is over 700 m deep (½ mile), and covers more than 7 sq km (2.8 sq miles). Second, the overburden (soil, subsoil, and unconsolidated earth and rocks) must be stored and then replaced systematically in their original order after the mineral is removed. Even under optimal conditions, which rarely occur, most restoration results in a soil that is less productive than it was prior to mining. Subsurface drainage is always disturbed and revegetation is often slow (Jordan, Gilpin, and Aber 1988). The restoration of land is always more difficult when toxic materials are involved or leaching of the overburden occurs during its storage. These conditions often occur in mining operations associated with coal, lead, copper, and gold, which together comprise a significant percent of surface mining operations.

Because of the critical role of minerals in development, mining operations are inevitable. How could any urban area exist without the quarried stone, sand, and gravels needed for construction or the minerals needed for industry and energy? But the destruction of the land resource, with soils blighted and water fouled, needs to be counted against the expediency of short-term economic gain (Simons 1992).

Nuclear Energy Production

One of the authors remembers from his childhood the commissioning of the first U.S. commercial nuclear power plant at Shippingsport, Pennsylvania. Among other advantages proclaimed for atomic power during this early stage of nuclear energy development was that it would provide a clean alternative fuel (environmentally benevolent) to the region's (western Pennsylvania) traditional polluting energy source, coal. Yet, the raw materials required for nuclear fuels result in the same disturbances of the landscape as other mined minerals. In addition, the hazardous nature of the spent fuels creates another set of problems with regard to finding a safe site to store them. During the infancy of atomic power, little public awareness existed to the "new" environmental problems that would be an inevitable by-product of this proclaimed clean energy.

Today, even if one only considers the issue from the perspective of land degradation, both the hazardous wastes associated with nuclear power generation and the radioactive impacts of nuclear accidents make nuclear energy environmentally risky. Because of these inherent risks, Denmark has never permitted atomic power plants to be constructed within its boundaries, and Sweden has a stated policy of decommissioning all of its existing plants. In the United States, it has become so costly due to required environmental safeguards and the inevitable litigation of nuclear opponents that plant construction has for all purposes ceased. Austria tried to get the Czech Republic to close an atomic power plant as a bargaining chip for entry into the EU. But it must be emphasized that nuclear power remains a major source of energy. For example, France relies very heavily on nuclear power and it remains an important source of energy in the United States and many other countries. In fact, pressures are building in the United States to begin to explore the idea of building new plants.

In March 1979, when the Three Mile Island (near Harrisburg, Pennsylvania) atomic power plant sustained a near disastrous event, the general public in the United States became more alert to the potential for acute environmental problems associated with energy generated from nuclear power. Fortunately, most radiological impacts of the Three Mile Island accident were confined to the power plant, and minimal if any land degradation was associated with this event (table 6.1).

Table 6.1. A Comparison of Estimated Releases (%) for the Chernobyl and Three Mile Island Nuclear Power Plant Accidents

| | Three Mile Island | | Chernobyl |
Isotopes	Containment Building	Released to the Environment	Released to the Environment
Noble gases	48	1	100
Iodine	25	0.00003	20
Cesium	53	not detected	10–15
Ruthenium	0.5	not detected	2.9
Cerium	none	not detected	2.3–2.8

Source: Adapted from Collier and Davies 1986.

In contrast to the mild Three Mile Island incident, on April 26, 1986, a nuclear power plant accident occurred at Chernobyl in Ukraine. This meltdown released radioactive isotopes, which have been detected almost worldwide in the northern hemisphere. A multitude of major environmental consequences occurred, including land degradation that continues to affect this area of the Ukraine. The radionuclides released from the Chernobyl plant were transported throughout the atmosphere. The former Soviet Union and Eastern Europe, Austria, Scandinavia, and northwestern Europe experienced significant fallout on their lands (OECD 1987). Crops and milk products were contaminated in many European countries for a period of time (OECD 1987; Mould 1988), requiring them to be destroyed. With regard to land degradation, the damage resulting from Chernobyl was confined to the Ukraine, Belarus, and Russia. Significant contamination was concentrated over an area of approximately 1,000 sq km (400 sq miles) radiating outward from the destroyed reactor. Contamination affected both plant and animal life within this zone, and continues to limit land-use options for the future. In the immediate surrounding areas of the plant, decontamination activities themselves resulted in significant land degradation. In the vicinity of Pripyat, the town closest to the plant, decontamination actions left only one tree standing (Mould 1988). The spared tree was a memorial to Ukrainians killed during World War II in this area. Workers carrying out decontamination strategies in close proximity to the nuclear plant cut down all of the trees in this previously forested area and buried them. The soil in the vicinity of the plant was removed to depths between 1 and 1.5 m and buried elsewhere (Mould 1988).

Because of the intensity of the radioactivity in proximity to the plant, a 30 km (19 mile) wide evacuation zone centered on the power plant has been closed to human activities, with the exception of the remaining power plant. This zone is now used for investigating the effects of the accident on the environment. By any criterion the contamination of the lands in the vicinity of the plant and the required decontamination procedures represent a new form of land degradation and sacrifice zone resulting from some of the technological advances of modern societies.

While less dramatic than acute nuclear accidents such as Chernobyl and Windscale (UK), the chronic accumulation of hazardous wastes from nuclear facilities also has land degradation impacts. In the United States, federal nuclear activities were not required to adhere to the same standards as civilian installations. This aspect of U.S. government nuclear policy is reflected in the Defense Department's current attempt to get the military exempted from many other environmental rules (Dowd 2002). Because of this difference in standards, significant land degradation occurs on many federal lands in the United States. Only occasionally do the negative impacts of governmental installations' management become part of the public record. One problem brought to wide public attention was associated with the Savannah River Nuclear Facility in South Carolina.

Sited next to the Savannah River, this facility included five nuclear reactors and several large processing installations. The activities carried out at this site resulted in the storage of large quantities of radioactive and nonradioactive wastes on the grounds in various locations on these government lands (USGAO 1984). The public became cognizant that there were hazardous wastes leaking into the groundwater from these wastes. If these wastes were to enter the nearby Tuscaloosa Aquifer, a major water supply source for the Southeast would be in jeopardy. Cleanup operations were initiated to rectify the problem, and the Tuscaloosa Aquifer appears to have been protected (USGAO 1984). The potential and actual physical impacts can be contained and the habitat restored, but other damaging social impacts are harder to deal with. Not all parts of the Savannah River area benefited from the jobs created by the Department of Energy's facilities. Moreover, agriculture in the district went into sharp decline, and the region became stigmatized as a waste-contaminated district replete with major environmental problems, which seems certain to hinder future economic productivity whatever objectively might be the actual land condition (Greenberg et al. 1998).

With radioactive wastes being stored on site throughout the United States, the potential for similar environmental problems elsewhere is widespread. It is for this reason that the Yucca Mountain storage facility in Nevada was established to remove wastes accumulated throughout the country to a hopefully more environmentally safe storage area. Even this decision remains controversial and contested (Kuletz 1998). Because the site is located only 100 miles from Las Vegas, the decision to implement the Yucca storage project has taken place over the strong opposition of the Nevada political establishment. From a national technoscientific perspective, Yucca Mountain is a barren, largely lifeless, low-productivity domain that can readily be sacrificed to larger social and economic needs (USGS 2000). Not so to the Shoshone and Paiute communities for whom Yucca Mountain is a sacred site and who regard the surrounding environment as a sustaining spiritual habitat of great beauty. In this instance, the region's very remoteness from national centers of population and power, which heretofore has helped Native Americans retain their claims to space unwanted by mainstream national interests, now works against their ability to defend interests of vital importance to their material and cultural well-being.

If radioactivity had entered the Dnieper River from Chernobyl or the Tuscaloosa Aquifer from the Savannah facility, the ramifications for the bioproductivity of the affected land and water systems would have been momentous. Land degradation problems associated with radioactivity are a negative ramification of progress in the industrial/technological sector. Because of the wide spectrum and the nature of negative impacts that can result from the misuse of nuclear energy, prevention of the degradation from ever occurring is clearly a superior option to decontamination of an area after an accident has transpired. In fact, one of the tenets of the Green Movement is zero nuclear power so as to reduce the likelihood of such potential problems to zero.

Wood Consumption and the Environment

Vegetal materials make significant contributions to the demands for energy and materials. Wood provides approximately 20 percent of all energy in Asia and Latin America. Almost 50 percent of energy in Africa is derived from either wood or charcoal (Arnold 1992). Unlike most minerals, vegetal materials should be treated as renewable resources in that new growth should replace the cut plants. In reality, management of growth and harvest cycles is often deficient. As a result, a degrading environment and a lower production in biomass results after a number of harvests—or even one forest clearing in some cases. Through a brief examination of two vegetal commodities, trees and cotton, some relations between land degradation and biomass utilization are explored.

Comparison of a natural vegetation map to one of actual groundcover reveals a major discrepancy between the two. Over 33 percent of the world's forests have been cleared for agriculture, grazing, or urbanization (Postel and Ryan 1991). Of the remaining 4.2 billion ha (10.4 billion acres) of woodlands and forests, less than 25 percent are primary forest, the vegetation type delimited on maps of natural vegetation. The remaining 75 percent are either secondary or managed forests that differ significantly from the primary tree cover they replaced. Most of the world's remaining primary forest is either tropical rainforest, especially under pressure in the Amazon, Africa, and Southern Asia, and taiga in Alaska, northern Canada, and Siberia. Almost all of Europe's natural forests are gone, while in the United States, excluding Alaska, less than 4 percent remain (table 6.2).

Trees play a multitude of environmental roles. They affect climate, are a carbon sink, minimize erosion, contribute to a complex biodiversity, and strongly influence groundwater/surface water flows. In many areas they are a source of animal feed and provide supplements to the human diet (Antonsson-Ogle 1990). By intercepting rainfall and providing litter, they affect groundwater recharge and hence have a significant effect on stream discharge (Pereira 1973). Additionally, trees may improve soil fertility by adding nitrogen to the soil and/or reducing salt concentrations. Under current demographic, economic, and political conditions, in many parts of the world current management practices are failing to renew existing forests, let alone restore those already sacrificed. Through the early 1980s forest cover was decreasing at an estimated rate of 11 million ha, or more than 27 million acres, per year (WHO 1992). More recent estimates indicate that deforestation occurs at a rate over 50 percent higher (World Resources Institute 1992 and table 6.2). Reforestation and afforestation at best offset these losses by 10 percent.

In the poorer regions of the world, landlessness and other contemporary pressures on the remaining tropical forests could result in the same magnitude of change as has occurred in the midlatitude areas. But because the environmental situation in many tropical areas is more fragile than in the former middle latitude areas, unlike the southern Appalachian area (chapter 5), these lands could experience land degradation that will be very difficult to reverse. Originally, widespread wood fuel shortages in the developing world were viewed as the primary cause of tropical deforestation and in semi-arid areas a cause of desertification

Table 6.2. Estimated Forest Cover (millions of hectares; 1 ha = 2.47105 acres)

Location	Original Forest	Current Forest	Existing Primary Forest
Europe	n.a.	157	<1
United States (excludes Alaska)	384	244	13
Brazil	286	220*	180
Other	5,530	3,623	1,320
Total	6,200	4,244	1,519

*total deforestation in 1996–1997 was 1,322,700 ha; 1997–1998 increased 27 percent to over 1,600,000 ha. INPE (Instituto National de Pesquisas Espaciais)

Source: Data adapted from Postel and Ryan (1991).

(Eckholm 1975). Later it was felt that widespread tree cutting was a result of agricultural expansion (O'Keefe, Raskin, and Bernow 1984) and urban energy demands (Munslow et al. 1987; Lewis and Berry 1988). Now, except in extremely local areas, fuel wood demands are not considered a catalyst for deforestation (Leach and Mearns 1990). Today, it is known that a multitude of factors cause deforestation in tropical areas (Geist and Lambin, 2001) among which are poverty, demand for tropical woods, conversion to pasture and others (chapter 5).

Only 10 percent of the midlatitude rainforest in the Pacific Northwest remains intact (Miller 1993). Forests in their natural condition once masked the terrain. This has been replaced by a checkered mosaic composed of plots of bare ground, recently replanted commercial saplings, and some mature forests. This type of land cover threatens the habitat of many plants and animals. Current conditions represent a chronically degrading environmental condition. The U.S. Forest Service, under public pressure and court orders, has altered the former approved clear-cutting strategy. Now some mature trees are left standing with the hope of encouraging a more natural regeneration of forest lands. Throughout the tropics, a pattern of replacing numerous species with a few favored ones is the pattern. In particular eucalyptus has been preferred over existing local species because it is fast growing and, when cut for coppicing, its shoots quickly develop into new tree growth. However, the oil in its leaves results in a ground litter that inhibits undergrowth with the result that soil erosion occurs. Hence, reforestation utilizing eucalyptus trees is not environmental friendly.

If the remaining primary forests are to survive, a significant effort needs to be undertaken to curtail the demand for wood. The U.S. government estimates that at least 10 percent of lumber and plywood/chipboard used in construction could be saved by changing standard, traditional house-building techniques without any sacrifice to quality. A simple change, such as increasing the distance between wall beams to 60 cm (24 in), is but one example. Curtailing waste related to convenience is another example. Over 25 billion pairs of disposable chopsticks are produced annually (World Resources Institute 1992). If reforestation expenses were included in wood costs, waste would likely be lowered and reforestation would contribute to preventing the forested areas from degrading.

Cotton Production's Environmental Impact

Cotton fabric is one of the most desired textile materials (Yafa 2005). The cotton plant requires a long, warm growing season and ample soil moisture for successful cultivation. Within the former Soviet Union, very few areas experienced the physical conditions needed for commercial cotton production. To meet the needs of their domestic market, to curtail imports, as well as to export cotton products, "cotton independence" was a goal of the USSR during the 1950s and 1960s. "Produce millions of tons of cotton at any cost, and fulfill The Plan, at any cost"(Precoda 1991, 114). To create the environmental conditions needed for successful cotton production and other agricultural crops in Central Asia, especially Uzbekistan, major water diversions were undertaken to increase irrigation (Rutkowski 1991). The diversion of water in the Aral Sea Basin for cotton irrigation continues to be one of the most significant contemporary ecological disasters. Widespread land degradation continues as a result of the curtailing of river discharges into the Aral Sea due to irrigation demands. The reduction in size of the Aral Sea degraded a complex land-water ecosystem. The changes set in motion by the expansion of irrigation not only resulted in unintentional destructive changes in the areas of cotton production and the Aral Sea, but also have set into motion a chain of events that is resulting in negative regional climatic change as well.

Through the 1960s, before river diversions reached a magnitude that affected the sea, the Aral Sea's surface level fluctuated between 50 and 53 m (164–174 ft) above sea level (Precoda 1991). Including its islands, the surface area was around 67,000 sq km (25,900 sq miles) and its volume was greater than 1,000 cu km (244 cu miles). Having no surface outlet, its waters were saline, but not excessively. These conditions resulted from a general equilibrium among freshwater surface and groundwater inflows, evaporation, and transpiration from the wetlands surrounding a large proportion of its coastline, and groundwater outflows (table 6.3).

These prediversion conditions resulted in a highly productive ecosystem within a climatologically dry area. A rich aquatic life existed within the sea. Highly commercial species such as sturgeon accounted for over 45,000 tons of fish annually caught from the sea and its tributary rivers. The wetlands along the coastline provided a rich habitat for wildlife. Immediately inland were over 250,000 ha (620,000 acres) of forest and shrubs. The Amu Dar'ya delta, along the southern shore (figure 6.9), provided highly productive year-round pasture (Precoda 1991). Much of this productivity in both the Amu Dar'ya and Syr Dar'ya deltas has been lost to a combination of water diversions for irrigated agriculture and overgrazing (Ogar 2001).

According to USSR plans, the Amu Dar'ya and Syr Dar'ya waters (the major water sources for the sea) used in the basin's irrigation projects were eventually to result in an increase of 1.5 million tons in annual cotton production (Precoda 1991). The massive expansion of irrigation to reach this goal began in 1956. By 1960, because of the increasing use of these rivers' discharges for irrigation, river flows into the sea decreased to levels that were less than the evapotranspiration and groundwater losses from it. At this time, the sea began to contract, and it continues to shrink to the present day. This reflects the continuing imbalance of excess annual evaporation over the decreased yearly inflows into the sea. By 1988

Table 6.3. Characteristics of the Aral Sea, 1960–1989

Year	Height above sea level (m)	Area (1,000 sq km)	Volume (cu km)	Mineral content (g/l)
1960	53.3	67.9	1,090	10.0
1965	52.5	63.9	1,030	10.5
1970	51.6	60.4	970	11.1
1975	49.4	57.2	840	13.7
1980	46.2	52.4	670	16.5
1985	42.0	44.4	470	23.5
1989	39.0	37.0	340	28.0
1995*			272	

Sources: Kotlyakov 1991; *The Aral Sea Homepage 2002.

Figure 6.9. Between 1960 (top) and 1995 (bottom) the Aral Sea shrank drastically as water from the rivers that fed it was diverted to nourish irrigation schemes. By 1995 the sea had divided into two separate entities separated and surrounded by extensive salt flats. (Photo by USGS and Landsat; see www.dfd.dlr.de/app/land/aralsee/)

less than half of the 35 cu km of river discharges needed to prevent further shrinkage of the sea were flowing into it (Ellis 1990). If current trends continue, some estimates say the sea will cease to exist by 2010 (Sigalov 1987). The desiccation of the sea has resulted in a multitude of deleterious changes. The changes are of such a magnitude that they threaten the production of cotton, the primary reason why the diversions occurred in the first place.

Destructive creation, resulting from the river diversions, has proliferated in the region. The sea itself has ceased to be a productive habitat. Of the more than 20 species of fish formerly found in the sea, increasing salinity and dried up coastal spawning grounds have caused the disappearance of almost all forms of higher aquatic life (Micklin 1991; Breckle et al. 2001). The sea's fishing industry is completely destroyed. With vast areas of former seabed now dry (the islands are or will be connected to the mainland), the exposed precipitated salts covering the former seabed are now a major source of windborne pollution. Over 100 million tons of fine dust and salts are transported annually by winds blowing over the exposed portions of the former seabed (Precoda 1991). The continuous shrinkage of the sea results in a never-ending supply of materials for the wind. Some of the windblown salts are being deposited in the irrigated agricultural areas. By raising the soil's pH, these deposits are lowering cotton yields. Estimates indicate that huge areas of irrigated land will eventually become saline from these deposits thereby decreasing the soils' fertility and preventing most forms of agriculture. Since the harvesting of cotton was the major reason why the region's river waters were diverted from the Aral, this negative, land-degradation feedback seems particularly ironic and destructive. The salt-laden air also is responsible for respiratory and other health problems for the inhabitants of the basin. In the Kara-Kalpak region of Uzbekistan (south of the Aral), the incidence of throat cancer and respiratory and eye diseases has soared. One contributing cause has been increased levels of airborne dust and salt. Other likely factors are pesticides and insecticides associated with the cotton production (Ellis 1990).

The exposed seabeds are not only a source for dust and salt pollution in the areas surrounding the sea, but also these materials are being carried to and deposited on the distant glaciers and snowfields along the Afghanistan/Tajikistan/Kyrgyzstan/China borders in the Pamirs. Some of these areas are within the Aral watershed. The deposition of the windborne deposits on the glaciers is causing acceleration in the ablation of the glaciers and snowfields (Precoda 1991). The long-term effect of these changes on the rivers flowing from these areas is unknown. Yet if any significant changes do occur, they will cause further alterations in the hydrologic conditions of the basin, as well as in other river systems that have headwaters in the Pamirs.

Prior to the shrinking of the Aral, the frost-free period (growing season) in the surrounding area was around 200 days. Through its maritime influence, the sea protected large areas of Uzbekistan from the harsh winter winds that blow from the northeast. As these winds blew over the sea, moisture evaporated and the air temperature moderated. A proportion of the sea's evaporated moisture eventually precipitated as snow in the distant mountains to the south. These are the very areas experiencing accelerated ablation due to the increase in dust and salt and the decrease in moisture carried by these winds. As the surface of the Aral becomes smaller, the ability of the sea to ameliorate the cold winds decreases. Frosts are increasingly coming earlier in the autumn and later in the spring. The frost-free period is now around 170 days (Kotlyakov 1991). Thus climatic conditions for cotton are becoming marginal due to the shorter growing season. "Climatologists think that for cotton-growing, the Central Asian region will be lost permanently" (Moroz 1988). If this occurs, it will be the final irony in this example of land degradation. In the attempt to modify nature by increasing the production of cotton, the Aral Sea Basin will have undergone significant land degradation and will no longer have climatic conditions that permit cotton cultivation, even if ample water exists and the soil resource can be rehabilitated.

In the case of the Aral, some of the degradation processes set into motion by the massive diversion schemes were predicted; but they were ignored due to the critical national goal of increasing cotton production (Mainguet and Létolle 1998). No matter what strategies are decided upon in the future, the Aral Basin will never recover to its former self. Even if cotton production is lowered and water waste is curtailed—Nicolai Orlovsky, Michael Glantz, and Leah Orlovsky estimate that as much as one-fifth of all the water diverted for irrigation currently is wasted (2001)—many of the degradation events that have been set into motion by the irrigation projects will be irreversible. The salinization of some of the irrigated lands, the decreased productivity of the sea itself, the climatic change to a more severe winter and hotter summer, and the changes in the distant snowfields and glaciers illustrate some of the complex linkages and interactions between the sea and its surrounding environment. These linkages have been drastically altered by the single-dimensional plan to increase production through the utilization of this region's water resources without considering the feedbacks among the water, land, and wind systems. With the breakup of the USSR, the degraded lands of the Aral Basin lie within two independent nations that must contend with the associated problems which resulted from a former political/economic system and nation.

Transportation and Land Degradation

As land, water, and air transport evolve, each improvement generally places increasing demands on the environment. As they become larger ships require deeper and larger harbors, which often requires the dredging of channels and harbors. Railroads, automobiles, and trucks have replaced beasts of burden; mechanized transport requires gentler gradients than trails used by animals. As these vehicles become faster and more powerful, curves in the roadbeds are reduced or eliminated, road dimensions increased, and natural surfaces replaced by a variety of all-weather surfaces. Airplanes, like their water and land-based counterparts, have increased in size and speed. Runways are longer, wider, and natural materials are replaced by either asphalt or concrete. Grass or dirt airstrips are largely anachronisms.

Changing Scale in Transport Systems

Older transport systems largely were situated in topographic settings that naturally satisfied the demands of the conveyance. With more powerful earth-moving equipment and explosives, topography is now "created" to meet the demands of specific transportation equipment. Harbors and navigable rivers were once restricted to sites where natural conditions were suitable. Rivers are now dredged and dammed and canals are built to expand shipping possibilities. River transport, as far inland on the Arkansas River to Tulsa, Oklahoma, is possible only because of significant human alteration of the natural flow and channel of the river. Because of dam and canal construction, river traffic can now navigate from the Danube to the Rhine. This permits movement of goods from the Black Sea to the heart of Western Europe. Likewise, the St. Lawrence waterway enables Buffalo, Toronto, and other inland cities to become gateways for ocean shipping. The Mississippi is a complexly managed river system with locks, meander cuts, and the shipping channel in the delta "stabilized" through continuous control operations preventing the Mississippi from shifting its main outlet 160 km (100 miles) west to the Atchafalaya River. Without massive engineering interventions, New Orleans would not be able to function as a seaport. Unlike Bruges, Belgium, where silting of its harbor prevented it from remaining a major port, and gave a critical advantage to Antwerp, New Orleans has avoided suffering the same fate because of major manipulations of the lower Mississippi. Modifying topography for surface transport is common. Where trucks and trains previously had to climb steep hills, laboriously, terrain gradients have been reduced through cut, fill, and tunneling operations. The

TGV (high speed train) rail network in Western Europe is continuously expanding as new gentler and straighter tracks are laid. The narrow and steep winding horseshoe curves of yesterday's highways and railroads are being replaced by modern multilane expressways and straight gentler rail beds that permit high speed even in hilly settings.

Airport locations, once restricted to flat and open terrain, are now less restricted. Hills are removed and wet areas filled, if needed, when airport construction priority is high. Boston's airport exists on a massive landfill within the harbor. The new Hong Kong airport required extensive alterations in topography for it to meet modern aircraft needs. From the perspective of satisfying the needs of transportation goals, modification of the Earth's surface is logical. Nevertheless, in the process of meeting these needs, the alterations often set into motion land degradation. From the bioproductivity perspective, just replacing the natural earth surface materials with concrete and asphalt for roads in the United States alone accounts for 130,000 sq km (50,200 sq miles) of land degradation (Barrow 1991). In the remainder of this section other reasons for causal relations between transportation and land degradation will be presented.

Highways

With modern earth-moving equipment and powerful explosives, highways provide fast linkages within and between urban areas and out-of-town locations, and in the process create anthropogenic environments. They alter the terrain by changing vegetative cover, land morphology, surface permeability, and surface and subsurface hydrology. Slope stability problems, in areas with relief, are common along roads due to alterations in slope and drainage characteristics. This is because topographic changes resulting from road construction usually create a less stable terrain. Thus erosion and mass movement potential increases (Reed 1920). In particular, three ramifications of road construction are the primary catalysts for the instability of lands juxtaposed to major roads and railroads. These are:

1. the creation of cut slope areas, where terrain is excavated
2. the establishment of fill slope areas to raise the preexisting ground to maintain road grade or, in areas susceptible to flooding, to raise the paved road above high-water
3. the immediate areas adjacent to the road directly affected by the drainage and other processes originating from the highway construction (Molinelli 1984; figure 6.10).

Worldwide, modern expressways crisscross through diverse landscapes as relatively straight ribbons with gentle to moderate gradients. Construction of this form, which permits consistently high vehicle speeds to be maintained regardless of the neighboring terrain, requires road cuts through hills and road fills along valley bottoms. As relief increases in the zone that the highway traverses, the cuts and fills become ever more massive to meet the specifications of modern roads. While the cuts and fills minimize steep road gradients, the land slopes on the cuts and fills themselves are often steeper than their natural angle of stability. There are both economic and environmental reasons for the steep slope characteristics of cut and fill. As the slope of a cut becomes steeper, the slope, the area of the preexisting terrain altered in road construction, decreases. This reduces the land that must be purchased for road

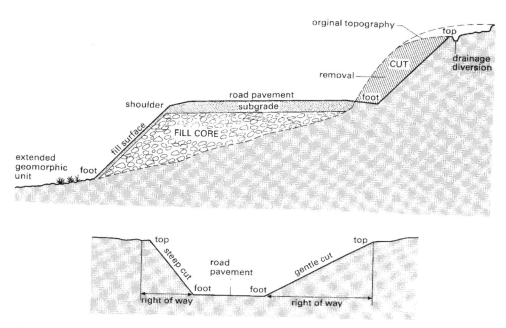

Figure 6.10. Removing steep grades from modern highways requires cutting and filling of existing terrain. When these cuts and fills are not constructed properly, excessive erosion and mass movements can occur. (From Molinelli-Freytes 1984, by permission.)

rights-of-way, and it directly disturbs a smaller area of the existing terrain. The flip side of this common engineering strategy of relatively steep slopes on the flanks of cuts and fills is that erosional processes and mass movements are more likely as slope increases. This is especially true when the materials involved are not consolidated. Road design considers the physical and engineering properties of the materials involved, but the fact that erosion and mass movement rates are generally higher in proximity to highways indicates that road slopes are often underdesigned (Skempton and Hutchinson 1969; Younkin 1974). The areas that experience landslides, slumps, or excessive erosion clearly become degraded. Roads that increase the accessibility of areas within the Alps and north and south of them are one reason for the occurrence of landslides and areas of degradation in parts of Switzerland.

Road construction affects not only the terrain directly altered by the presence of the road, but also areas immediately beyond the road. In particular, water runoff is a major culprit. Poorly designed drainage outlets, constructed to quickly move precipitation from the road surface, often result in accelerated erosion and sometimes gully formation on lands in the immediate area of the culprit. Road construction in parts of the Kenyan Highlands resulted in the formation of gullies 2 m (6 ft) deep, over 1 m (3 ft) wide, and over 100 m long (328 ft) at road drainage outlets. Portions of farmers' fields were eroded and hence bioproductivity was reduced. Drainage works along roads in particular are potentially disastrous in areas where the runoff is diverted to flow over deep, unconsolidated earth materials, such as in Central Province, Kenya. When bedrock is close to the surface, as in many parts of New England, such degradation is minimized. Problems similar to those just presented are associated with both railroad and airport construction.

Harbors and Coasts

With regard to physical processes, coastal areas are extremely dynamic areas. Morphological changes are the norm. Energy released during storms and waves along with longshore currents continuously move materials along coasts and from the land to the sea. In the UK, along the Yorkshire coast in northeast England near the mouth of the River Humber, the coast retreat is over 4 km (2.5 miles) since Roman times. Twenty-nine villages and their surrounding lands have been lost to the sea by normal geomorphic processes (Sheppard 1912). On Cape Cod, the central (Wellfleet) east-facing shoreline retreats 1–3 m (3–10 ft) annually due to normal winter storms and occasional hurricanes. In this zone, the Marconi Station, the site of the first transatlantic wireless transmission, has already been lost to the sea. Rapid coastal erosion in parts of the Cape results in destruction of built-up areas (figure 6.11). Likewise, these eroded materials produce deposition along the outer Cape's extreme northern and southern portions (Provincetown and Chatham).

Many sea harbors likewise are situated in dynamic environments. Two phenomena, sedimentation and storm waves, often require human interventions to keep a port's waters

Figure 6.11. Dwellings along the coast in Chatham, Massachusetts, are threatened by beach erosion. Residents often take expensive, and often only partially effective, steps to harden the beachfront of their property in an effort to escape damage. (Photograph by L. Lewis, autumn 1994)

from becoming too shallow for ships to enter and to protect ships that are anchored. By definition, a harbor is a sheltered water body protected from strong currents and waves. Thus the characteristic that makes these areas a haven for ships, namely quiet water, makes them ideal as sediment traps. Rivers or offshore currents with ample sediment have the potential to transport materials into harbors. When they do, the harbor silts up and becomes too shallow for shipping. Ephesus, Ostia, and Bruges are examples of ancient ports that lost their harbor status due to siltation. Many contemporary ports require constant dredging operations to remain open. New Orleans and Buenos Aires are but two such examples. Without dredging, these and many other ports would suffer the same fate as their ancient counterparts.

Even when siltation is not a problem, the depth requirements of modern ships mean that most ports have to undertake dredging operations. The deepened channels create unstable conditions that require constant maintenance to prevent them from becoming too shallow for the deep-draft ships. In New York Harbor alone, 4.6 million KL (6 million cu yd) of sediment are dredged annually (Fisher 1992). Today the competition among seaports to receive the largest and most economical ships requires channels up to 15 m (50 ft) in depth. This has initiated a new cycle of channel deepening in a multitude of major harbors. For example, along the U.S. East Coast, while New York (including all shipping berths such as Newark and Staten Island) remains the largest port, its main rivals, Baltimore and Norfolk, already have deepened their channels to 50 ft. This has set into motion new dredging in the New York area since, without dredging, the shipping channels would only be 5.5 m (18–19 ft) in many places (Ehrenman 2003).

By preventing siltation in coastal situations where new highly productive land areas would otherwise form naturally, dredging curtails some aspects of bioproductivity. By itself, dredging cannot be considered a cause of land degradation in most instances. But as harborsides are often the scene of intense industrial activity and large urban conglomerations, many of their waters and sediments have become polluted. This has degraded the lands and waters in the vicinity of these activities. For example, Boston Harbor had essentially become a dead body of water by the 1980s (figure 6.12). The result was that many activities had been curtailed. Fishing ceased, shellfish beds were often closed, and recreational activities were limited; beaches especially were often closed. The magnitude of the problem was so severe in Boston, that major human interventions costing hundreds of millions of dollars were required to reverse the ecological situation. A massive wastewater treatment plant requiring the creation of an artificial island in the harbor and an underwater outlet pipe for the treated waters, miles in length, has not only reversed the pollution and land degradation in the harbor, but today the beaches are open, and fishing and shellfish beds again are operational. Of course, Boston Harbor as a shipping port today is relatively small, making these improvements easier to accomplish.

For years pollutants have found their way into New York Harbor waters and contaminated its sediments. This situation exists in many ports. With dredging under way to deepen its channels, toxins such as dioxin have been uncovered (Fisher 1992). If buried elsewhere in the region, these toxic sediments will degrade outlying landfill areas. If the 460,000 contaminated KL (600,000 cu yd) of silt are dumped in the ocean, the toxins could enter the food chain and degrade a productive offshore area. Thus while dredging by itself does not result

Figure 6.12. Pollution in Marblehead Harbor on the north shore of Massachusetts Bay closed many popular swimming and shell-fishing areas in the 1990s. With the construction of a state-of-the-art water treatment facility on Deer Island, Boston Harbor's waters have improved significantly, allowing both recreational and commercial use of the bay. (Photograph by L. Lewis, March 1992)

in land degradation, getting rid of the sediments can result in a degraded environment where they are deposited.

Ports need to be havens for ships during stormy weather. To improve the natural sheltering properties of harbors, as well as to prevent siltation in many cases, engineering works such as jetties, groins, breakwaters, and seawalls have been constructed. Some ports, because of a lack of natural harbors in the vicinity, are largely the result of engineering works. Vera Cruz, a major Mexican port, is an example of an artificial anchorage. Because of the inherent dynamic nature of coastal and harbor areas situated at the interface of terrestrial and aquatic environments, anthropocentric changes in either milieu usually result in modifications in both. Equilibria along coastal areas are inherently dynamic. Any structure that alters littoral drifts often upsets many properties along a coastal segment. Jetties and groins built to protect harbors from waves and siltation trap materials moving against these barriers and alter their flow away from the harbor toward deeper waters. In response to the decrease of sediment transport in the currents from these protrusions, the shore currents will erode sediments down current from the jetties (Pilkey and Neal 1992). The accelerated erosion of undefended coastal lands, as a result of the engineering works intended to protect another segment of a shore, illustrates the concept of process and response in the physical world. In this case, the response is the unintended degradation of land experiencing accelerated coastal erosion due to the starving of sediments in the currents—the result of the constructed jetties and groins. Modern transportation systems, no matter if they are land, water, or air-based, usually require alterations in the topographic areas directly impacted by them. Harbor improvement is no exception to this rule. Because feedbacks exist in nature, the previous balance that existed in the area is usually upset by construction activities. Change almost always is set into motion due to permeability, slope, vegetation, and any other change that humans initiate. In the extreme, land degradation can directly occur from the "improvement" such as a gulley due to increased runoff from a road. However, most land degradation problems associated with transportation systems are usually the result of the indirect affects of the transportation system, namely the activities. The presence of dioxin in harbor sediments and the removal of these sediments for channel improvement is one example of the concomitant responses among land resources, activities, and transportation systems.

The Global Water Crisis

In the United States and Canada, initially most governmental policies and human behavior are not constrained by limitations of water supply. The overpumping of the Ogallala Aquifer in the Great Plains (chapter 5) is illustrative of this characteristic. "We'd like to believe there's an infinite supply of freshwater on the planet" (Barlow 1999). This perception is far from the truth. Furthermore, the limited supply of freshwater available for human use is decreasing due to pollution. The World Bank states that lack of available freshwater is one of the major factors that will limit economic development in the future (Serageldin 1995). While 70 percent of the Earth's surface is water; less than 2.5 percent is fresh and less than one-third of this freshwater is liquid. Most is locked in glaciers and ice caps, hence, unavailable for use (World Commission on Dams 2000). Only freshwater found in lakes, rivers, and accessible groundwater is renewed by precipitation and can be considered available on a sustainable basis (UNCSD 1997). Considering accessibility (location) of the world's freshwater, only about one-one hundredth of 1 percent of the world's total water is practicably available for human use (Lefort 1996). According to UNEP, besides global warming, water shortage is the second most crucial environmental problem for this millennium. There are several reasons for this water crisis, and these reasons are related to the land degradation and resulting environmental problems that result from this crisis. The four primary reasons for increasing shortages of water are:

1. population growth—"Population growth alone will push an estimated 17 more countries, with a projected population of 2.1 billion, into the water-short categories within the next 30 years" (Hinrichsen, Robey, and Upadhyay 1998, 3)
2. better living standards for greater numbers—Associated with increased living standards is increased water usage which includes not only direct uses, but indirect ones such as increased consumption of irrigated crops
3. inefficiencies in water use—Urban areas and irrigation systems often lose large quantities of water through leakage in the distribution system, while wastage in irrigation is profligate including watering of urban lawns and commercial agriculture
4. pollution—Increasingly, the limited water that is available is becoming polluted, making it unfit for use unless expensive water treatment first occurs.

Inter- and Intrabasin Water Transfers

Most freshwater is not found where humans need it. More than 50 percent of all runoff occurs in Asia and South America. Of that 51 percent, 20 percent alone occurs in the Amazon Basin, an area having a population of only about 10 million. From the human perspective, just as in the Nile Valley before the construction of the Aswan High Dam, most of the Amazon's waters are not utilized before it discharges into the Atlantic. Besides the location aspect affecting human usage, the spatial variability of water availability differs greatly at the national scales and within nations. As but some examples, at the national scale, The Netherlands is completely humid; Niger is almost completely arid. Available renewable freshwater on a per capita basis ranges from over 600,000 cu m (785 cu yd) in Iceland to less than 75 cu m (98 cu yd) in arid Kuwait (Hinrichsen, Robey, and Upadhyay 1998). Most countries have spatial variability of water within their borders. In Nigeria, most runoff occurs in the humid southern and western portions of the nation; in the United States, the eastern portion of the country, while smaller in area than the west, has the greatest runoff; and within Mexico, less than 10 percent of its national territory experiences over 50 percent of its runoff. Worldwide, the temporal variability of freshwater also is not what humankind desires in most cases. In Asia, including India, most precipitation occurs in the summer months, in some places over 90 percent. The majority of the massive runoff during the rainy season results in floods and large moisture surpluses that are difficult to capture for future use during the dry eight or nine months (for example, the Luni Block discussed in chapter 5). Thus human systems under these conditions must learn to cope with the extremes of huge moisture surpluses and major periods of severe moisture deficit. The amount and variability of precipitation under a wide range of environmental and livelihood systems are some of the most important determinants affecting the land resource.

Because of the natural temporal and spatial variability of freshwater, humans in water-deficient areas often utilize engineering works of various degrees of sophistication to move water from where it is available to where it is needed. Dams, the major strategy utilized in water transfers, have major impacts on the environment. In both the source areas and consumptive locales, the alteration of topography for storage and transportation as well as the actual use of the water in the consumptive areas often result in environmental problems and, at times, major land degradation. Dam construction over the last one hundred years created thousands of reservoirs. In magnitude they run the whole gambit in size from small farm and local water supply ponds less than 0.4 ha (an acre) in area to massive lakes such as Volta in Ghana covering 8,480 sq km (3,275 sq miles) with a 450 km (280 mile) length upstream from the Akosombo Dam. This single reservoir (Lake Volta) floods approximately 10 percent of the country's national territory (Lewis and Berry 1988). Today over 45,000 dams taller than four stories (about 15 m/50 ft) exist worldwide. Approximately one-half of all rivers have dams on them (World Commission on Dams 2000). The alteration of the earth's surface and its reservoirs are one of the few human features visible clearly from space. Dams, while decreasing discharge variability, alter ecosystems by drowning terrestrial areas. Although sediment trapped in reservoirs alters downstream sediment flows, threatening deltas (for instance, the Colorado and Nile), it also initiates downstream erosion and deposition affecting river geometry and habitats. Likewise, reservoirs often destroy the river valley's most fertile lands by permanently flooding former river floodplains upstream from the dam. Because of the demands of urban areas and agriculturalists during dry years, river waters are often overallocated. In these years, the river flows do not reach the oceans, as the waters trapped behind dams are not sufficient to meet the needs of humans and nature. This is true not only for small rivers, but also for large ones. Included in the latter group are the Nile, Yellow, Colorado, and Murray-Darling. While every river system is unique, there are some general environmental impacts associated with dams and their water diversions. To illustrate some of these impacts of moving water for human use, the Snowy River Diversion in Australia is examined. Unintended negative impacts, both in the basin supplying the moisture and the basin receiving the additional flows, illustrate some aspects of land degradation associated with humankind's insatiable appetite for water and the constraints of nature.

Australia, the driest continent, has always looked for ways of increasing the use of its drier interior where many areas have high agricultural potential if they could be irrigated. One strategy to increase irrigation in Australia is to transfer water from its limited humid areas near the coast to the drier areas in the interior. The lands within the Murray Basin, juxtaposed to the Snowy River Basin, have a lower relief and a greater agricultural potential than most areas within the rugged humid lands found along the Snowy River. Thus increasing the Murray's water supply by capturing a proportion of the Snowy's makes sense if viewed solely from the agricultural perspective and ignoring environmental concerns. This was the driving force behind the 1960 decision to divert water from the south-flowing Snowy River into the Murray River Basin. In the formulation of the plans for this diversion, "only riparian users were considered, with no account taken of amount of water flow required to maintain ecosystems (*in the Snowy*)" (Bergmann 1999). From 1949 through the early 1970s, a number of dams were built, tunnels were bored through mountains, and pipelines were constructed in the Snowy Mountains Hydroelectric Scheme. Besides the generation of electricity, this project's main purpose was to divert some of the Snowy River's waters to supplement the flows of the Murray River, a river system on the northern divide of the Snowy Basin. The Murray's lands, while drier than those in the Snowy, were relatively flat and had a high agricultural potential. Irrigated agriculture, expanded as a result of the Snowy River Scheme, made the Murray area the "food basket" of Australia.

The diversion of the Snowy River's headwater tributaries in the vicinity of Mt. Kosciusko was so extreme that the river's annual flow in the upper portions of the basin was only about 1 percent of its predam discharge (Bergmann 1999). The net affect of the diversion of the waters into the Murray River system is that discharges in the lower portion of the Snowy River are reduced about 50 percent. Equally important from the environmental perspective to the reduction in river flow is the complete change in the river regime. Snowmelt in the upper portion of the basin resulted in spring flooding under natural conditions. The scheme's implementation and its controlled discharges result in the elimination of these high waters and the resulting scour of sediments from the riverbed during high flows. Floodwater is no longer "wasted" but captured for later release into the Murray for use in irrigation.

Whenever humans alter natural systems, feedback mechanisms always result in changes within the affected river channels. The lack of any environmental safeguards in the planning and implementation stages of the scheme resulted in a series of deleterious impacts from the environmental perspective. The planned releases of water from the dams to maintain a minimal-constant flow altered the streambed. Without the high velocities associated with the high flows during spring, silts have filled in most pools within the channel, creating a generally shallow, flat riverbed with resulting negative effects on the biodiversity of the floodplain. Also, the 50 percent decrease in discharge altered the coastal water balances at the mouth of the river. This resulted in the migration of a saline wedge several kilometers upstream in the lower portion of the river as the decreased river flows no longer could compensate against the salt flows. The result is that the local groundwater in the extreme lower portion of the basin has increased in salinity and approximately 550 ha (1,360 acres) of agricultural land in this humid area of Australia were degraded due to excess salinity (Bergmann 1999). These negative environmental changes in the Snowy River Basin were not considered a problem from the perspective of the planning goals of the scheme, since the aim of the interbasin transfer of water was to add more irrigated land to the Murray River areas. Economic and environmental costs in the Snowy River Basin were not factored into the project. This illustrates how not considering the environmental costs beyond the boundary of a scheme, which is clearly only an artifact of human accounting and one that nature does not respect, results in land degradation.

Just as the decrease in flows initiated environmental feedbacks in the Snowy River Basin, the diversion increased water-initiated feedbacks in the Murray Basin. In the case of negative feedbacks, most of them can be traced to poor management of the irrigated lands. While the Murray system is overallocated, just as is the Colorado, much of its waters are lost in transmission and through evaporation. Forty percent of the Murray's waters are allocated for irrigation; 14 percent is lost in distribution to the irrigators. In addition, the Murray-Darling Basin Commission estimates that irrigation water use could be reduced by 20 percent if better on-farm water management was practiced. One effect of the overuse of water in the irrigation process without proper drainage has been that water tables in many of the irrigated areas have risen. This rise in the water table has brought quantities of salt from dissolved minerals in the subsoil closer to the surface, degrading some of the areas' farmland as its increased salinity curtails yields. Currently, 200,000 ha (494,000 acres) are currently affected with up to 1 million ha (2.47 million acres) at risk (Brewsher Consulting 1999). Thus, not only has the waters from the Snowy River helped increase the development of this agricultural area, it has contributed to wasteful water practices resulting in salinization of some of the areas, and hence destructive land degradation.

The diversion of waters from one system to another clearly was conceptualized as a "nation-building" project. From the perspective of economic development it clearly has been an important catalyst. Since environmental considerations, except erosional control, were not integrated into the project, numerous negative feedbacks have occurred both in the supply portion of the project (Snowy River) and the receptor portion (the Murray Basin). Today, in an attempt to arrest the degradation in the Snowy River, water flows have been restored to it by closing a valve on the Mowamba Aqueduct which diverted the mountain waters to a holding dam (O'Conner 2002). How this action to restore water to the Snowy River as well as to deliver water to the irrigated farming in the Murray Basin works remains to be seen. Just as intrabasin transfers, exemplified by the Colorado and Nile rivers, result in both expected and unexpected environmental impacts, so do interbasin transfers. The land degradation that has resulted in southeastern Australia illustrates the need to integrate environmental concerns during the initial stages of a project. Even then, strategies must be developed that will allow corrective measures to be implemented prior to major unexpected environmental damage.

Invasions by Plants and Animals

Invasive plants and animals often affect biological productivity and land degradation. Many of these species are deliberately promoted by humans for specific economic or aesthetic objectives. Numerous others are introduced into an area for experimental reasons in an effort to test their suitability for naturalization under local conditions. Even when not deemed suitable for the purposes for which they were imported, escaped or released individuals often are able to establish themselves in the wild and survive. Occasionally members of exploited domesticated species are able to escape and breed successfully without human support and nurture. Such feral populations can often become pests that affect land productivity in agricultural settings. Occasionally animals that are neither wanted nor helped by people manage to hook a ride with humans and reach areas that obstacles of distance, climate, and physical barriers previously prevented them from attaining. These fellow travelers are accidental introductions, a by-product of the increasing globalization of human contact, economy, and travel. All of the plants and animals that have invaded local ecologies have had a serious impact on the quality of the ecosystems into which they have advanced. In many instances the expectation of those who helped introduce the new species was that productivity would increase. In practice, the outcome was often more destructive than creative in nature. This section briefly reviews the recent record of plant and animal invasions and the impact that these introductions have had on the worlds they entered and created.

Plant Invasions

Agricultural crops are the most common invasive plants. To grow them, humans have massively altered local landscapes. The grasslands of North America have largely disappeared in favor of cultivated fields of wheat, irrigated cotton, and sorghum. This triumph of mechanized farming has certainly resulted in greater productivity for humankind, although it is doubtful that the prairie dog, now on the brink of extinction (Long 1998), would regard the drastic reduction in biodiversity and habitat represented by this grassland transformation as a positive development. Neither would the bison, whose demise under excessive hunting pressure paved the way for an invasion of cattle and sheep, and whose ideal adaptability to grassland conditions continues to spark calls for its reestablishment (Callenbach 1996). In this section we discuss three plants, water hyacinth, kudzu, and African grasses (buffelgrass and lovegrasses), whose introductions were intended to enhance the quality of local ecosystems but whose invasive behavior has resulted in more controversial and problematic consequences.

The saga of water hyacinth is well known. Native to tropical riverine South America, water hyacinth (*Eichhornia crassipes*) first attracted attention because of its beautiful flower. Brought as an ornamental to a trade fair in New Orleans in 1884, its bloom resulted in widespread use as a water garden beautifier in the tropics and subtropics. Inadvertent releases of the plant into uncontrolled lakes and rivers occurred wherever it was introduced for ornamental purposes. Because water hyacinth reproduces vegetatively, mechanical control is difficult since broken pieces left in the water during removal are the basis for rapid reestablishment. Once let loose, water hyacinth can readily become a major management problem for irrigation systems. Not only is the plant a physical impediment to water movement and small craft transport on main canals, but also massive amounts of water are lost to the atmosphere by evapotranspiration. In the absence of insect enemies and other control agents outside its native habitat, substantial amounts of energy resources, time, and labor must be invested in control that is invariably partial. Chemical control at levels that promise to be effective is undesirable because the impact on other components of the system is severe and potentially persistent, and the plant has a considerable innate ability to resist toxins. Indeed, water hyacinth's ability to flourish in contaminated water has made it a potential natural filtering system for sewage treatment systems. Finding ways to transform water hyacinth from a pest to an asset is an innovative idea. Experiments in Sudan indicate that harvested water hyacinth can be chopped into a cattle fodder that is palatable and nutritious (Wolverton and McDonald 1979, 7). Widespread use of the plant as a source of supplemental feed for livestock would go a long way toward establishing an economic use with real meaning in the local economy. Use of the tough, fibrous vines of this water plant as the base material for wicker furniture is another development that has begun to find favor with furniture and household supply stores in the industrial world. Converting a pest into a productive resource is a promising way to transform a destructive invader into a creative source of long-term sustenance for people.

Travelers in the deep south of the United States often observe the striking sight of a climbing vine reaching up into the canopy of roadside trees. This is kudzu (*Pueraria montana*), an East Asian native first brought to North America as an ornamental shade plant at the Japanese exhibit in Philadelphia's Centennial Exposition in 1876 (Stewart 2000, 66). From use as an occasional veranda shade ornamental, kudzu was transformed in the 1930s into a miraculous solution to two agricultural problems: soil erosion and livestock feed. The U.S. Soil Conservation Service promoted the plant as a quick fix to the gully erosion instituted by poor land-use practices. At the same time, poor farmers were encouraged to use the vine as a source of fodder for their livestock, seedlings and seeds were widely distributed by government extension agents, and "valuable pastures of kudzu" (Bennett 1943, 182) were seen as an integral part of conservation farming programs. The combination worked well as long as a farmer's cows, sheep, or goats were able to move across an open landscape, browsing on kudzu as they went. Once fencing became more widespread and land management practices changed, control over the vine broke down (Alderman 2004). On lots from which animals are barred, but in which kudzu becomes established, the vine proliferates rapidly. So dense, entangling, and impenetrable can massed kudzu become that it poses a mobility problem for both infantry and armor on many military bases in the southern United States (Baskervill 2002). While many native insects consume kudzu seeds and act as a brake on its expansion, vegetative reproduction promoted, often inadvertently, by human action spreads the plant (Stewart 2000, 68).

Road building is a particular culprit. To minimize slope instability, and hence degradation along roads, in humid areas, one method is to plant a complete groundcover and then allow bushes and trees to colonize the area. Kudzu is a fast-growing broadleaf creeper that was considered an ideal plant to provide a dense groundcover. For this reason, it was planted along many road cuts and fills in the southeastern United States. Its rapid growth did provide good groundcover on highway rights-of-way and encouraged slope stability. But this "miracle" vine became a scourge. The vine not only covers the ground, but it climbs and covers trees paralleling many southern highways. When infestations are severe, tree growth is curtailed and, in extreme cases, kudzu can kill the trees (Forests Industries 1984; Watson 1989). While this is not strictly a form of land degradation, it illustrates how an attempt to minimize negative environmental impacts resulted in other negative feedbacks.

Invasion by alien grasses into the drylands of the southwestern United States and northern Mexico is a classic example of the generic and ambiguous issue of the ecological and economic impact of exotic species (Tellman 2002). Three African grasses, buffelgrass (*Pennisetum ciliare*) from East Africa and Lehmann lovegrass (*Eragrostis lehmanniana*) and Boer lovegrass (*Eragrostic curvula*), were introduced into the Sonoran Desert region in the middle of the twentieth century. These species were selected after a lengthy international search process whose goal was to ameliorate the adverse impact of cattle grazing on native grasslands in the region. Seriously overstocked rangeland, combined with a severe drought in 1891–1892, resulted in both devastated grassland and the death of three-quarters of the herd animals in southeastern Arizona (Bock and Bock 2000). From this coup de grace the already-stressed and degraded rangeland proved unable to recover, and not unreasonably people set out to rehabilitate the environment.

When local plant species proved unable to or too tardy in recolonizing disturbed areas, researchers sought plants from analogous habitats that might cope better with harsh local conditions. The U.S. Soil Conservation Service experimented with lovegrass as a soil stabilization groundcover in the 1930s and then, "following the philosophy that if a green grass survives, it is useful. . . ." (Bock and Bock 2002, 154), encouraged planting lovegrasses outside experimental plots. In practice, lovegrasses soon needed little assistance. They colonize aggressively on disturbed sites with bare soil. Roadsides are particularly good sites where seed transport to new locations is easy, disturbance is frequent, and water, drained from road surfaces and adjacent slopes and culverts, is relatively abundant. From lodgments along roads, the newcomers then spread into adjacent areas where local species found it difficult to compete with the prolific invaders.

Several features of the new situation favored the invaders. Inserted into a new habitat because they represent particularly vigorous species in their native setting, lovegrasses encounter few pathogens or species-specific predators in their new environment. Thus they have a comparative advantage not granted to native species. Drought appears to favor the newcomers, opening up holes in the native plant cover that can be exploited successfully by the nonnatives when better moisture conditions return. Combined with the pressure of grazing cattle, sufficient disturbance zones exist to promote the expansion of African grasses. Even on the Appleton-Whittell Research Ranch Sanctuary near Elgin, Arizona, where grazing pressure has been removed completely for thirty years, stopping the spread of lovegrasses into the remnants of native grasslands has proven impossible (Bock and Bock 2000). Originally promoted as a major new fodder source, available when native grasses are not accessible, lovegrasses have proven to be lower in quality than anticipated. What the introduced species do manage better is resist fire. Indeed, their greater density provides an important fuel source for fire, which seriously harms local varieties. The more prolific seeding of the alien invaders allows them to spread aggressively into areas that fire opens up, while their more abundant fuel encourages a greater incidence of fire in the first place (Esque and Schwalbe 2002, 175). This is bad news for native plants and the animals and birds adapted to life among them, and the biodiversity of local habitats seems to decline as the introduced species expand (Bock and Bock 2000, 143). Ranchers in northern Mexico hardly harbor such concerns, and have promoted vigorously the invasion of buffelgrass at the expense of local plants. Dispersal is aided by barbs that catch onto skin, fir, clothes, and vehicles, while plumes promote carriage by wind. Not only is buffelgrass dominant in large sections of dryland Mexico, but also greater density than native fodder species, whatever might be its poorer nutritional qualities, has permitted ranchers to triple their stocking ratio (Búrquez-Montijo, Miller, and Martínez-Yrízar 2002, 134). Few ranchers are willing to forgo the income enhancement that more livestock represent in favor of a contribution to biodiversity! Given that Mexican governmental policy was precisely to encourage greater productivity for humans in the drier portions of its national space, the more simplified environment that results is regarded as a sacrifice well worth making. In the Organ Pipe Cactus National Monument just across the frontier in Arizona, the official view of buffelgrass was the opposite of sympathetic. Determined to preserve native habitat inviolate, and concerned that an increased incidence of fire (promoted by buffelgrass' high dead fuel loading) would have a negative impact on the signature cactus that the monument was created to preserve, park personnel and volunteers mechanically eradicated buffelgrass from roadside, parking lot, and riparian "hot spots" of invasion and dispersal (Rutman and Dickson 2002). Whether two completely different policies on opposite sides of an international boundary can result in permanent exclusion is an open question.

Animal Invasions

Animal invaders exhibit the same basic characteristics as their floral counterparts (Elton 2000). Often introduced accidentally, escaping from captivity, and beginning from very small numbers, alien animals are seldom noticed initially. How invasive animals are viewed is often a perceptual issue that changes over time. Whether such animals degrade environmental quality is often as much a point of view as an empirical reality.

Pigeons are a good example of this ambiguity. While the vast flocks of passenger pigeons that congregated in rural North America were slaughtered with impunity and driven to extinction in the nineteenth century, their feral, urban cousins expanded rapidly, largely unmolested, in the twentieth century. Descended from the rock pigeon (*Columba livia*), bred for centuries by the aristocracy as an ornamental and a culinary delicacy, selected assiduously for its navigational abilities, the pigeon has in the last one hundred years become a major pest in urban areas (Todd 2001). Few natural enemies remain in large urban environments to limit pigeon populations. An ability to reproduce in stark spaces with limited nesting infrastructure, coupled with an ability to scavenge food morsels dropped by humans, makes the pigeon an ideal urban denizen. Land is not degraded by their presence, although concentrations of pigeon droppings can harbor disease and certainly public spaces experience aesthetic deterioration, flowers and other ornamentals are eaten, and native birds are driven from parks when concentrations of pigeons swarm over people and infrastructure. New York City officials have found it necessary to hire falconers to unleash Harris hawks upon the resident pigeon population in an effort to drive the birds from Times Square and other popular tourist venues (Fine 2003). Yet pigeon sympathizers, in contradiction to public policy, do not hesitate to chuck fodder before the hard-pressed avian invaders!

Any number of animals was introduced into new habitats by humans for economic reasons. Often this was done with a view to profiting from the fur of the new species. Once present in a new place, it has always proven difficult to keep all of the animals in captivity. The muskrat (*Ondatra zibithica*), introduced into many parts of Europe for its fur, can be traced to a group of five individuals brought to Czechoslovakia in 1905 (Elton 2000, 24). Nutria (*Myocastor coypu*), large aquatic rodents from South America, were brought to Louisiana for their pelts. When the fur market declined in the 1980s, many nutria were released by farmers and augmented an existing population of fugitive cousins. The consequences of this rapid expansion of wild nutria were deemed potentially significant, but when measured against the impacts of hurricanes and other natural processes they were considered minor, largely unknown, and relatively unimportant (Walker et al. 1987, 200). A burrowing animal in levees and bayou shorelines, nutria impact the habitat by eroding banks and consuming large amounts of vegetation. Mark Ford and James Grace (1998) point out that there is a direct relationship between the number of pelts taken by trapping in a given year and the rate of wetland loss. With trapping active between 1956 and 1983, nutria grazing diminished and biomass increased. This not only provided more protective organic litter on the soil surface but also helped to augment soil mass through the development of roots. The result was less soil leaving the local system. When pelt prices crashed, the nutria harvest dropped from 1.5 million in 1983 to less than 300,000 annually today (Ford and Grace 1998, 978). The resulting increase in grazing pressure not only can seriously impact the natural vegetation but also can weaken the levees that control water levels in rice fields. Not all impacts of nutria are regarded as bad; because problems controlling aquatic vegetation often were severe without resort to the use of chemicals, the nutria were introduced into eastern Texas as a natural mowing machine (Swank and Petrides 1954). With fears rampant that the Louisiana coast is disappearing at a rate of 35 sq miles a year and that the nutria is augmenting the impact of global processes, the state government is offering four-dollar per tail bounties for nutria (TG 2002). This incidentally provides an important source of cash for rural populations who have otherwise lost income due to the decline in fashion fur. And it provides some relief for the native muskrat, who finds the nutria to be an aggressive competitor for essentially the same niche.

Other animals are introduced as objects of sport. Rabbits first appeared in Australia as live targets on a commercial shooting establishment. Their escape into the wild, in an environment without natural predators to limit their numbers, introduced a major new grazing competitor for rangeland forage. This was hardly appreciated by ranchers and farmers, who demanded and received substantial governmental assistance in campaigns against the bunny. Efforts to control the spread of the rabbit resulted in the construction of hundreds of miles of barrier fencing (McKnight 1969), which became a prominent landscape feature in western Australia. The release of mountain goats into the Olympic Mountains was the product of a sportsmen club's desire to increase the supply of targets. When the area became a national park in 1938, hunting policy changed (Sleeper 1990). The protected goat population then began to have a major impact on the park's alpine habitats. How to conserve a sensitive habitat from a nonnative species that is itself protected by the park and is itself endangered in other areas is a management problem of considerable proportions.

No such ambiguity surrounds the story of insects accidentally introduced into novel habitats. Until recently, the full range of lethal tools has been employed against these invaders without qualms. The gypsy moth (*Lymantria dispar*) was brought to the United States in 1869 in an effort to breed a silkworm that would launch a local silk industry (Elton 2000, 111). Some of the insects or eggs escaped confinement, and dispersal from the Boston, Massachusetts, area began, which continues today. Episodically severe outbreaks occur in which massive infestations of caterpillars defoliate trees, sometimes several times in one growing season. The defoliation of millions of acres and considerable income losses are the result. Efforts to control the gypsy moth have included mechanical removal of egg sacks, which is very labor intensive and never complete, and the use of chemical agents, which pose threats to the environment. The use of Bt (*Bacillus thuringiensis*), which is an effective natural agent, is undesirable as a frequent, large-scale control agent because this organism is a major resource of organic farmers. Too frequent and widespread use of Bt only reduces its effectiveness against insects that are a particular plague on commercial crops. Interest now centers on a fungus (*Entomophaga maimaiga*), whose airborne spores give it wide distribution and whose ingestion by the caterpillar result in rapid death (Adler 1994). Unlike chemical pesticides, which almost invariably persist in the environment and kill more than the intended victims, targeted natural killers usually are species specific, although there are concerns that Entomophaga may attack other moths as well. Learning to control an introduced animal, whose often explosive growth occurs because specific natural predators are absent in the new place, is a lengthy process. Concern that the cure introduced as a control may itself escape control and cause unforeseen impacts is very real.

Conclusions

Why do exotic invaders so often do better than their native counterparts? The reasons are numerous. Where they are consciously picked, the aliens are the cream of the crop, the best that nature and scientific selection can choose. Frequently they enter an open niche, one where few if any competitors are found. Where native species do exist, the more aggressive foreigners frequently take their nesting sites, consume their food, and create conditions of harassment and disturbance that tip the ecological balance in favor of the invaders. Few natural enemies exist to control the population size of the new species; pathogens and predators that limit population growth in the territory of origin are left behind. Usually the newcomers are a type of weed, accustomed to living in disturbed sites, able to reproduce rapidly, and adapted to aggressive dispersal. In many cases the invaders are unseen, transported accidentally to a new site as hidden fellow travelers aboard ship, stuck on the undercarriage of vehicles, or hidden in a shipment of food or building materials. From small beginnings of a breeding pair or a pregnant beetle can come a population explosion if local conditions are ripe for success. But so insignificant does the first lodgment often seem that notice is seldom taken. When control officials move to eradicate an invasive colony, for example of mute swans, public opposition to elimination of an aesthetic resource or object of sympathy creates public relations problems and deliberate efforts to flout the regulations.

Summary

Regional forces acting as catalysts for land degradation clearly are growing in importance during this age of globalization. With ever increasing urbanization, land conversion from vegetative covered surfaces to cleared and paved lands result in decreasing bioproductivity in almost all environmental and economic settings. The demand for water, materials, and energy that are required for the successful functioning of urban centers exacts a high environmental cost in areas often distant from cities. In addition, waste products resulting from the high inherent population concentrations and economic activities associated with cities often results in pollution problems affecting both land and water resources.

Increasingly tourism, itself partially a by-product of urbanization, initiates land degrading pressures as city dwellers search for less crowded ambiences, new and exotic recreational activities, and more natural settings. Often in this quest, major land cover changes result. From the clearing of forest cover to meet skiing needs to major stresses placed on coastal/beach settings, successful resort areas must meet the demands of the temporary populations that arrive in their areas during the high seasons. The visitor crush often overwhelms local ecosystems, and an unfortunate side effect is often land degradation.

Transportation needs likewise are ever-increasing to meet the needs of the contemporary world. Construction of highways, airports, seaports, and rapid mass transit often requires major modifications in existing topography. With these modifications, land and water systems are modified and all too often land degradation results. Because of the demands generated by and the intensity of change associated with contemporary conditions, these modifications are no longer local in scale but often regional in nature. For example, agricultural production today is global in scale. Products are transported often to distant markets. To meet the needs of agriculture, regional alterations often are required. Water diversions and removal of groundwater to meet irrigation needs and the mining operations required to meet fertilizer demands are but two examples that can result in major regional-scale land degradation. Tighter and more frequent global linkages produce more rapid exchanges of plant and animal species with serious impacts on biodiversity and environmental health. With an ever-increasing world population and steadily more powerful technology, regional interactions are likely to grow in the future and to affect the environment in a multitude of ways.

But not all contacts, exchanges, and environmental changes produce land degradation. In many instances people have created sustainable, if dramatically altered, agroecosystems. It is to examples of such creatively destructive modifications of Planet Earth that we now turn in chapter 7.

Creative Destruction

LIKE THE ancient Roman god Janus, represented by two opposite faces of dramatically different character, human use of the Earth has a dual visage. One face is creative destruction, the process by which the natural world is modified and sustainable land-use systems are developed. The opposite process is destructive creation, a condition characterized by the failure to achieve long-term sustainability and by the initiation of progressively more serious patterns of land degradation. Creation and destruction are closely linked for, in the interest of sustaining human population, one must often first destroy in order to create. Cultivation of most crops desired by humans, for example, requires the removal of preexisting natural vegetation in order to create the conditions that are most conducive to crop growth. Despite the evident risks involved in this strategy, from a reduction in species diversity to increased prospects for land degradation, simplification of the natural world in the pursuit of human sustenance is the resource management approach most frequently employed by humankind.

A successful process of *creative destruction* is characterized by several factors. Of primary importance is the subtle application of the *genius loci principle*. This principle stresses the special conditions found in every locale and the need to know these system states in detail. Each place is characterized by particular attributes of soil, climate, vegetation, culture, and system behavior. Without an intimate knowledge of these conditions, it is relatively easy to attempt to erect stable resource-use systems but, instead, to generate systems that are vulnerable to collapse or that are poorly adjusted to local resource constraints and opportunities. Time is required in order to comprehend the basic dynamics of any place and to develop livelihood systems that are closely attuned to these rhythms. This is why many traditional resource-use systems provide important insights into how to erect sustainable systems.

Intimate knowledge of place makes it possible to identify the *critical zones* that are essential for system stability. These crucial resources must be maintained and enhanced if a sustainable system is to be generated. Increasing the productivity of a critical zone often is linked to the corresponding decay of a *sacrifice zone*. Sacrifice zones are resource areas that are degraded in order to transfer the productivity of that sacrificial space to the critical zone. The movement of soil and water from hillslope catchments and their concentration in valley bottom terraces by the Nabataeans, cited in chapter 2, is an example of the use of sacrifice zones to improve the productivity of critical resources. In other instances, there is no direct transfer of productivity to the critical zone from a sacrifice area. Rather, the changes in the critical zone can involve linked developments in other areas that reduce the productivity of those districts and make them sacrificial offerings, the degradation of which is accepted as a necessary cost of promoting progress and productivity elsewhere. Enduring creative destruction is sensitive to the changes that take place in both types of sacrifice zones, since otherwise degradation in the sacrifice zone can have negative feedbacks that diminish productivity in the critical zone.

In addition, there exist two other attributes of creative destruction. The first is *waste capture*, which converts a potentially negative output from one source into a positive resource. The more that wastes are internalized within a system and become input resources for other parts of the system, the more likely it is that the system will attain long-term stability. Chinese fishponds, discussed in greater detail later in this chapter, are an apt illustration of this principle, since manure and other organic wastes from terrestrial activities are a basic resource for the carp cultivated in the pond catchment. Equally important is *cost export avoidance*, an objective that strives to prevent imposing negative impacts upon distant areas and converting those affected districts into sacrifice zones. This is a vital concern, because failure to diminish or prevent distant negative impacts can produce *unintended sacrifice zones*. Unexpected impacts, spatially connected to distant activities, are an important contributing factor to many contemporary land degradation problems. Because causation in these instances is far removed from the immediate setting of the negative developments, it is usually difficult to rectify the situation until often irreversible damage has occurred. It is one thing to sacrifice deliberately the productivity of one zone for the gain of another when this has been carefully considered and, insofar as it is possible, the consequences have been meticulously evaluated. It is quite another matter to trash an environment heedlessly without compensating gains in stable productivity elsewhere. It is particularly in settings in which cost export avoidance is ignored that the often-delicate balance between creation and destruction is upset and degradation occurs. Only when critical zones are preserved and the larger ecumene is protected from the export of undesirable costs can the creative destruction process bring into existence resource-use systems that possess the stability to recover quickly from disturbance and the resilience to absorb change without negative consequences.

In chapters 4 through 6 we examined some of the myriad ways in which destructive creation occurs in land, water, vegetation, common property, and urban systems. This chapter explores the operation of creative destruction in human systems. Examples are grouped into low-, moderate-, and high-intensity systems depending on the capital, labor, land, and technological investment that each system makes in extracting sustenance from its habitat.

Low-Intensity Systems

Resource-use systems that are spatially extensive, employ a long fallow cycle, and invest relatively little human labor, capital, and technology in exploiting the total area to which they have access are low-intensity systems. Hunting and gathering, swidden (slash and burn) cultivation, and nomadic pastoralism exemplify low-intensity systems.

In low-intensity systems, a fundamental factor in their success is mobility. This characteristic allows a band, village, or herding group to shift to new locations whenever seasonal conditions or local resource scarcity require. A linked variable is the size of the territory exploited by a particular group, which is often very substantial compared to the population supported. Thus low population densities are typical in low-intensity systems. Long lag times characterize the periodicity of resource use, and it may be years before a specific site exploited in one year is revisited. At the same time, there are often seasonal periods of quite intense use of spatially limited resources, and these episodes of concentrated use, coupled with alternating periods of dispersion (Ingold 1987, 184–87), establish both the rhythm of seasonal activities and the base constraints that condition population dynamics and resource use. Population growth in excess of what can be maintained during the most constrained period requires either a change in location or an altered resource exploitation strategy, and often signals a shift to more intensive systems of resource use.

In a significant fashion, low-intensity resource-use systems exhibit few of the characteristics of creative destruction. In particular, they are able to avoid frequent use of long-term sacrifice zones because the overall intensity of resource use is sufficiently diffuse to make a sacrifice zone unnecessary. Moreover, low-intensity systems are able to protect the critical zones and resources essential to sustained exploitation. Departures from the practices employed by these systems result either in intensification and creative destruction or land degradation. In this sense, low-intensity systems are the base setting from which more intensive, creative destruction modes of livelihood may develop. A brief examination of hunting, shifting cultivation, and herding follow in subsequent sections as a prelude to examination of the more ubiquitous intensive examples of creative destruction that exist in most populated regions of the contemporary world.

Hunting and Gathering

The oldest human mode of food production is the combination of hunting and gathering activities (Lee and DeVore 1968). This system of resource management encompasses all but the past ten thousand years of human existence. It is the ultimate base from which all subsequent livelihood systems evolved. Although hunter-gathers possessed a sophisticated hunting technology and had a deep knowledge of both plant and animal characteristics in their local area, they were not able to support large human population densities. Generally, hunter-gatherers utilize a variety of population control devices (prolonged lactation; infanticide; post-partum taboos) that discourage population growth (Neel 1970). Today's hunter-gatherer cultures only survive in the most marginal and remote settings, environments not able to support many people. In the past, hunter-gatherers would have lived in more abundant habitats that were able to sustain more people. Undoubtedly, it was from these better locations that experimentation leading to agricultural domestication took place. It is from better-endowed sites that hunter-gatherers were gradually pushed by more intensive livelihood systems until they were restricted to their present isolated locations. Even in these limited habitats, replete with constraints, most hunter-gatherer groups are able to enjoy an adequate diet and experience significant leisure (Lee 1968). They are able to accomplish this by virtue of values such as sharing, a gender-based division of labor in which men hunt animals and women collect plant material, and cultural practices that emphasize seasonal mobility. The success of hunter-gatherer cultures in sustaining themselves throughout most of human history is a product of two factors: the modest concentrated pressure that they brought to bear on the resources in their territory, and the modicum of creative destruction that figured in their livelihood system.

The bulk of the hunter-gatherer diet is composed of plant material, not meat. Women are primarily responsible for these food resources, although there are some communities in which the women also figure as prominent hunters (Estioko-Griffin and Griffin 1981). Depending upon the abundance of the habitat, the range of plant resources collected can be quite large. Among the !Kung, for example, two hundred plants are identified, and some eighty-five edible plants are gathered seasonally in an environment that can hardly be considered lush (Lee 1969, 59). The palatability of these items varies, and there is a seasonal cycle in their collection. As the dry season increases, the !Kung shift from the more desirable species and consume the less tasty. As collecting continues, it also becomes increasingly difficult to find food within easy walking distance of the base camp. Lee (1969, 31) indicates that a 10 km (6 mile) collecting radius is the maximum that people can conveniently go from their base camp in a day in order to hunt and collect. This "day prism" sets a limit to the time-space resources of a band (Carlstein 1982). When available resources begin to decline, a band must either shift to poor resources or move on to another site. In drier places there is always a movement away from permanent water sources during the wet season in order to conserve the resources of this more bountiful area for the dry season. In more moist habitats, bands must still move in order to avoid overexploiting local resources. Failure to move, combined with excessive local extraction, would invariably result in the starvation of the band, a result that has undoubtedly occurred on occasion. Such localized population crashes allowed for recovery of an overstressed resource base before any future cycle of human use. Movement was also mandated because food storage technology was insufficient to permit permanent settlement. Without stored reserves to carry the population through a long "empty" period, movement to new districts and/or the consumption of less palatable species was mandated.

Within this framework of selective and relatively short-term use of resources, hunter-gatherers had relatively little impact on their environment. There were three ways in which gatherers may have had an impact. First, by creating disturbed conditions around their campsites and along their favorite trails, gatherers may have produced conditions that were suitable for the growth of some of the species that they collected. Fire was the major tool for environmental modification employed by hunter-gatherer communities. It was the second mechanism used by gatherers, probably as much by accident as by design, to favor some of the food plants that played a role in their diet. Use of fire by the !Kung collectors to enhance the growth of plant foods is reported by Wiessner (1982, 65), although overall it is not likely to have had a great deal of impact on productivity. By virtue of their collecting activities, gatherers expanded the range and frequency of the food plants they preferred. Except for the use of fire to alter species composition in favor of plants they regularly consume and at the expense of others, gatherers have little creative destruction impact upon their environment. Indeed, even among the !Kung, who derive more than one-half of their vegetable food from one source, the mongongo nut (*Ricinodendron rautanenii*), millions of nuts are left uncollected each year (Lee 1969, 59). Even with a basic dietary staple, the !Kung did not begin to interfere with the reproduction of the mongongo by natural means; no efforts to achieve domestication occurred.

As is the case with most collectors, the !Kung's best creative destruction activity is a negative one. It is found in the mobility that characterizes their seasonal pattern and the social network that enables them to leave their home territory and to stay with friends and relatives in distant bands during periods of scarcity. By temporarily reducing carrying capacity under adverse conditions, hunter-gatherers avoid doing serious harm to their environment. This permits their habitat to recover so that it can once again sustain their population and its extractive activities.

If the evidence for deliberate creative destruction is weak among gatherers, it is also problematic among hunters. By far the most important tool used by hunters is fire, which is employed in two ways. In the first instance, fire is used as an aid in hunts in order to drive animals over cliffs, into marshes, or into settings where they can be killed more easily. In the second case, fire is used to favor the growth of grass or browse that would both attract game animals to sites closer to human settlement and provide more abundant and nutritious fodder for animals. Simmons (1989, 78–79) believes that fire, especially in wooded areas, can improve the quality of fodder by a factor of ten. This occurs because fire mineralizes the surface litter and makes nutrients more readily available to the soil and ultimately to the fodder plants that grow in it. Benefits not only include more desirable browse with higher leaf protein content, but also greater concentrations of game animals in smaller areas, so that hunters have to invest less energy in their capture. Given the attentiveness that characterizes hunter and gatherer relations with their home territory, it is likely that in many instances they readily recognize the utility of fire. Consciously employing fire to modify vegetation in a direction that is conducive to human sustenance is the essence of creative destruction. At the same time that some species will find burnt areas attractive, other species will discover that the altered environment is unsuitable; their numbers will decline, at least locally, in consequence. Fire used in hunting would have that same general effect, although the purpose of its use is not always directly aimed at environmental manipulation. Fire was used by !Kung to attract game (Wiessner 1982, 65) and burning was a common practice in the woodlands of eastern North America for many reasons, including to provide browse for deer (Day 1935; Cronon 1983, 147–51). It seems clear that some creative destruction was practiced by hunters to increase both grassland and browse, although it is unclear how extensive this practice was.

Hunters could have initiated more destructive creation directly by predation on animals themselves. By concentrating their efforts upon specific species, and by failing to spare pregnant females and adolescents, hunters could place specific species and age sets at risk (Simmons 1989). These activities may gradually alter the population structure of preferred animals and reduce the number of older individuals present. The result would then be a larger and much younger population with a rapid growth to hunting size, which would use less energy in support of aged individuals. With a population pyramid much closer to a herd of domesticated cattle, the antelope, seals, or deer being hunted would channel more energy into the diet of the hunting community. Greater demands than this made upon animal populations by their human predators are generally a product of contact between contemporary hunters and modern markets, with dangerously excessive off-take rates the result (Simmons 1989, 71). It is the demands of an urban market that commercialized the hunting of green turtle by the Misquito Indians, rather than their own consumptive demands, resulting in drastic overuse of this important marine resource (Parsons 1962; Carr 1967; Nietschmann 1972, 1973).

One hypothesis suggests that hunters entering North America at the end of the last glaciation were responsible for driving to extinction a number of genera of megafauna (Martin 1973). The arrival of hunters synchronous with the loss of a number of species of large game was a suggestive coincidence. Yet there are few examples of hunters running amok and overexploiting their prey in pre-European contact conditions. It is unlikely that human predation was the primary causal mechanism in the loss of North America's large animals. Nonetheless, it is possible that heavy hunting combined with the very severe environmental changes that accompanied the end of glaciation could have had a catastrophic impact on fauna that were already under very severe stress (Butzer 1971, 503–15). In at least one case, that of Maori immigrants to New Zealand, a hunting culture was capable of driving a number of flightless birds to extinction (Simmons 1989, 79). Such instances appear to be isolated and the product of very special circumstances. Traditional hunting technology was limited sufficiently, despite its long evolutionary development and considerable sophistication (Oswalt 1973), that hunters lacked the ability to decimate the animals that were their preferred food species.

Most hunter-gatherer groups possessed a value system that promoted respect for the animals that were an important part of their subsistence. This respect for and sense of kinship with the main object of predation is quite different from the attitude that develops when physical and psychic distance between humans and wild animals increases. In the classical world of Rome and Greece (Anderson 1985), this turned hunting into a sport for the elite, a spectacle of ritualized mass killing in the arena for the enjoyment of the mob, and a setting for the display of manly courage.

A striking example of the differing impact of hunters upon their prey is afforded by Kemp's (1971) analysis of the Baffin Island Inuit, whose adoption of the rifle and motorboat changed their relationship with the seals that they hunted (figure 7.1). Under traditional conditions, the hunter took only as many animals as were needed, and captured those animals only after a long, arduous wait near the seal's air holes or by difficult stalking efforts by boat on the open water and by foot on the floe ice. Believing that Eskimo and their prey were joined in a common realm of spirituality, whenever possible the hunter tried to avoid offending the spirit of the animals that he captured. He did this by killing no more than were needed and could be consumed. Almost every seal harpooned was retrieved because a harpoon formed a firm attachment that prevented the carcass from sinking. With the introduction of the gun, the ease and range of capture increased. The price of this increase in labor efficiency is a dramatic increase in the number of seals killed but lost by sinking before they can be retrieved by the hunter. Kemp observed losses of up to 60 percent of the animals shot. Moreover, the seals became cautious about approaching people, a disruption in intimacy, rooted in a technological change, that has profound potential implications for the wise maintenance of the basic animal resource.

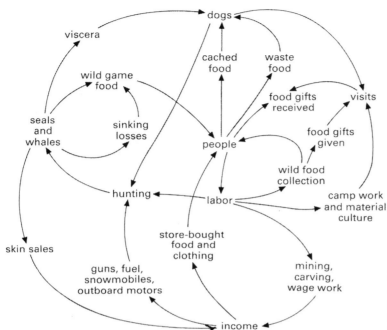

Figure 7.1. The structure of Arctic hunting societies changed dramatically when guns, snowmobiles, and outboard motor boats were introduced.

Hunter-gatherer bands exhibit little conscious creative destruction behavior. But their livelihood reflects a stability that is rooted in the basic principles of creative destruction, a sustainability that contrasts sharply with many modern systems. While there is little firm evidence to indicate a strong tendency for hunter-gatherers to overexploit their food sources, there is evidence that in special circumstances this may have occurred. Certainly some habitat and species modification was promoted, especially through the use of fire, to the benefit of human sustenance without initiating significant land degradation.

Shifting Cultivation

Shifting cultivation is an agricultural system in which crop production moves to new locations whenever yields decrease to a point where continued labor investment in the crop and field environment is no longer worthwhile. Fields so abandoned are usually left fallow for a generation or more rather than for a few years (Richards 1985, 50). The objective is to allow natural vegetation to recolonize the cleared field. The roots of these plants tap deeper mineral layers and leaf litter (at least to some degree) contributes nutrients and organic matter to the soil without humans having to engage in expensive and time-consuming rehabilitation efforts. In most instances, practitioners of shifting cultivation return to plots that they cleared on a previous occasion. Diminished clearing effort in new-growth forest is the primary reason for this practice. Carter (1969), in studying the slash-and-burn cultivation systems of lowland Guatemala, estimated that clearing an old-growth forest required 368 hours of labor to prepare a 2.8 ha (7 acre) field. In contrast, a similar size plot on which saplings were growing required only 113 hours to clear, while but eighty-four hours had to be invested in clearing dense low-growth vegetation. Thus a cyclical return to formerly utilized plots avoids the arduous work involved in clearing old-growth timber. But the temptation to return to previously cleared areas too frequently must be avoided at all costs. Returning too soon reduces the accumulation of nutrients in the soil and can shorten the period within which farmers can anticipate receiving adequate yields. For this reason, maintaining the length of the fallow cycle is essential to the success of shifting cultivation. This is also why extremely long-fallow agricultural systems such as shifting cultivation are land extensive, support low populations densities, and make little investment in infrastructure.

Creative destruction is an integral part of any shifting cultivation system. The primary creative destructive act is the removal of forest cover in the area in which cultivation is to take place. Large trees are girdled, small trees are cut, and material not needed for housing or fencing is piled in heaps. Often branches and small saplings are collected from a larger forest area, transferred to the field site, and added to the vegetation piles (Manshard 1974, 58). This transfer of primary productivity in the form of burnable trash treats the uncut forest as a mild form of sacrifice zone, since organic material that should be available via decomposition for plant growth in one area is moved to and concentrated upon an adjacent zone. Once dried, the vegetation mass is burned and the ash residue functions as a fertilizer source for the newly established field. This nutrient flush plays an important role in the initial yields generated from the cultivated plot. The use of fire to liberate nutrients in the ash also heats the upper soil horizon and plays an important role in destroying the seed of some weeds (Manshard 1974, 58). Reduction in the local availability of competing weedy vegetation is important to the initial high yields of fields, just as the increase in weeds over time contributes to greater maintenance costs and ultimately to plot abandonment.

Vegetation removed and altered by cutting and burning is essential for the success of the shifting agricultural enterprise; the forest in both its morphology and its species composition is sacrificed for the benefit of the agricultural population. Distant sacrifice zones are not created; rather, one component of a given unit of space, the forest vegetation, is sacrificed in order to create conditions in the same space that are conducive to the crops that people desire to grow. In most instances, these crops would not be able to grow in sufficient abundance under natural conditions to support a small, semi-sedentary village population living for several years in the same place. The presence of a closed canopy forest would shade many potential corps sufficiently to inhibit growth. While light holes would always exist in the forest as a result of natural events, edible plants growing in these locations would be likely to sustain only mobile hunters and gatherers. Forest removal is the essential creative act mandated for the success of shifting agriculture. Clearing the forest is a human device for enlarging and enhancing the conditions that occur naturally, but which are found too sporadically to sustain an agricultural, semisedentary existence.

The success of shifting cultivation depends on four factors. The first is the design of the gardens themselves. Particularly in the tropics, gardens are not monocultural; rather they contain an abundance of different plants, the "high diversity index" of which mirrors the variety of plants found in the forest (Ruddle 1974, 5). Plants cultivated in shifting cultivation gardens are intercropped. Not only are only small numbers of any one plant present at any particular point in the field, but also plants of different heights are intermixed. The leaf archi- tecture of this complex structure insures that raindrops do not fall directly onto the soil sur- face. Instead, raindrops cascade more gently toward the surface through an intermediary series of layers. This greatly reduces soil splash and erosion is diminished (Richards 1985, 60). Richards also points out that soil conservation is encouraged by employing minimal tillage in high-rainfall areas in order to minimize soil disturbance, and by keeping some uncut tall trees and the stumps of cut trees within the field. The cohesive action of this intact root mass enhances soil stability. To a degree these isolated tall trees, together with the diverse heights of the densely packed cultivated vegetation, replicate the structure of the forest that the crop habitat replaces (Geertz 1966, 16–25). The small, cultivated clearings also reproduce an opening analogous in scale to the natural light holes produced by blowdowns and other nat- ural processes, and encourage rapid crop growth by permitting sunlight to reach the vege- tation. Small-scale fields such as these also contribute to erosion control (figure 7.2; Watters 1971, 7). By mimicking the same organizational principles as the forest that it replaces, the shifting cultivation field represents a stable act of creative destruction.

A second adaptive feature of shifting cultivation is the way in which forest recolo- nization is encouraged. When fields are cleared, individual trees are always left standing in the fields. These trees are never so large or numerous as to cause shade problems for the crops. But the existence of a number of trees in the field with well-established root systems means that some of the minerals leached from the surface soil layers will be captured by the trees. The presence of individual trees is also a secure source of seeds for forest regrowth once the plot is abandoned (figure 7.3) and so too are trees along the forest-field edge, pro- vided that the plot is not too large. Thus, built into forest clearance is the mechanism for forest reestablishment.

Nonetheless, periodic forest removal has an influence on the species composition of the regrowth forest. Because tropical rainforest trees in particular are widely scattered, there is no guarantee that the regrowth forest plot will reflect the same species composition as did the forest before it was originally cleared. An altered regrowth forest is even more likely when one considers the tendency for shifting cultivators to retain within plots those trees that have some useful purpose in their dietary regime. Regardless of species, very large speci- mens are readily ignored if there is no pressing reason for removing them, since this prac- tice minimizes work effort. It always makes sense to retain within fields scattered individual trees that contribute fiber, medicinal products, and beverages to the local larder. In parts of Latin American not only are valuable preclearance crop trees retained and protected within agricultural clearings, but also fruit trees are planted when fields begin to lose the fertility needed to sustain field crops. These sites can be identified long after agricultural use has ended by virtue of their unusual concentration of beneficial trees (Watters 1971).

As a result of this selective pressure, the structure, composition, and morphology of the forest that is the raw material for the shifting cultivation system can be expected to undergo change over time. Hecht, Anderson, and May (1988) have described a successional forest in the intermediate ecological conditions found between the Amazonian rainforest and the semiarid northeast region of Brazil. This forest is a *babassu* (*Orbignya phalerata*) palm for- est, which is the basis for much more than the shifting cultivation system of the region's inhabitants (figure 7.4). From this particular new-growth forest are extracted items such as

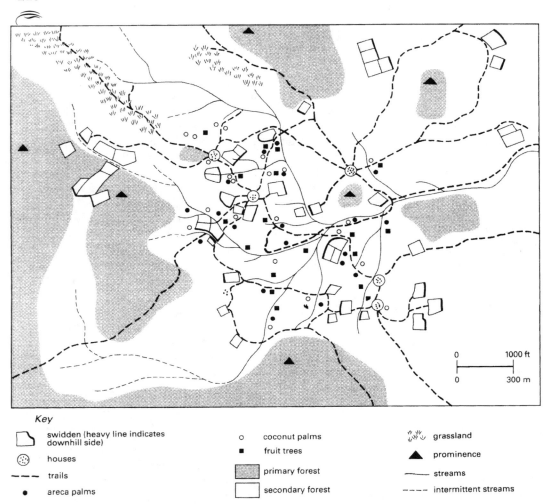

Figure 7.2.   Land use in a shifting agricultural society in the Philippines. (From Conklin 1953. Reproduced courtesy of the New York Academy of Sciences, United States)

Key

⬜	swidden (heavy line indicates downhill side)	○	coconut palms	⅏	grassland
⊙	houses	■	fruit trees	▲	prominence
– – –	trails	▨	primary forest	——	streams
•	areca palms	⬜	secondary forest	- - -	intermittent streams

thatch and other construction materials, fiber, animal feed, palmetto, and charcoal. All items are of vital importance to the income of the region's impoverished population, and all of them come from an early successional stage in forest development. This is a developmental phase that local folk would probably be happy to see permanently maintained were it not for the need to remove the babassu and other species episodically in order to plant crops. Most of the shifting cultivators who collect the products of the babassu palm do not have legal tenure to the areas from which they derive their livelihood. Therein lies a major problem. The expansion of commercial development into shifting cultivation zones, in particular ranching, mechanized cultivation, and plantation production of babassu fruit for

Figure 7.3. Shifting cultivation fields in East Kalamatan: (top) Deforestation for agricultural expansion; (bottom) Crops growing on recently cleared plot. (Photographs by David Kummer, 1993)

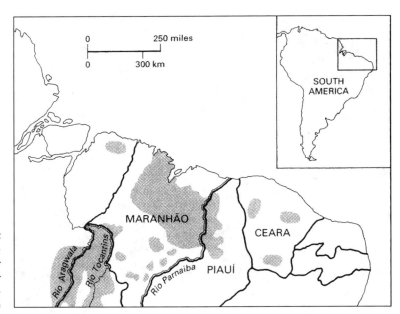

Figure 7.4. Secondary forest in northeast Brazil is dominated by *babassu* palm. (After Hecht, Anderson, and May 1988, by permission of the Society for Applied Anthropology.)

industrial uses, produces progressive impoverishment for already poverty-stricken people who have few development alternatives other than shifting cultivation and babassu exploitation (Kates and Haarmann 1992). Yet the existence of the babassu palm forest upon which their livelihood depends is itself an artifact, a result of an altered successional sequence initiated by shifting cultivation.

Fertility maintenance is a third feature of shifting cultivation's creative structure. This is primarily achieved through the ash produced by burning vegetation accumulated on the field. Burning increases the phosphorus available to cultivated plants and the ash raises the soil pH into the intermediate range (Richards 1985, 56). While these effects are temporary, they are critically important for decent crop yields. Beckerman (1987, 81) reports that the Bari of Venezuela obtain yields of 18 tons/ha (7 tons/acre) per year from their swidden fields, a yield that is twice what more sedentary farmers are able to obtain from the same habitat. Beckerman regards this figure as the norm for shifting agriculturalists throughout South America's tropical forests. The productivity of swidden in Southeast Asia is more variable and is generally lower, ranging from 500–3,000 kg/ha, although yields of 1,000 kg/ha are common (Kunstadter 1987, 146). In the latter case, these yields compare favorably with rain-fed fields in the same habitat, but are modest in comparison with the yields obtained from irrigated fields. Nonetheless, the productivity of swidden fields in the short-use/long-fallow system is considerable, and the fertility of soils and the productivity of the system is a direct result of the long fallow cycle that allows for an adequate period of recovery from use. Attempts are also made by swidden agriculturalists to extract the greatest possible return from fields and to prolong their use to the maximum extent once the effort has been made to clear the plot. Paul Richards (1985, 57–58) reports widespread use of intercropping as a fertility and pest control management technique. In these situations, farmers develop pragmatic combinations of crops in order to fix nitrogen for nitrogen-demanding plants as well as to maximize returns from microscale variations in soil type. Crude rotation schemes in which hardier crops are planted in successive years are another fertility management device that is designed to extend the life of cleared fields as much as possible.

Episodic mobility is the fourth and final factor that contributes to the stability of mature shifting cultivation systems. It is particularly important that the period of field use is a short one. For this reason, the uncleared portion of a community's territory is as much a part of the total resource base as is the cleared field. The situation is analogous to that of the large runoff catchment area attached to each small Nabataean field described in chapter 2. Without the extensive forest fallow area, the small slash-and-burn plots would not be productive. Moving the field site, and often the village as well, is essential in order to permit forest regrowth, which restores local fertility. Without this natural regenerative process, and without the use of fire to unlock the nitrates and phosphorus contained in the vegetation, a rapid shift into land degradation would occur. But as long as the fallow cycle is maintained, the shifting cultivation system continues to employ constructively the principles of creative destruction.

Pastoral Nomadism

A third type of low-intensity system that has traditionally employed creative destruction is pastoral nomadism. Pastoral nomadism is an extensive land-use system that is based on the herding of animals. This economy is most commonly found in areas that lack sufficient reliable water to sustain significant agricultural activity. Instead, people move animals to seasonal pastures where adequate grass and water can be found. This especially well-suited adaptation permits nomads to take full advantage of the highly variable nature of rainfall in dry places. To accomplish this, pastoral nomads must develop a lifestyle that emphasizes mobility and that subordinates their material culture to the need to change location many times each year. A tent, rather than a permanent house, is their main dwelling, and their most valuable possessions (in addition to their animals) are easily moveable goods such as rugs and jewelry. Despite their focus on animals and their mobile lifestyle, pastoral nomads are hardly self-sufficient. Whenever practical, they plant crops themselves, and their diet is very much dependent on agricultural products. Philip Salzman (2000) describes the multiplicity of nonherding activities engaged in by the Yarahmadzai Baluch of southeastern Iran, who engage in gathering activities, raising of tree crops, trading and smuggling, wage labor, and service as guides depending on local conditions and opportunities. In most parts of the world, this combination of opportunism and deficits has meant that nomads and farmers engaged in a number of economic interactions in which their respective products are exchanged to their mutual benefit.

Like hunting and gathering and shifting cultivation, prior to modern conditions, traditional pastoral nomads had a very limited impact on their physical environment. This is due to the large territories within which nomads traditionally operate, the small population densities supported by pastoral nomadism, and the frequent movements in which pastoralists engage. The pattern of movement is a very extensive rotational grazing system that involves oscillating between summer and winter pastures (Johnson 1969; Ingold 1987, 179–95). This movement pattern is an ecologically astute practice that is adjusted to the seasonal availability of water and grass in a given district at a specific time of the year (figure 7.5). The objective is to protect critical pasture zones from overuse by moving to new areas seasonally. Movement to high-altitude summer pastures has the advantage not only of reaching fodder and water resources that cannot be exploited in the winter, but also of reducing pressure on the crucial lowland dry season pastures. In terrain in which an altitudinal variation in pasture resources does not exist, herders move their animals to distant districts that receive only occasional rainfall in the wetter season in order to conserve grazing possibilities close to reliable water resources in the dry season (Galaty and Johnson 1990, 33).

Little adverse impact is brought to bear on the pastoral environment by these mobile herds, because under premodern conditions animal numbers seldom reached sufficient size to place local resources under prolonged pressure. While herd numbers would increase during a sequence of good years, the corresponding decrease in herd size under extended drought conditions, or as a result of exposure to an epizootic epidemic, would result in a dramatic decline in grazing pressure. As long as herd numbers did not fall below a critical minimum, the viability of the pastoral nomadic community was not threatened. Every effort was made during a run of good rainfall years to increase the herd size to a point at which it could withstand severe reduction under adverse conditions and still provide minimal support for the dependent human population. In effect, traditional herders used their animals to capture positive pulses in the environment and carry those captured resources over into deficit periods. If nothing else, the captured energy transfers could always be eaten to sustain human life at a time when sustenance was in short supply. In effect, mobility and livestock created conditions of increased reliability in ecosystems characterized by uncertainty and risk (Roe, Huntsinger, and Labnow 1998). As a consequence, catastrophic failure seldom occurred or affected more than localized groups.

The only sacrifice zone that ever temporarily emerged in traditional pastoral nomadic conditions was found around dry season water sources. O'Leary (1984), in discussing the pastoral patterns of the contemporary Rendille of northern Kenya, argues that herders take the easy option in determining their dry season grazing location and in providing security for themselves and their herds. This leads the Rendille to locate near springs, shallow water holes where the effort required to raise water to the surface is minimal, and settlements with secure water supplies. Security concerns encourage herders to locate in areas in which raiding is unlikely, and only after alternative pastures are depleted will they move to more vulnerable sites. Extrapolation into the past of the present-day pastoralist's preference for secure water source sites, where modern technology makes it possible for deep groundwater resources to be exploited and adjacent vegetation resources to be overused, is unreasonable. Around surface springs and shallow groundwater sources that could be reached by traditional technology, there is a more reasonable case to be made for possible overgrazing, although this commonly was confined to a zone a couple of kilometers in diameter around the watering point (Valentin 1985). Repetitive, concentrated grazing in dry season sites tended to reduce the frequency of the most favored palatable forage species as animals selectively rummaged in their habitat. This selective pressure, in turn, encouraged the expansion of species that possess thorns, unpalatable leaves, or harmful exudes (Simmons 1989, 146). Despite this selective pressure,

Figure 7.5. Seasonal grazing by pastoral communities: (top) In Morocco, migratory herders still bring their sheep to summer pastures in the Middle Atlas. (Photograph by D. L. Johnson, June 1987) (bottom) Kababish camel nomads bring their animals to a seasonally flooded area in central Kordofan, Sudan. (Photograph by D. L. Johnson, September 1982)

the limited water resources of most shallow, hand-dug wells make it very unlikely that unpalatable vegetation was substituted for desirable fodder species over extensive areas. Overgrazing did occur in specific places, as large herds were temporarily attracted to available water resources, but the geographic extent of the disturbance always was small.

Local tribal authorities used their political power in order to expel lower status client groups when deteriorating resources developed, and in this way were able to prevent severe damage at threatened sites (Peters 1968; Bernus 1990). Differential access rights to water and grazing resources were mechanisms that enabled nomadic pastoralists to lighten pressure on critical fodder resources and to manipulate the stocking ratio in a given habitat in order to insure the survival of some herders and their animals. This stratagem was always a last resort, and a variety of institutions existed in order to distribute pressure on land resources more equitably before attempts were made to drive low-status groups away from resources. For example, in periods of abundance when herds were growing, surplus animals were transferred to the care of "stock friends" living in adjacent districts, as a device to spread the owner's risk and lower local stocking levels to manageable proportions. Unlucky friends, unfortunate kinsmen, and neighbors with inadequate herds were lent animals in order to spread stock more widely in the social system, and at the same time disperse livestock more widely in space. In this way, ties of kinship and marriage were maintained with other groups in a complex web of social relations. This network of social debts and obligations enabled herders to request access to the temporarily more favored areas of other groups should natural conditions on their own tribal turf prove unfavorable in any given year. In these ways excessive concentrations of animals and people that might threaten the viability of critical grazing zones were avoided and the intensity of use was diminished. Detailed local knowledge of the carrying capacity of the local resource base made it possible for herders to use the local environment intelligently (Hobbs 1989). Compared to modern pastoral conditions, traditional pastoral nomads used their environment lightly and had only a modest adverse impact upon their resource base.

Fire was the major tool of creative destruction employed by traditional nomadic pastoralists, just as most contemporary pastoral groups use fire at some point in their seasonal cycle for a multiplicity of purposes. Burning grassland just before the beginning of the rainy season encourages many grasses to sprout. These new shoots provide a welcome source of fresh fodder at the end of a long period of poor grazing. Regular burning creates a more open habitat—a setting in which pastoralists can more easily see their flocks over large distances—and eliminates protective cover that might shelter potential predators. Fire also manipulates the species composition of the vegetation. The primary impact of fire is upon woody vegetation, since the heat of the fire adversely affects trees and shrubs that lack a thick insulating bark.

Grasses are favored in these circumstances. The removal of the shade dominance exerted by trees and shrubs creates more niche openings for other plants, thus increasing the diversity of the grassland flora that is of primary interest to the herder (Ruddle and Manshard 1981, 123). A denser sod cover also reduces erosion. When combined with light grazing, fire removes dead stems and old growth and promotes an increase in the vigor and growth rate of many desirable plants (Warren and Maizels 1977). Increased availability of grass means that more livestock can be supported on a given area. More livestock means greater manure deposition and a higher rate of nitrogen return to the soil. Since domestic stock have definite consumption preferences, the plants they favor tend to become more widely distributed as grazing animals spread the seed of preferred plants about the landscape as an accidental by-product of defecation. The key ingredient in this process of coevolution is the combination of burning and grazing at intensity levels that are not excessive. Light, episodic impacts have profound beneficial long-term effects.

Not only does fire improve the quality and quantity of the forage available to domesticated livestock, but also in many places the removal of moist, shady habitat reduces the breeding opportunities for insects that carry disease. In sub-Saharan Africa, the tsetse fly, the intermediate vector of trypanosomiasis (sleeping sickness), is controlled by burning its breeding sites. Stopping managed burning leads to an increase in brush and a decrease in the safe habitat within which nomads can herd their animals. Lacking sophisticated chemicals to use in an antitsetse campaign, materials that pose their own set of environmental problems when employed in large concentrations (Linear 1985), traditional pastoral nomads have had fire as their only tool for instituting large-scale environmental modification. That tsetse abatement has been an accidental result of burning for other purposes in no way diminishes its importance as an act of creative destruction (Lewis and Berry 1988). By burning, nomads have created a grassland setting that is conducive to their grazing animals at the expense of forest and shrub lands.

Despite the reality that traditional pastoral nomads cause almost no land degradation (McCabe 1990), herders have a bad reputation for following destructive land management practices (Wikjman and Timberlake 1985; Hjort af Ornas 1990). This image is a product of failure on the part of central planners to understand the dynamic nature of nomadic pastoralism (Ruddle and Manshard 1981, 124) and the tendency of settled farming societies to export environmental costs onto pastoral areas (Simmons 1989, 264–71). Nomadic groups that once were able to control their own destiny have lost power and territory to an expanding agricultural economy. This agricultural expansion invariably takes the best pastureland, and leaves the herder with the less desirable land; the more arid the district in which this process takes place, the more destructive and the more rapid is the impact on the surrounding habitat (Falloux and Mukendi 1987). Confined to increasingly more marginal locations, nomadic herders have had to maintain their animals for longer periods in rangeland that once was only used seasonally. The inevitable result is land degradation in the excessively pressured districts (Olsson and Rapp 1991).

In many river floodplains, nomads were able to graze their animals on fallow land, riparian forest, and postharvest crop residues once the agricultural season was completed. Conversion of these seasonal agricultural zones to perennial irrigation has cut the nomad off from vital dry season pasture (Adams 1989). In many African floodplains and wetlands, development displaces huge numbers of animals permanently to more distant locations (Adams 1995). Not only does the farmer lose the manure formerly deposited by the nomads' herds, thus rupturing the manuring cycle that once tied farmer to herder and captured waste resources, but also herders lose seasonal use rights that reduced pressure on rainy season pasture. In effect, part of the cost of floodplain irrigation development is exported onto the rangeland. The remaining grazing areas are made the sacrifice zone for prosperity in the agricultural sector.

Many pastoralists have become sedentarized, both on their own initiative and in response to pressure brought to bear by nation-states for political and ideological reasons. The result is an excessive concentration of people and animals, usually around sites with decent water resources, and the outcome is degradation (figure 7.6) in the surrounding vegetation and soil resources (Bedrani 1983; Janzen 1983; al-Ibrahim 1991). War and civil conflict have also restricted nomadic movements and placed more pressure on local habitats (Bascom 1990; Arkell 1991; Dahl 1991). Nationalization of the rangeland (Beck 1981; Bedrani 1991) has meant that nomads have lost the ability to control their resource base. The shift of political power to governmental agencies from traditional tribal leadership has usually proven ineffective in managing range resources, because these institutions seldom possess the staff or the local support to implement management decisions (Agrawal 1999; Humphrey and Sneath 1999). Land degradation is the inevitable consequence of the resulting scramble to extract something useful from the rangeland resource base before access to the rangeland is forever lost to the pastoralist, although herders often show remarkable resilience in adapting old patterns to altered economic, political, and environmental conditions.

In a spatially extensive, long-fallow, low-intensity ecological system such as nomadic pastoralism, few, if any, sacrifice zones exist. When they do occur, these sacrifice zones are limited in spatial scale and temporal duration. Creative destruction in traditional nomadic pastoral systems is a common process, but it takes place slowly through the gradual expansion of rangeland habitats. Fire is the primary tool employed in this positive process of habitat modification. When substantial land degradation occurs in pastoral environments, it is almost invariably linked to the export of costs from other sectors of economy and society onto the pastoral community. This process of land degradation impacts vulnerable pastoral people who are much more victims than they are villains in the development of rangeland degradation. For it is when cost export avoidance fails in other habitats and economic sectors, largely because linked cultural ecological systems are treated in disaggregated fashion, that serious land degradation occurs in rangeland exploited by pastoral nomadism.

Moderate-Intensity Systems

The Agricultural Revolution, whereby humans began consciously to manipulate and control plants and animals, altered the relationship of humankind to its environment. Unlike the low-intensity systems described in the previous section, which today are relic livelihood systems existing on the margins of sedentary civilization, moderate-intensity systems represent a fundamental change in the nature-society relationship. Moderate-intensity systems demonstrate an enhanced ability to produce significant impacts that transform concentrated spaces, last for a substantial time, and affect distant ecosystems. More sophisticated and numerous technological inputs characterize moderate-intensity systems compared to low-intensity ones. A major difference between the two is the greater ability of moderate-intensity systems to shorten, and in some cases even eliminate, the fallow period.

Positive use of creative destruction underlies this ability to use land resources more intensively. Much larger populations are supported in a smaller territory than is the case in low-intensity systems. As a result, human labor can be invested in greater amounts to

Figure 7.6.   Grasslands in Sudan exhibit differing degrees of human impact: (top) Between El Obeid and El Khuwei in Sudan's Kordofan Province, vegetation is used only seasonally and remains largely intact, but (bottom) around the important bore-well site and village of El Khuwei, large numbers of animals attracted to reliable water supplies place heavy pressure on the local vegetation, particularly during drought years. (Photographs by D. L. Johnson, September 1982)

transform local space. It is equally significant that, unlike low-intensity systems, the actions of moderate-intensity systems begin to affect areas far beyond the their immediate vicinity.

By massive use of human labor in some cases and/or the selective application of technology—including the use of machinery—in other situations, constraints of the natural environment that restrict production are ameliorated. As is typical of most successful examples of creative destruction, moderate-intensity systems usually evolve slowly and are rooted in an intimate understanding of local habitat conditions. New areas of sustainable use on lands previously avoided due to environmental constraints, and drastically shortened fallow cycles on established lands that require careful management of system inputs and outputs, are two hallmarks of moderate-intensity systems. Both improvements result in the ability to sustain large populations on limited areas for unlimited periods of time.

An important difference between low- and moderate-intensity systems is exemplified by the ability of moderate-intensity systems to expand food-producing activities despite frequently poor natural conditions. This expansion is the result of direct modification of the land, initiated to meet the desired production goal. This characteristic is completely unlike low-intensity systems in which vegetation cover is almost exclusively the product of the direct alteration that occurs due to human activities and in which conscious modification of terrain features is unknown. Low-intensity systems modify their habitat very little, and then only slowly over long periods. Moderate-intensity systems accelerate this process often through direct alterations of critical components in the existing topographic and/or hydrologic systems.

Prior to the development of hybrids and bioengineered plant varieties, expansion of plant varieties beyond their native environmental settings only took place if constraining components at the local site could be altered. In particular, drainage and/or terrain morphological changes often were undertaken to meet the specific demands of those crops the inhabitants wanted to grow. This ability to transform nature radically, convert the environment into habitats that are the product of human conceptualization, and create habitats that are utterly dependent upon humans for their continuation is the unique contribution of moderate-intensity systems. In these settings, nature and society no longer exist as separate categories, but rather are integral parts of a common system. Moderate-intensity systems take a giant stride down the road toward the domestication of nature, a process that succeeds only when sustainable agroecologies can be created.

In this section we consider several moderate-intensity systems that have successfully achieved a sustainable agroecology by the use of the principles of creative destruction. We begin with two examples where traditionally the use of large amounts of labor was essential in order to create—over centuries of evolutionary development and elaboration—sustainable and productive cultural ecological systems. These are the terrace systems that transform steep slopes into fields, and the integrated agricultural-aquacultural systems that have converted coastal and deltaic lowlands in many parts of Asia from forest and marsh into highly productive farmland.

Our second set of examples explores the American Midwest, a region that today is one of the world's premier rain-fed agricultural areas. Throughout this agricultural heartland, technology, applied at varying levels of scale and importance, has played a critical role in creative destruction. This example looks at the role of drainage in converting the region's large expanses of poorly drained, but good-quality soils into prime agricultural terrain. We look at how a simple technology, land tiles, made it possible to open to cultivation vast areas of the Midwest whose soils were wet enough to impede agriculture but not wet enough to be considered swamp or marsh. We examine how farmers adjusted the needs of land drainage to an existing arbitrary land division system, the township/section survey system of the United States.

In the second part of this example, we explore the role played by larger-scale, machine-dominated drainage technology in partnership with labor, which transformed a substantial swamp in the heart of the American Midwest. The Kankakee Marsh, which remained a poorly settled island surrounded by productive farmland in the Chicago hinterland until the turn of the twentieth century, was altered from a diverse wetland ecosystem into productive farmland that better met the needs of the local population and the nation. Today this area of northwestern Indiana and northeastern Illinois has evolved from lands where moderate-intensity agricultural interventions were applied at the turn of the century into a successful zone within the American Corn Belt. Here farmers utilize high-intensity interventions that permit large agricultural yields in a nonrotational, nonfallow, high-energy input, sustainable agricultural system.

Terracing and Creative Destruction

In chapter 5, terracing was introduced largely within the context of Rwanda where poor design resulted in infertile soils being exposed and land degraded. In Central Province, Kenya, many of the steep slopes have been terraced for the production of coffee and tea. In Kiambu District slopes as steep as 22 percent have been terraced for coffee (Lewis 1985). Unlike the wide terrace benches common in Rwanda, the bench portion on the steeper slopes in Kiambu are only approximately 1 m (1 yd) wide and no infertile subsoils are ever exposed. To maximize yield, farmers in addition irrigate the crop during a portion of the year. The result of proper terrace construction, continuous maintenance, and the growth of soil-conserving crops on the terraces (coffee) is very low soil losses, on the order of 1.6 tons/ha or less per year (Lewis 1985).

When combined with a planning process that is organized around catchment areas and involves a high level of interactive participation by local residents, integrated soil conservation including terracing has experienced considerable success (Pretty, Thompson, and Kiara 1995). Farmers who engage in terrace agriculture employ a technology that intensifies the agricultural system and requires considerably more time and labor from the farmer than traditional farming systems. But at least in Machacos District, this effort has paid off in less erosion, higher levels of productivity, and greater environmental stability in areas that were viewed three decades ago as agriculturally marginal and environmentally problem-plagued (English, Tiffen, and Mortimore 1994; Tiffen, Mortimore, and Gichuki 1994). Thus when terracing is done correctly and maintained, it is possible to create a sustainable agriculture on steep slopes. But the potential problem of poor maintenance and resulting terrace instability requires that the terracing strategy only be implemented with caution. When economic conditions change (for example, southern western Europe) or political instability occurs (for instance, Rwanda and Lebanon), poor terrace maintenance can result in wall deterioration, accelerated soil loss, and land degradation.

Creative Destruction in the Chinese Pond-Dike System

Sites at transition zones between major ecological systems often are productive habitats for humankind. Coastal locations offer access to a greater diversity of ecosystems than almost any other site. Mountain habitats would perhaps show greater diversity along an ecological gradient (elevation), but they would not be as close together. In coastal zones both terrestrial and aquatic ecosystems exist in close proximity and can be exploited readily without engaging in substantial movement. Within the littoral zone, these advantages are maximized in delta environments where rivers empty into the ocean. Historically, these sites have attracted human settlement by virtue of their abundant plant, animal, and aquatic life. In the past two centuries, government-sponsored settlement schemes have transformed these habitats in dramatic fashion, and in no region more so than in coastal districts of south and eastern Asia (Richards 1990). Here most of the coastal wetland districts have been transformed into zones of peasant agriculture by forest clearance, drainage, and flood protection embankments.

Spurred by the desire to enhance agricultural productivity in support of a growing population, this process of land development destroys coastal vegetation that contains many valuable resources, particularly mangrove ecosystems (Hellier 1988). These terrestrial ecosystems are linked to offshore sea grass ecosystems that are important food and habitat resources for marine animals. Many species, such as the spiny lobster, pink shrimp, and various grunts and snappers, migrate between these land and sea systems during their life cycle (Fortes 1988, 211). Any reduction in the coastal vegetation results in an increase in erosion that smothers the sea grass beds with sediments. These grassy marine habitats provide shelter and food for marine animals. At the same time, habitat destruction eliminates breeding sites for many species and damages the feeding setting that supports their juvenile stage. Thus terrestrial modifications constitute a threat to the primary productivity of aquatic ecosystems. This produces a classic case of creative destruction in which the transformation of mangrove and other coastal ecosystems into agricultural land turns the marine ecosystem into a sacrifice zone. Increased terrestrial output is gained at the expense of diminished marine productivity.

Although accelerated by systematic government action in the past two centuries, the process of deltaic transformation has ancient roots that began with the development of paddy cultivation in the region (Lo 1990, 405; figure 7.7). In its most advanced state, coastal and delta modification follows the pattern described by Ruddle and his colleagues in their analysis of a rural commune in the Zhujiang (Pearl River) Delta in South China (Ruddle et al. 1983; Ruddle and Zhong 1988). Here conversion of the wetland soils of the delta environment has not halted with paddy rice development. Instead, there has evolved an agricultural system that tightly integrates aquaculture with crop production on the dikes that separate the fishponds. In this system, waste products from one activity become the inputs to other segments (figure 7.8). Livestock, crops, and fish are treated as integral parts of one system rather than as separate entities managed by different production units. The result is sufficiently robust to be sustainable for centuries, yet resilient and flexible enough to accommodate change.

It has taken the pond-dike system over one thousand years to evolve to its present state (Ruddle et al. 1983, 49). This long evolutionary development is important in determining

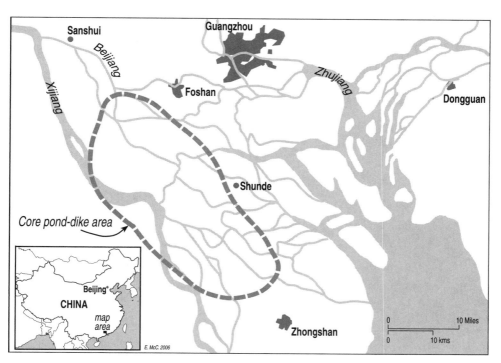

Figure 7.7.   The Zhujiang (Pearl River) Delta in South China has been the setting for pond-dike agriculture for centuries. (Cartography by Eliza McClennen, Mapworks)

the character of the system. It means that it has grown organically from a rich base of empirical knowledge. This bedrock of indigenous knowledge is enormously sensitive to the peculiarities of the local environment, and as such demonstrates how the genius loci principle is applied. Because it is so rooted in place, the pond-dike system is not readily transferred elsewhere, although its management principles can inform land management in other locales.

The foundation of the system is the pond and dike subsystems and the complex manner in which they are woven together. The pond is typically an east-west-aligned rectangular water body, 2.5–3 m (8–10 ft) deep and covering less than 0.5 ha (1.2 acres). The ponds are shaped in a ratio of 6:4, an evolved configuration that minimizes bank erosion due to wave action. The same rough balance exists between area devoted to water and that devoted to land. The depth of the pond also reflects empirically derived folk knowledge, since digging deeper risks reducing sunlight penetration to the lower pond depths and diminishing the productivity of the bottom-feeding fish varieties.

The beauty of the pond structure is that it is conceived as a total volume that can be exploited by polycultural techniques. This is achieved by raising different varieties of a freshwater fish, the carp (*Cyprinus carpo*) in the pond (Lo 1990, 410; Korn 1996, 7). Indigenous to China, carp have evolved specific feeding preferences. Some consume detritus on the

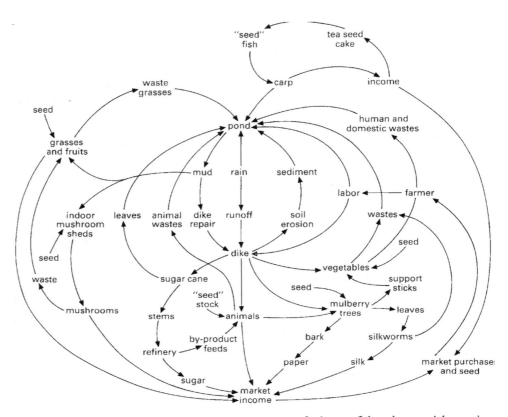

Figure 7.8. The pond-dike production system integrates freshwater fish and terrestrial crops in a very tight fashion that uses the wastes from one component as the inputs for another component.

pond bottom, some eat grass, others specialize in snails, and still others concentrate on algae and plankton. As a result, the entire pond volume can be filled with different varieties and ages of fish in varying stages of development, each of which utilize a different segment of the pond. The key to the entire system is the grass carp. This is an herbivorous species that lives in the top layer of the pond and consumes both aquatic plants in the pond and the grass clippings and crop residues that are grown on the dike and dumped into the pond. Grass carp are inefficient feeders and probably do not digest 50 percent of the food they eat (Ruddle and Zhong 1988, 30). Partly digested grass carp excrement becomes the raw material for plankton growth. Silver and bighead carp, phytoplankton and zooplankton consumers respectively, benefit most from this fertilizer input. Invertebrates that live on the bottom of the pond feed on the plankton and unused detritus that rains down from the levels above. These creatures in turn are the food source for the black, mud, and common carp that prowl the pond floor. Common carp in particular are aggressive rooters in the bottom mud since they also feed directly on detritus. By their actions they stir up material that enters the water volume and becomes available as food to other carp species at higher levels in the

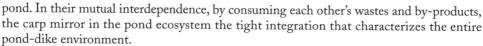

pond. In their mutual interdependence, by consuming each other's wastes and by-products, the carp mirror in the pond ecosystem the tight integration that characterizes the entire pond-dike environment.

The pond itself is kept constantly fertilized at a high level in order to produce the maximum amount of food for the fish. This high level of fertilization requires constant vigilance on the part of the pond operator in order to insure that the dissolved oxygen levels in the pond do not sink too low. Were this to happen, fish would become lethargic, growth would slow, and, if dissolved oxygen levels were to drop too low, fish survival would be threatened. The first sign of the emergence of these problems is the movement of the bottom feeders, the common and mud carp, to the surface (Ruddle et al. 1983, 55). Attention must also be paid to the turbidity of the water, since overfertilization can so cloud the water that sunlight penetration is reduced. Any loss of solar radiation reduces photosynthesis, and thus strikes at the primary productivity of the plankton and algae upon which the aquatic food chain depends. The source of the organic wastes used to fertilize the pond is partly the undigested wastes of the fish themselves, but primarily is derived from terrestrial sources. Basically the more the fish are fed, the greater the yield in consumable fish biomass they return to the farmer (Korn 1996, 8). Terrestrial manure is produced from the excrement of animals—especially pigs, ducks, cattle, humans, and silkworms—and the organic residue from agricultural activities carried out on the dikes. Pigsties and latrines are often located close to the pond so that manure will drop directly into the water, because these wastes are most valuable when fresh (Ruddle and Zhong 1988, 33). Wind and wave motion then redistributes the wastes more evenly through the pond. It is the bacteria and the protozoa in the manure that are the prime food source for bottom-feeding carp, and the fish feces returned to the bottom are also available for recolonization by bacteria and subsequent reconsumption by fish in a continuous cycle (Ruddle and Zong 1988, 32). Once established, the fishpond is capable of sustaining high stocking rates as long as manure inputs, water turbidity, and dissolved oxygen levels are continuously monitored in order to insure system stability.

Until recently, juvenile fish (fry) used to stock the ponds were captured from adjacent streams. This was necessary because carp did not breed in captivity. Today river pollution places these natural sources of fry at risk. Fortunately, at approximately the same time as river pollution has become a serious problem, dramatic improvements in nursery breeding and rearing of carp have occurred. To prevent inbreeding, however, fry and larvae are still extracted from natural sources (Ruddle and Zhong 1988, 34). Whether artificially bred or captured in nearby streams, fry are raised in specially prepared nursery ponds for about one month. When the fry reach fingerling size (ca. 3 cm [1 in] long), they are transferred to separate ponds where they are "grown out" in a two-stage process (each stage using separate ponds) lasting about one month per stage. When the fingerlings reach approximately 12 cm (3 in) in length, they are shifted to the production ponds.

The second major component of the system is the dike. Dikes are artificial creations built up by the systematic excavation of the adjoining ponds. In this process of pond and dike creation, any semblance of natural vegetation and topography is destroyed and a completely humanized landscape is created. Despite the effort in pond design to minimize the impact of bank erosion, a gradual collapse of the dike bank is inevitable, because none of these edge environments are constructed with hardened materials. The pond never fills in because two or three times each year mud is excavated from the pond bottom and placed on the adjoining bank (Ruddle et al. 1983, 57). Since the mud is rich in organic matter, it serves as the most important fertilizer source for the crops grown on the bank. The need to remove mud from the pond several times during the year is just one reason why the pond-dike system is very labor-dependent. It also demonstrates how dynamic the system is within a general structure of stability, and how important that basic dynamism is to the overall design and function of the pond-dike system.

Dikes are carefully calibrated structures and they can only exist within a narrow range of tolerances. The dike cannot be too narrow or it is impractical to cultivate; yet no dike can be too wide or there is not sufficient mud fertilizer to sustain year-round cultivation. Dikes that range in width between 6 and 20 m (20–66 ft) are typical (Ruddle and Zhong 1988, 55). Dikes cannot be too low or they become waterlogged. Conversely, if the dike is too high, its soils and plants are more readily affected adversely by drought. Thus the dike exists within a rather narrow tolerance limit that must be carefully monitored and maintained.

The major crops grown on the dikes are sugar cane and mulberries. Less numerous are fruits, especially the litchi and longan for which the region is famous, vegetables, and oilseeds (mostly peanuts), which are intercultivated with the main crops or are cropped in rotation with them. Fruit trees were actually the first major commercial crop in the region when a production shift from paddy rice cultivation to commercial production began. After AD 1620 mulberry cultivation began to replace fruit as the price of silk rose substantially and as the benefits of integrating pond and dike more closely began to be widely recognized (Ruddle et al. 1983, 49–50). In particular, farmers learned empirically that direct application of animal wastes to the plants harmed the mulberry leaves. The same manure dumped into the pond for the benefit of the managed aquatic food chain was perfectly benign when withdrawn after a period of pond fertilization and applied directly to the fields. The use of the pond as an intermediate fermenter converted harmful wastes into beneficial inputs and made the integrated land-water system possible.

Mulberry leaves are harvested seven or eight times between April and November and are fed to silkworms housed in sheds. Silkworm excrement, cocoon waste, wastewater, dead larvae, and scraps of unconsumed mulberry leaves find their way into nearby ponds as fertilizer and fish food. Similarly, waste from silk-spinning factories, whose wastes are bought by farmers, becomes fertilizer for the fishpond. Feedback loops such as this silkworm to fishpond linkage are vital elements that complete the cycling of nutrients and waste products in the total system.

In the winter, when sheds can no longer be used to raise silkworms, they are used to cultivate mushrooms. Mud from the pond bottom is used to prepare the mushroom bed on the silkworm-shed floor. Seed is purchased from off-farm sources, and most of the output of mushroom buttons is sold off-farm as well. When the mushrooms are harvested and the next cycle of silkworm cultivation is about to begin, the nutrient-rich mushroom mud is cleared from the floor and used to fertilize the vegetable gardens (Ruddle et al. 1983, 59). The sheds are then cleaned and disinfected, and the next cycle of silkworm cultivation begins.

Sugar cane is another major dike crop, and has increased in importance during the past several decades to the point where it now surpasses mulberry trees in total acreage. Like all crops grown on the dike, the primary source of fertilizer for the cane is the mud dredged from the adjacent pond and deposited on the dike. The cane has several uses within the farm. In particular, the young leaves are a source of fish feed, while the old leaves provide shade for the gardens. The cane's major use is as a commercial crop. Stems, when harvested, are processed in off-farm refineries to produce sugar, which is both marketed nationally and consumed locally. Bagasse and other wastes from the refinery are returned as inputs to the farm operation and are important sources of animal feed.

Animals are a major component of the integrated pond-dike farm. Pigs are arguably the single most important animal, and are regarded as walking manure factories, the excrement from which is a vital input into the pond subsystem. In addition to their role as producers of manure for the fishponds, large amounts of animal protein and cash income are derived from the farmers' pigs. Small numbers of water buffalo are also kept for traction power. Chickens, ducks, geese, and in some cases eels are raised in addition to the ubiquitous carp. All of these animals contribute to the subsistence and cash income of the farmer, as well as return excrement to the pond environment. The cycle is enriched by the growth of aquatic plants, particularly water hyacinth, in separate ponds as a food source for the animal population, especially the pigs. Supplemented by crop residues, kitchen wastes, and food material scavenged from accidental spills, these water plants provide the bulk of the food supply for the domestic animal population.

The pond-dike system is a tightly integrated example of creative destruction; it is a system joined together by a complex pattern of exchanges that only can be suggested in diagrammatic form. The integrated agriculture-aquaculture system leads to complete removal of the native vegetation and its replacement with a much simplified, human manipulated complex of cultigens. Part of the success of the system is its ability to cut costs dramatically by finding most of its major inputs within the local environment (Ruddle and Zhong 1988, 152). Expensive imported inputs are kept to an essential minimum. This is accomplished by capturing as many wastes as possible from each subsystem and using these wastes as major inputs into other components. Thus mud, which as a waste product fills the pond and threatens it with extinction, becomes a major source of fertilizer and a construction material for the dike. Similarly, the dung of animals based on the dike becomes the primary energy input contributing to the controlled eutrophication of the fishpond. Carefully monitored and regulated, this process, which potentially could result in the death of the aquatic environment if dissolved oxygen levels were to decline too much, creates ideal growth conditions for the varieties of carp that inhabit the pond. Release of eutrophied pond water is infrequent and only occurs when fish are harvested and when as part of pond maintenance mud is extracted from the pond and used as fertilizer. The natural ecosystem of the delta and at least some of the productivity of offshore fisheries are sacrificed for the human-oriented productivity gains of the pond-dike farm. Because the pond-dike farm possesses high levels of stability and resilience, and has survived for centuries, this sacrifice seems defensible and a classic example of creative destruction.

But what has endured and developed for centuries is not immune to change. Although the pond-dike system functions well as a largely self-contained unit, emitting few pollutants to the surrounding environment and requiring few inputs from off-farm sources, it is today seriously threatened by industrial development in nearby regions (Korn 1996; Yee 1999). Urban and industrial growth has taken some dike-pond land out of production.

Fishpond monocultures that stress higher value species, such as eel, turtles, and prawns, and increase the ratio of water to land surface are becoming more common. Water to land ratios in these intensely capitalized and profit-driven enterprises approaching four to one are increasingly common; these new relationships neglect the water-land nutrient exchanges that are the basis for stability in the more traditional system (Yee 1999, 531). Dikes collapse more frequently, and inputs from outside the local system are increasingly required to maintain and fertilize the pond environment. What was a remarkable example of creative destruction runs the risk of spinning into an emerging instance of destructive creation.

Drainage and Creative Destruction in the American Midwest

The American Midwest, the nation's granary, reflects its recent glacial ice sheet heritage. Large areas of the region were poorly drained, as the river and stream systems had insufficient time to develop an integrated drainage by the time the major influx of European migrants settled the area in the 1900s. To convert this region from the natural vegetation of woodlands, wetlands, and grasslands into prime agricultural land required major alterations in the drainage of vast areas as well as the removal of the prevailing forests. The alteration in the vegetation cover and drainage that rapidly changed this area from a region where hunting and trapping were replaced by farms is a prime example of creative destruction on a regional scale. The transformation of the land from natural vegetation to a myriad of crops resulted in significant changes in the flora, fauna, and erosion/drainage characteristics of this immense region. Once valued as a source of meat or skins from muskrat, mink, opossum, and raccoon, wildfowl such as ducks and geese, carp, black bass, and other freshwater fish, mussels from which to make pearl buttons, and timber, the advent of drainage districts and steam-powered dredging machinery transformed the landscape and drastically decreased the spaces available for anything other than agriculture (Thompson 2002). The contemporary landscape, with farmland stretching from horizon to horizon, appears to the casual observer to represent a simple, inevitable outcome that humans have created in a natural setting just waiting to be cultivated. In actuality, it required major alterations in the natural environment. Today it requires continuing technological inputs such as drainage maintenance, pesticides, fungicides, and fertilizers to keep sustainable the agricultural systems that replaced the natural ecosystem.

The thousands of square kilometers comprising the American Midwest, the country's agricultural heartland, possess numerous characteristics ideal for modern, high-technology farming. With a large percentage of its lands having slopes of less than two degrees, most Midwest topography can be described as ranging between flat and very gentle. Where relief is evident, it generally results from short and not particularly steep slopes. These terrain characteristics are ideally suited for modern farm implements. The soils that veneer most of this region are derived primarily from either glacial or alluvial deposits. They are relatively young, generally high in soluble mineral plant food, and most of the soils' nutrient deficiencies can be easily countered through application of fertilizer supplements, a critical requirement in the intense monoculture agriculture practiced today. From the agricultural perspective, the climate of long, hot summer days and sufficient moisture complements the area's topographic and pedological advantages. Today in many areas, highly productive farm fields stretch as far as the eye can see, from Ohio in the east to Iowa in the west. In the northern, cooler portions of this region, wheat, potatoes, barley, oats, sugar beets, or soybeans cover the flat checkerboard fields, while dairying is important in many of the hilly areas (Borchert 1989). In the southern, warmer sections of the Midwest, corn (maize) is king. In Illinois alone, 400,000 ha (1×10^6 acres) generally are planted in corn (Horsley 1986). Soybean, wheat, oats, and livestock production follow in dominance.

That this bountiful area is the heartland of America's rain-fed agricultural production is partially the result of its natural endowment. Of paramount importance to its agricultural success have been the changes that its settlers have initiated and implemented on this landscape over the past 120 years. Farmers occupying former forest, prairie, and wetland areas reduced and eradicated multitudinous plant and animal species when creating the conditions needed for the successful introduction of livestock, pasture grasses, wheat, and other exotic species brought from Europe and more recently Asia. Without the creative destruction of this region's enormously large areas of natural wetlands, a significant proportion of today's rich farmlands would not be cultivated. The early pioneers largely avoided the wet prairies, swamps, and woodlands during the first waves of settlement in favor of the healthier, drier lands. Today these former wetland areas contain some of the most productive farmland found anywhere within the Midwest. The implementation of vast artificial drainage schemes by both individuals and government organizations converted the thousands of square kilometers of poorly drained bogs, bottomlands, meadows, and woods into the region's ubiquitous, highly productive, well-drained farmlands (Meyer 1936; Hewes 1951; Kaatz 1955; Johnson 1976), the hallmark of the contemporary landscape. Today a dramatic shift in attitudes toward drainage has occurred (Prince 1997). Support for new drainage schemes is largely absent, and emphasis instead is on the preservation of wetland wildlife habitat, flood protection, recreation, and aesthetic landscapes. But this attitudinal transformation belies the reality of the Midwest landscape's transformation from wetland to fields of soy and corn and, except for particular unusual local niches, is unlikely to reverse nearly a century of drainage improvements.

Drainage throughout this area is not obvious to the casual observer. Rarely are open drainage ditches utilized. Furthermore, the majority of the drainage ditches constructed in this region do not follow the region's natural gradient, where you would expect them to have been built. Their locations are determined not by topography but by land ownership. Property ownership and field pattern throughout this region are related to the location of the rectangular township and section lines, which are determined by survey, independent of topography. It is this survey system that gives the Midwest its checkerboard appearance (Pattison 1957; Johnson 1975). If, under this survey system, artificial drainage had been developed mirroring the land's subtle topography, it often would cut diagonally across fields, a hindrance to both field preparation and harvesting. Furthermore, outlets to the ditches would have required many farmers to secure permission from their neighbors to cut their ditches through their property (Johnson 1976). This was something that most farmers would not appreciate. To overcome these shortfalls, tiling became the dominant method utilized for farmland drainage.

Tiling of the American Midwest began in the 1880s (Hewes and Frandson 1952). By 1910, thousands of kilometers of tiles had been laid (Johnson 1976). Drainage by underground tiles required farmers to dig a system of integrated ditches in their fields, place and connect the tiles in the ditches, and then to bury them. One rule of thumb used in a poorly drained soil was that the tile would draw water from one rod (5.03 m) on each side of the pipe for every foot (0.13 m) that it is buried (Hewes and Frandson 1952). Tiles throughout the Midwest are generally buried at depths of at least 1 m (3 ft). This depth permits both sufficient drainage at an economical cost for most soils as well as the minimization of land loss to drainage practices. A depth of 1 m is usually sufficient to permit annual crops to be cultivated over the tiles without field preparation and harvesting activities damaging the buried tiles. Tiling was so successful in extending agriculture on lands previously too poorly drained for cultivation that it is common for 40–60 percent of the land area in many Midwest counties to be tiled (Hewes and Frandson 1952) and is the fundamental reason for the region's agricultural prominence.

The Kankakee Marsh in northwestern Indiana and northeastern Illinois is illustrative of the role that artificial drainage and creative destruction has played in the significant expansion of farmland throughout the Midwest. Stretching for 135 km (85 miles) in northwestern Indiana and northeastern Illinois, the Kankakee wetlands were a poorly drained intermorainal area prior to the reclamation of its swamps and marshes for agricultural use. In its natural condition, the Kankakee River, with an average gradient of 779 mm/km (5 in/mile), meandered extensively and continuously shifted its course throughout the entire length of the wetland. This created a complex landscape with an intricate maze of poorly drained, abandoned channels, including oxbow lakes and sloughs (Meyer 1936). Generally, for three-quarters of each year, water up to 1 m (3 ft) in depth covered a 5–8 km (3–5 mile) wide zone on each side of the river. Swamp vegetation occupied a 1.5–3 km (1–2 mile) wide band along the entire course of the river, with marsh vegetation dominating the other portions of this wetland. Both the Native Americans and the early pioneers who settled in this area largely viewed it as a common property resource, hunting and trapping the Kankakee's abundant wildlife. Thus the early periods of use of this wetland were characterized by human adjustment to the natural environmental conditions of poor drainage (Meyer 1936).

Chicago, the major livestock and grain center of the United States at the turn of the century, is situated only 70 km (43 miles) north of the Kankakee area. With most lands in close proximity to Chicago already occupied by the middle of the nineteenth century, the lightly utilized Kankakee area was one of the last large blocks of land in the region available for agricultural expansion. Beginning around 1880, land scarcity resulted in large herds of cattle being introduced into the "empty" Kankakee area despite its generally poor environmental situation. Initially cattle ranged on the wild hay marshes that occupied the drier, higher lands along the marsh's margins. But shortly after the introduction of livestock (1884), steam dredging of the first ditch began in an effort to extend the cattle ranges and permit livestock grazing throughout the whole year (Reed 1920). The creative destruction of the "natural" Kankakee had begun. Engineering works, including dredging, ditching, and tiling, altered the environment to better meet human needs, first for grazing and eventually for farming. These landscape modifications ultimately resulted in over 203,000 ha (500,000 acres) of good farmland being added to this portion of the "corn belt" (Meyer 1936). Today the transformation of the Kankakee from a poorly drained wetland fit primarily for wildlife to well-drained, highly productive modern farmland, albeit with most of the original wildlife and its habitat exterminated, is largely complete.

The meandering and wandering river has been replaced by a permanently dredged Kankakee River that is straight, with abrupt angular bends. By substituting a straight channel for the sinuous river, the river's gradient has been increased. The result is that today the river's velocity is swift compared to the slow current characteristic of the previous meandering stream. This change in the geometry of the channel contributed to the draining of the Kankakee wetland by moving water through the area in a shorter time. The straight ditches and tiled fields that form artificial rectilinear drainage networks now drain former marshes. These drainage works follow the surveyed boundaries of the township and range system, since these boundaries are the basis for property lines in the Midwest. Farmland has replaced the previously dominant pasture, swamp, and marsh vegetation, which is largely restricted to areas of former meander bends and sloughs that remain close to the water table even with the artificial drainage (figure 7.9).

Channelization of rivers and streams and the construction of tiled and other forms of drainage were common strategies throughout the Midwest (Thompson 2002). This was a principal tactic to convert the region's poorly drained lands to the farmland found throughout this area today. Clearly, some negative aspects of this widespread alteration of drainage exist besides the immense changes that took place in the region's flora and fauna. The increased instability of some river channels due to the rapid runoff of precipitation and resulting land degradation is one such example (Simon and Hupp 1986). Likewise, areas of excessive soil erosion occur where good farming practices are not observed. However, in general, the conversion of the Midwest from its natural state of forest, swamp, marsh, and wet prairie to productive farmland is an example of creative destruction.

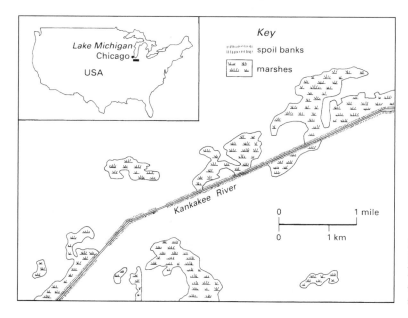

Figure 7.9. Drainage of the Kankakee Marsh near Hebron, Indiana, made it possible to convert wetland to productive agricultural land.

The Midwest's contemporary farming systems have evolved from the sustainable, but less intense, rotational systems of the first half of the twentieth century to the specialized and high energy demanding systems of today (Hart 1972). This trend toward greater crop intensity has evolved while largely maintaining sustainable patterns of land use. From the human perspective, the croplands that replaced the native vegetation and wildlife better meet both individual and national needs than if the land system had been left in its natural condition. Today, due to creative destruction in this region carried out by both governmental and private initiative, the area is characterized by some of the highest agricultural yields found anywhere in the world. The creation of the conditions needed for successful crop production in the poorly drained areas of the Midwest required widespread destruction—or at least massive alteration—of nature, with consequent changes in species diversity and habitat quality. The critical variable, the erection of a system that can withstand the test of time, appears to be met on these artificially drained lands. Corn (maize) yields averaging around 8.8 ton/ha (140 bushels per acre) are common in northern Illinois and have remained high over a period of years (Illinois Department of Agriculture 1998). Agricultural yields from farms on these drained lands remain among the highest in the world (Horsley 1986). Its agricultural surpluses, largely the result of the transformation of this zone into a well-drained landscape, contribute significantly to the world production of grains. In the years to come, farming practices will continue to evolve, and there is good reason to assume that land degradation need not occur and that sustainable agriculture will continue to be practiced throughout this region.

The moderate-intensity systems discussed in this section possess six common characteristics. First, all use labor in spatially much more concentrated forms than is the case in low-intensity systems. The existence of increasingly abundant and skilled labor, in turn, makes it both necessary and possible to engage in permanent modifications of the Earth's natural habitats. The increased use of technology to assist human activities is a second characteristic. This technology increasingly features larger and more powerful machines that initially supplement and increasingly replace human labor. Third, in moderate-intensity systems humankind first begins to alter topography in a significant way. This begins with very simple, small-scale changes and progresses to the modification of slope, hydrology, and surface relief. In almost all such instances, the systems and infrastructure created are dependent upon human vigilance and support for their continued operation. Fourth, an artificial vegetation is substituted for the flora found in nature. In theory, this habitat modification is permanent, although withdrawal of the human energy subsidy would undoubtedly result in the collapse of the artificial system and the emergence of new ecosystems dramatically different from the premodified habitat. A shift from long-fallow cycles to much shortened fallow cycles, or even patterns of continuous use, is the fifth distinctive feature of moderate-intensity systems. It is this ability to generate more production from more intensive use of the same plots that sustains the large labor supply needed to erect and maintain sustainable agricultural systems. Finally, with moderate-intensity systems the first examples emerge of sustained impacts upon the environment that have more than local implications. This change more than any other, because it often operates beyond the knowledge and control of local actors, increases the difficulty of creating sustainable agroecologies.

High-Intensity Systems

Although the origins of high-intensity systems are found in the less intense production systems described in previous sections, in their use of capital, labor, and technology high-intensity systems differ from other production modes in more than just degree. For the change in degree represented by the emergence of high-intensity systems is so great that it becomes a difference in kind. Technological innovation is the primary feature of high-intensity systems. The lavish use of technology is focused on machines, which begin increasingly to substitute for rather than simply supplement human labor. Massive draglines used in strip mining, for example, make it possible to remove a volume of overburden in a matter of days that would take thousands of hand laborers years to accomplish. In its most profound state, a high-intensity system also may be able to substitute for the land component. This is accomplished, for example, in closed environment agricultural systems such as greenhouses, in which the soil, climate, nutrient, and temperature conditions are all controlled and manipulated in an absolutely artificial creation. Such systems can function without reference to soil quality and weather conditions because they can make up for nature's local deficiencies and variability by controlling for the deficit ingredients or episodic extremes. Such systems are very dependent on inputs from outside their controlled environment and for this reason are inherently unstable. Because of this systemic instability feature, we do not consider these controlled environment systems to meet our sustainability criteria.

The use of machine technology as a labor substitute makes it possible to support large population densities in places distant from the primary production process. The worldwide shift of rural populations to urban centers is an artifact of high-intensity systems and the machines that sustain them. These systems also have profound impacts upon the natural world at great distances from the site of the intervention. As a consequence, managing high-intensity systems in order to avoid land degradation is difficult, since it is easy to initiate changes in one place that have negative, but unobserved or unassociated, consequences elsewhere. Thus despite (or perhaps because of) its powerful technology, high-intensity systems have greater potential to induce destructive creation than do low- and moderate-intensity systems. Worse yet, the pace at which these changes occur is very rapid and makes it difficult to respond to adverse change in a time frame that promises successful results at an economically feasible cost.

Nonetheless, examples of creative destruction in high-intensity systems can be found. In this section, we examine four examples in which the forces of creative destruction have operated in high-intensity systems. Our first example deals with one case, the Broads of eastern England, in which an accident of history and environment transformed a wasted sacrifice zone into a critical resource zone that is itself now threatened by destructive creation. An examination of the drained landscapes of the Zuider Zee polders, representative of similar systems worldwide, and the Scheldt estuary storm surge barriers that have met the needs of the Netherland's population for agricultural land, flood protection, and recreation follow. Finally, we consider how the introduction of irrigation technology into the Sonoran Desert created both the critical agricultural zones of the Imperial Valley and the sacrifice zones of the Salton Sea and the Colorado River estuary.

The Broads and Accidental Creative Destruction

The Broads of Norfolk and Suffolk in eastern England are an example of inadvertent creative destruction. The Broads are a group of freshwater lakes of varying size that are linked in complex fashion to the region's partly tidal rivers through a system of lock-free channels (called dykes locally). These water bodies are connected to adjacent wetland, alder woodland, pasture, and arable fields to comprise an attractive rural landscape. Although evidence exists to support the belief that the general geomorphic and fluvial features of the region are the product of postglacial deposition and marine transgressions (Jennings 1952), an accumulating body of historical evidence indicates that many of the region's distinctive features are the product of human activity (Lambert 1960; Moss 1979). Before AD 1450, peat mining met much of the local population's domestic energy needs. The pits were relatively deep, since the deeper, more consolidated peat layers constituted the most valuable source of energy. A rise in sea level, perhaps linked to climate change, led eventually to the flooding and abandonment of peat-mining operations.

In this instance, creative destruction, that is the formation of the Broads, is the accidental by-product of destructive creation. In the rural medieval mining operations, the peat mine pits were sacrifice zones. Extraction of energy resources from the peat pit was essential to the well-being of the rural peasantry. A portion of the wildlife habitat of small game animals and fish was sacrificed in order to provide people with fuel. These sacrifice zones scarred a significant proportion of the rural landscape. The flooding of the peat pits, which initiated a new set of ecological processes, masked this landscape disturbance.

Within the lakes and along the dykes (channels) that were cut to connect them and to provide easy cross-country transportation there evolved new plant and animal communities. Reeds grew along the banks and protected them from erosion. Bottom-growing water plants developed in the clear, unpolluted, sunlit, shallow river and lake waters. These plant communities provided the habitat for numerous fish and eels, the capture of which represented an important dietary addition to the local population's larder. Wildfowl flourished in the marsh and swamp habitats adjacent to the Broads and frequently found their way into local cooking pots. Many fish species seasonally spawned and spent part of their life cycle in these wetlands before returning to the main water bodies. Over time, the aquatic and terrestrial habitats that evolved in and around the Broads came to be viewed as natural systems, valuable for the ecological diversity they brought to the region and the resources they represented. The fact that they were largely the accidental consequence of human activity, the wasted and cast-aside sacrifice zones of a bygone era, was forgotten. After centuries, the memory of the Broads as artificial landscapes has faded and they are regarded today as an immutable part of the natural order. This evolutionary development, despite its accidental nature, must be regarded as an outstanding example of creative destruction both by virtue of longevity and stability.

Today this artificial natural habitat is once again faced with destructive pressures, albeit from new sources. Digging for peat is no longer a productive occupation, but other changes threaten the viability of the Broads environment (O'Riordan 1979; George 1992). Many of the wetlands located near the Broads and dykes have been drained and converted into agricultural land. Protection of these agricultural lands by flood barriers and levees cuts off fish from access to the seasonal wetland habitat and confines them to the major waterways and lakes. As long as the drainage schemes were maintained as grazing dykes, with limited use of fertilizers that are rich in phosphorus and nitrogen, the impact on the aquatic environment of the Broads was minimal. But the actual and impending conversion of much of this land to winter wheat threatens to increase nutrient loading in the Broads (Shaw 1986; Colman 1989). Sewer discharges from municipal sewage systems and rural septic systems are also major contributors to eutrophication, which fosters algal blooms (Stansfield, Moss, and Irvine 1989). These microphytic plants absorb sunlight, increase the turbidity of the water, shade bottom-growing plants, and lead to the loss of habitat that is essential to fish (Balls, Moss, and Irvine 1989). A contributing factor in this shading problem, which was not found in the region when sailboats dominated river traffic, is the increased turbidity generated by motorboats (Garrad and Hey 1988).

Because the Broads also function as a major recreation area, the region attracts many tourists with an interest in boating. Increasingly, this industry uses motorboats, as opposed to the traditional and environmentally benign sailboat. The wash created by both commercial and private powerboats has a serious impact on the banks of dykes and Broads as well as upon bottom sediments. Boat wash has an enhanced impact for three reasons:

1. decline in the water weeds and bottom plants due to pollution and turbidity removes a shock absorber to undiminished wave energy buffeting the bank and bottom;
2. indiscriminate mooring along the waterways breaks down the reed beds that protected the bank; and
3. the invasion of the accidentally introduced coypu, a bank-burrowing vegetarian rodent, has lead to heavy grazing of much of the bank and near shore vegetation.

A lack of clear policy guidelines and political jurisdiction compounds the problems of management and control (O'Riordan 1990; Shaw 1990). As a result, an act of accidental creative destruction, which transformed medieval peat pits via flooding into productive lake, marsh, and swamp environments and has been sustainable for five centuries, is so threatened by contemporary destructive-creation processes that it is in danger of being changed into a different, and less desirable, state. What this case suggests is that if one can take a time perspective of centuries, destructive creation can be converted into creative destruction—and vice versa!

Drainage, Flood Protection, and Creative Destruction

By necessity, the low-lying country of the Netherlands has created an environment secure from the ravages of North Sea flooding while at the same time areally expanding the nation through reclamation of wetlands and lands below sea level. While both reclamation and flood control projects have a long history in Holland, it was through the massive use of technology in the twentieth century that the Netherlanders finally created a safe haven for their industrial, recreational, and agricultural endeavors. The attainment of this stable condition has required major changes both in the terrestrial and coastal systems of the nation. The Zuider Zee polders are a superb example of this millennia-long process.

There is a widespread saying, dating back at least to Voltaire, that God created all of the world, but that the Netherlanders built their own country. This expression refers first to the fact that approximately one-half of the country is composed of land reclaimed from shallow sea floors just offshore from its natural coastline and wetlands, and second, that significant areas, especially in Holland, would be highly susceptible to coastal flooding if it were not for the multitude of engineering works, ranging in size from small to massive, that have been constructed since the sixteenth century (figure 7.10; Lambert 1971; van de Ven 1987). These works both have created and protected this lowland area from the damaging North Sea storms that have historically flooded large areas of this country. Despite centuries of hydraulic interventions, a massive 1953 winter gale resulted in 2,000 km² (772 sq miles) of farmland and populated areas being flooded in northwestern Holland. Over eighteen hundred inhabitants were drowned as the North Sea's waters, pushed southward by the gale, inundated the low-lying lands of Zeeland (Smiles and Huiskes 1981). This disaster acted

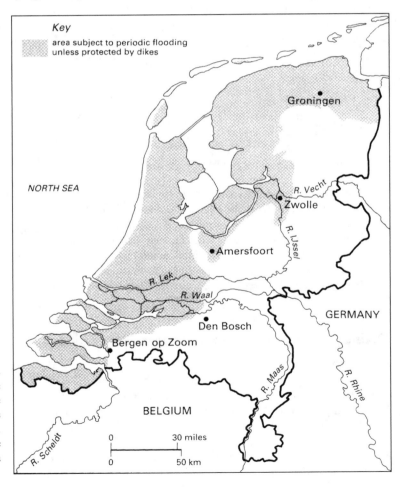

Figure 7.10. Low-lying areas vulnerable to flooding in the Netherlands comprise more than 50 percent of the nation's territory. (From Smits 1988. Reproduced courtesy of the Ministerie van Verkeer en Waterstaat, Netherlands)

as the catalyst for the Delta Plan, which, utilizing the technological advancements of the twentieth century would culminate in the fortification of the country's coastline that had begun with small local projects centuries ago. Now not only would damage from moderate-intensity storms be curtailed, but with the massive construction of hydraulic storm surge dams across the estuaries of the delta, potential flooding from the rare but high-intensity storm would be avoided or greatly diminished.

Over hundreds of years, the natural landscape of the Netherlands has been transformed by human actions (figure 7.11). For defensive purposes, by means of flood control, improved drainage, better transportation, and reclamation of the shallow sea floors, the Dutch have built thousands of kilometers of dikes and canals and have installed thousands of pumps throughout the nation to create the lands that comprise the modern Netherlands (Bardet 1987; van Meijgaard 1987; Schmal 1987). In this northwest corner of Europe, the

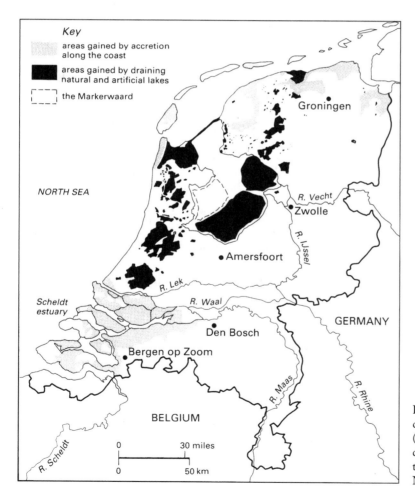

Figure 7.11. Lands gained by drainage in the Netherlands. (From Smits 1988. Reproduced courtesy of the Ministerie van Verkeer en Waterstaat, Netherlands)

end product of massive hydraulic engineering has been the creation of a complex set of artificial conditions necessary to sustain human life according to the desires of the inhabitants of this area (Volker 1982). Thus much of the Netherlands is a classic example of creative destruction. This includes even some of the country's "higher" eastern regions where former woodlands and marshes are today replaced by farmland, industrial parks, and urban and recreational areas that bear little resemblance to their natural state. The massive drainage of wetlands throughout this low-lying nation, the removal of most of the original forests—and their replacement by farmland, pasture, and urban land use—and the protection of the Meuse, Rhine, and Scheldt's deltas from the North Sea's storm floods have all contributed to the economic well-being of the Netherlands (Lambert 1971; Smiles and Huiskes 1981). The creation of the Dutch landscape that exists today, in response to the hundreds of years of land reclamation and coastal projects, is woven together by means of major environmental alteration and destruction of natural ecosystems (tidal flats, marshes, and heaths).

By the twentieth century the reclamation projects and coastal protection projects had evolved from the small incremental improvements of prior times. The pre-1900 projects primarily utilized animal and human labor for earthmoving operations and wind power for pumping. The massive landscape alterations and coastal controls initiated after the Zuider Zee Act of 1918 required modern technology in their reshaping of the landscape. Both the polder projects in the former Zuider Zee and the Eastern Scheldt Estuary Barrier Scheme altered the existing hydrologic conditions to a degree that was impossible prior to the technological advancements of the contemporary period. By creating 166,000 ha (410,000 acres) of arable land (table 7.1) on below sea-level polders that bordered on or were located within the Zuider Zee, now Lake IJssel (figure 7.12), as well as by protecting these new lands, along with all of the low-lying areas of central Netherlands, from North Sea flooding, the former salty and brackish Zuider Zee ecosystems were destroyed. In their place these areas became the largely human-managed polder and freshwater systems of the Flevoland (van Duin and de Kaste 1984).

Polder construction, the major method utilized in the Netherlands to create highly productive rural and urban areas from lands that formerly were below sea level, meets the criteria of creative destruction. In all cases, it alters the hydrologic systems associated with prepolder conditions to such a degree as to result in the destruction of the original ecosystems and their replacement with systems that the inhabitants desire. Polders are a very complicated form of land reclamation. Their construction is "carried out on lands with a high

Table 7.1.   Land Reclamation of the Zuider Zee/IJsselmeer

Polder	Year Drained	Area (hectares)
Wieringermeer	1930	20,000
North Eastern Polder	1942	48,000
Eastern Flevoland	1957	54,000
South Flevoland	1968	44,000
Markenwaard	(Proposed)	60,000

Source: Based on data from Graves (1990).

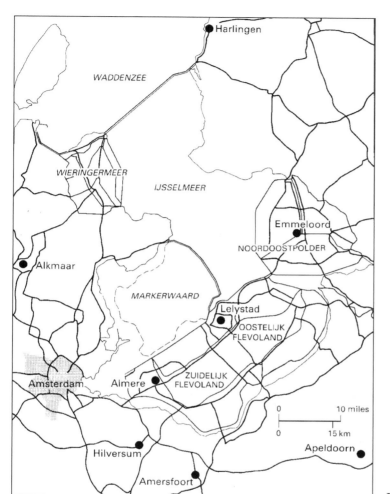

Figure 7.12. Lands recovered from the former Zuider Zee. (From Smits 1988. Reproduced courtesy of the Ministerie van Verkeer en Waterstaat, Netherlands)

potential for agricultural development, and when impoldering has taken place they generally show a high productivity. . . . Polders are found in deltas, low lying coastal areas, river valleys, marshes, swampy areas, and under former lakes, tidal embayments and gulf areas." (Volker 1982, 2). Potential polderlands have a high level of flooding, normal agricultural practices are impossible, and many other activities are greatly curtailed.

Impoldering includes a combination of strategies. In almost all cases, dikes need to be built to protect the lands from flooding. Second, diversion of local water systems from polder areas is needed. According to Volker (1982, 2) polders are defined as ". . . reclaimed level areas having a naturally high water table but where the surface and groundwater levels can be controlled." According to this definition, polders are found both near and below sea level worldwide. In various forms, they exist in Belgium, England, Denmark, Germany, Italy, Surinam, the United States, Egypt, Japan, India, Bangladesh, Southeast Asia, and elsewhere (Darby 1940; Biswas 1970; Fukuda 1976). In all cases, impolderment results in the destruction of natural conditions by altering the hydrologic, pedologic, vegetation, and animal domains. In their place is substituted a set of conditions that meet the needs of humankind.

Because of advances in technology, by the turn of the twentieth century, it was possible for the Dutch to believe that the flooding that had caused havoc to the nation from its earliest history could be curtailed through massive public works. According to the 1918 Zuider Zee Act, it was decreed that the Zuider Zee should be cut off from the Wadden Zee by the construction of an enclosing 30 km (19 mile) long dam, stretching from Friesland to North Holland. The aims of the decree primarily were to:

1. provide flood protection to the 300 km (190 mile) coastline, which was susceptible to disastrous floodings when breaks in the dikes occurred, primarily during winter storms
2. create new farmland for the country's agricultural overpopulation
3. increase food production; and
4. improve water management in the central Netherlands by replacing the salt and brackish water of the Zuider Zee with the freshwater of Lake IJssel (Smits 1970).

Through the impoldering process, four large polders were created on former seafloor lands (table 7.1). With the construction of the enclosing dike in 1932, the dredging of canals, the emplacement of subterranean pipe drains (similar to the tiles of the American Midwest), and continuous pumping of the polder's waters, both new land and an artificial hydrologic system was created. Soil properties were altered and new agricultural, recreational, and urban lands established. A complex cropping strategy increased the permeability of the soils and reduced their pH to levels that satisfy the requirements of the crops traditionally grown in the Netherlands (Smits 1970). New accessible recreational areas, both on Lake IJssel and in surrounding land through afforestation and landscaping projects, have been created in the densely populated central Netherlands (Smits 1988). This landscape creation was accomplished at the expense of ecological damage. The coastal ecosystems of the former Zuider Zee are gone, replaced by the new terrestrial polder ecosystems and the freshwater ecosystems of Lake IJssel. Because of changing economic conditions (farm surpluses), recreational needs, and environmental concerns about the ecosystem changes inherent in polder construction, the Markerwaard Polder was never built. Thus another area of creative destruction has not yet occurred, but could should the need arise.

The Scheldt Estuary Barrier is designed to protect the low-lying coastal and marshy lands of western Holland, which have been particularly susceptible to storm surge disasters throughout their modern history. Storm surges, in general, result from a combination of at least three major factors. First, a strong slow-moving storm that is over a large body of water is a prerequisite. A minimum fetch of 645 km seems necessary for sufficient wind energy to be transferred to the water to produce a possible surge (Thurman 1985). Under these conditions of high, steady winds and a large body of water, strong wind-generated waves are produced. Second, when the storm waves reach a coastal area during high tidal conditions, they interact with the tidal conditions to produce a higher storm surge. During high spring tidal conditions (new and full stages of the moon) the above average high tides are ideal for producing the highest storm surges. Third, the storm generating the waves must be moving toward the coastal area. This means that the wind-generated waves will pile up on the shoreline. If all of these conditions are met, the individual effects of each factor will reinforce each other and produce an exceptional storm surge. When these conditions occur south of the Faeroe Islands, large quantities of North Atlantic Ocean water can be forced into the narrow confines of the North Sea and flood the low-lying coastal regions.

From January 31 to February 1, 1953, during particularly ideal storm surge conditions, over 4 billion m^3 (5.25 billion cu yd) of North Atlantic Ocean water were pushed into the North Sea (Robinson 1953). A devastating storm surge, along with the high tide, produced waves over 5.5 mm (18 ft) in height that blasted the Dutch coastline. The surging North Sea's waters covered more than 25,000 km^2 (10,000 sq miles) of the Netherlands' land. This resulted in immense damage throughout western Holland. In response to this storm, the government of the Netherlands embarked on a massive public works program to protect this portion of the country in the same manner that the barrier dam now protected the coastal lands and polders of the former Zuider Zee.

Learning from their previous experience, and because of the importance of ecological criteria in decision making in the Netherlands in the 1970s, the construction of the Scheldt Estuary Barrier Scheme, unlike the Zuider Zee strategy, did not lead to the building of permanent dams to close off completely the estuaries of western Holland. On this occasion, a marine ecosystem was not converted to a lacustrine one. Through the construction of costly storm surge barriers, the estuarine conditions of the delta area and the wetlands of western Holland were largely preserved, while protecting the low-lying southwestern area.

The Oosterschelde barrier, the last component in a thirty-year project to prevent flooding in the Oosterscheldes estuary and western Holland, is comprised of a complex of dams, dikes, and channels. In contrast to the permanent barrier dam separating the Wadden Zee from Lake IJssel, the Oosterschelde barrier is made up of sixty-two hydraulic gates that can be lowered to the sea bed during storm conditions (figure 7.13). Most of the time the gates are in their raised position, which permits the waters from the Oosterschelde estuary and other western coastal waters to continue to mix and flow into the North Sea. When storm conditions occur, the gates are lowered. In their lowered format, the Oosterschelde barrier acts as a dam and prevents the North Sea's waters from flooding the low-lying areas inland from the barrier. When normal conditions return, the gates are raised, and once again the normal tidal rhythms of the estuary occur. To protect the rich estuarine environment, the project required an additional eight years to complete; the equivalent of over U.S. $800 million was contributed to the project compared to the original proposed closed dam (Kohl 1986).

At least in this case, the massive degradation costs of one type of creative destruction strategy were considered to be too high, even though it would have been economically less expensive than the final adopted scheme of hydraulic storm surge barriers. The chosen strategy is still one of creative destruction in that significant disruptions in the estuary's

Figure 7.13. The hydraulic gates on the Scheldt Estuary, when open, permit normal tidal flow to enter the estuary, but when closed they prevent storm surges from threatening to inundate productive land. (Photograph by L. Lewis, February 2003)

ecosystems took place during construction. Furthermore, even with the open barrier dam strategy, there was a 60 percent reduction in the area's salt marshes and a 45 percent reduction in tidal flats. Thus there were both large environmental as well as economic costs associated with the need to protect the lowlands of western Holland from future North Sea storm surges (Smiles and Huiskes 1981).

Irrigation: Creative Destruction in the Imperial Valley

The Imperial Valley, located in southeastern California, has been an important irrigated agricultural area since the early 1900s (figure 7.14). The transformation of this very arid and hot corner of the United States from desert to irrigated agriculture was marked by major alterations in the surrounding lands. Imperial Valley agriculture remains not only sustainable, but also very highly productive after almost a century of irrigation (U. S. Census Bureau 1987). This is a story of making the desert bloom through irrigation that is widely repeated in the arid portions of North America (Fiege 1999). The success of the Imperial Valley has required the creation of major sacrifice zones in the general vicinity of the irrigation perime-

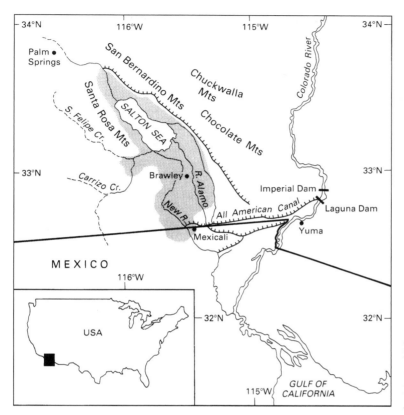

Figure 7.14. The Imperial Valley, situated in the Salton Sink, receives the Colorado River waters through the All America Canal.

ter as well as along the lower Colorado River from the Hoover Dam to the Gulf of California (Bergman 2002). Along with other users of the Colorado's waters, the Imperial Valley's pursuit of sustainability has contributed to large off-site alterations in the Gulf of California ecosystem, the Colorado River Delta, the lower Colorado, and portions of the Salton Sink—the topographic area in which the Imperial Valley is situated.

The development of Imperial Valley irrigated agriculture represents another example of creative destruction, albeit one that is in a delicate adjustment. Imperial Valley agriculture—both its creation and its maintenance—required high-intensity technological interventions. Especially critical are the series of dams built to control both the Colorado's highly variable flows and high rates of sedimentation. An increasingly delicate balance endures from some of these interventions, especially with regard to the interaction between the escalating salinity of the lower Colorado and the pH value of the valley's irrigated soils. Today, with expanding demands on the Colorado's limited water supply, especially to meet urban demands, major improvements are being implemented to increase the efficiency of irrigation in this area (Committee on Western Water Management 1992). The ramifications of these changes will continue the processes of creative destruction that were set into motion when individuals decided to convert this bleak and inhospitable area into highly productive agricultural lands. In the remainder of this section we explore some of the environmental alterations that have taken place to permit agriculture to prosper in the Imperial Valley.

Occupying a bowl-shaped depression, surrounded by a desolate desert area, and characterized by a furnacelike arid climate, the Salton Sink in which the Imperial Valley is located was referred to as the Valley of the Dead prior to the twentieth century (Worster 1985). This name harshly reflected the paucity of water and high summer temperatures experienced in this hot desert area. Average annual rainfall is less than 7 cm (3 in) per year and the whole depression is devoid of any significant nonsaline water (Seckler 1971). Temperatures over 40°C (100°F) are common during the region's summer months. The Imperial Valley is a rifting depression lying below sea level that, except in the south, is enclosed on all sides by low mountains. Its southern border is protected from the sea by the Colorado River's thick deltaic deposits. These deposits created a fanlike plug that cut off the Salton Sink depression from the Gulf. It is these deltaic deposits, comprised of fertile silts and clays (Robinson and Luthin 1976) and covering thousands of hectares, that formed the foundation for the area's agricultural development in 1901 (Henderson 1999, 175–95). Before this could occur, because of the aridity of the area, the scarcity of useful water for irrigation had to be overcome. Fortunately, the Colorado River, flowing only 80 km (50 miles) to the east, had a large enough discharge to meet the water needs of the Imperial Valley (figure 7.15). As the Colorado flows along a levee elevated above the Imperial Valley, its waters, when diverted, could flow by gravity into the area. The diversion of Colorado water into the Salton Sink, at the beginning of the twentieth century, initiated creative destruction processes, some of which continue to this date.

In May 1901, under the aegis of the California Development Company, a diversion channel was cut into the riverbank of the Colorado across from Yuma (Seckler 1971, 194; Committee on Western Water Management 1992, 235). This diverted some of the river's water through abandoned delta distributaries toward the Salton Sink (Reisner 1993). Using these waters, irrigated agriculture began to expand rapidly in the former Valley of the Dead. With its fertile soil, an apparent ample water supply, and the potential for a year-round growing season, two thousand settlers were soon on the scene and 40,500 ha (100,000 acres) of land were already under cultivation on formerly barren lands by January 1902 (Reisner 1993). Settlement and irrigation continued to increase rapidly after the initial success of providing water to the area. By 1904, seven thousand settlers had moved into the area, and water demands for irrigation expansion grew continuously during this early period (Worster 1985).

Due to the heavy sediment load found in the river's waters, the diversion channel from the river immediately experienced heavy deposition throughout its length. Maintenance procedures were inadequate, as they could not keep pace with the deposition taking place within the channel. This threatened the water supply to the valley and hence the success of the newly established agriculture. In 1904, to offset the effects of sedimentation, a new diversion channel was begun further downstream from the original cut. Before it could be completed, a series of major floods occurred along the Colorado River. In early 1905, overbank flows reached the new diversion channel before the crucial permanent controlling gate was constructed!

Upon reaching this channel in February, the floodwaters quickly cut back toward the river's main channel and breached its levee. By the summer of 1905, the Colorado's major

Figure 7.15. The All America Canal diverts water from the Colorado River to the Imperial Valley. Irrigated and nonirrigated lands present a stark contrast at the edge of the cultivated area. (Photograph by L. Lewis, June 2002)

flow of 2,520 cu m per second (90,000 ft^3/sec) had been diverted from the Gulf of California toward the Salton Sink. Now out of control, the Colorado flooded and destroyed large areas of the recently developed lands. The Salton Sea, occupying the lowest portion of this interior area, was created during this flood as an undesired by-product of the Imperial Valley irrigation development. From the land degradation perspective, all of the valley's lands permanently flooded by this 80 km long (50 miles) by 24 km wide (15 miles) interior sea were destroyed (Reisner 1986). From this perspective, the Salton Sea became an unintended sacrifice zone for Imperial Valley agriculture. Unlike the Aral Sea, which is today largely destroyed by irrigation follies, the Salton Sea was created by a different series of management mishaps and problems.

In 1907, through a massive engineering and construction project, along with a rather tranquil period of river flows, the Colorado was returned to its former channel and agricultural development of the Valley once again was possible. However, by this time thousands of acres of farmland had been destroyed, more than four times the volume of earth excavated for the Panama Canal had been eroded and deposited in the sink, and the Salton Sea, which permanently flooded some of the former agricultural lands, had been created (Worster

1985; Henderson 1999). The sea's future was guaranteed by the continuous inflow of drainage waters from the surrounding irrigated lands. This regular input was required to prevent the salinization of the valley's soils.

Even before the Colorado's waters were used for irrigation, the lower Colorado had a naturally high concentration of total dissolved solids (TDS) compared to other major rivers. By 1968, because of the heavy utilization of its waters throughout the whole river basin, it had become even more saline. In addition to the natural inputs of soluble minerals from waters running off the region's alkaline soils and mineral springs, the Colorado received additional mineral inputs from drainage of irrigated lands as well as from mining, manufacturing, and municipal wastes. From its sources in the Rocky Mountains of Colorado and the Green River Basin in Utah and Wyoming until the Imperial Dam near the Mexican border, a twenty-one-fold increase in its TDS from 38 to 809 mg/l occurs. Almost two-thirds of this increase is due to natural sources (Committee on Water 1968). By 1976, due to increases primarily linked to irrigation, the TDS level in the Colorado had increased to 877 mg/l. In the mid-1970s, some estimates indicated that with increasing water use and diversions the TDS level could reach a value of around 1,300 mg/l by the year 2000 (Robinson and Luthin 1976). As salinity readings approached these levels throughout the decade, the United States attempted to meet its treaty obligations on the quality of water delivered to Mexico by diverting agricultural wastes from the Gila River irrigation systems toward the Gulf of Mexico. The consequence of this diversion was the accidental development of an entire marshland ecosystem—whose preservation is now the object of intense concern for environmentalists—in the mudflats of the delta north of the Gulf of California (Bergman 2002, 47–50). Efforts more broadly to reduce the Colorado's TDS load has meant that the alarmist concerns in the 1970s have had a positive impact; today TDS levels at the Imperial Dam on the lower Colorado remain at about the level they had attained three decades earlier.

To prevent salinization of the Imperial Valley's soils through its Colorado River irrigation water, the accepted strategy has been to apply more water than was required to meet the evapotranspiration demands of the plants and climate. This excess of water prevents salts from accumulating in the top layers of soils by leaching the soil salts downward through the soil profile and out of the root zone of plants. A system of 2,240 km (1,400 miles) of pipelines throughout the Imperial Valley carries the salty drainage that results from this irrigation strategy to the saline rivers flowing into the Salton Sea. The increasing salinity of the Salton Sea resulting from these discharges, along with the inputs of untreated sewage, pesticide residues, and heavy metal contamination from Baja, California (flowing in the New River), threatens the viability of the Salton Sea ecosystem (Committee on Western Water Management 1992; Bergman 2002). Already, high levels of toxins are found in the fish caught in this sea, making them unfit for consumption. The Salton Sea, while a sacrifice zone for the Imperial Valley's irrigation, has led to recreational benefits, and a new marine ecosystem developed after its creation in 1904. These positive facets of the region's creative destruction are now being threatened by contemporary activities occurring within the Salton Sink and Mexico.

Because of increasing demands on the limited water supply of the Colorado, most notably urban growth in California and water transfers to Arizona, new strategies are required to utilize water more efficiently. Through water conservation, such as canal lining, building new regulating reservoirs, and canal spill interceptors, it will be possible to divert 13.1×10^6 m^3 (106,100 acre-feet) annually from irrigation to the Metropolitan Water District, which supplies water to southern California's urban areas, without reducing agricultural activities. But this diversion will result in lowering the Salton Sea by 0.67 m (2 ft) and will likely exacerbate salinity and pollution problems in the sea (Committee on Western Water Management 1992).

Today about 203,000 ha (500,000 acres) of irrigated land in the Imperial Valley generate $1 billion a year in agricultural products. Among these products are an array of vegetables, fruits, grains, and sugar beet (U.S. Census Bureau 1987). While some crops, such as lettuce, onions, carrots, and snap beans, might be reduced in the future due to increasingly saline irrigation waters, most crops will continue to prosper under the environmental conditions created by water transfers into the Imperial Valley. Technological improvements, such as desalination of drainage water, which are already under construction or in the planning stage, should insure the continued sustainability of the area.

To keep the Imperial Valley in operation over the past ninety years, a series of activities have been required, many of which have destroyed previously productive ecosystems. In order to assure a reliable water supply, land was flooded behind the dams built on the Colorado. Impoundment of water along the Colorado and its diversion to the Imperial Valley and other areas in southern California starved the delta of the Colorado of water and sediment. Riparian and delta ecosystems were impoverished as a result. Efforts to rehabilitate these areas by diverting drainage waters from the Wellton-Mohawk Irrigation District have been assisted by larger channel flows in the Colorado during El Niño periods, although recovery has been only partial (Glenn et al. 2001). While these pulse flood inputs have succeeded in establishing riparian trees along the banks of some of the Delta's watercourses, doubts remain about whether these events, either from wet years or by releases from the Colorado's dams, can reestablish more than a fraction of the Delta's former extent (Varady et al. 2001).

More recently, the agricultural chemicals and fertilizers that enter the Salton Sea, with their toxic residuals, as well as the copious salts in drainage water, threaten the viability of the accidentally created aquatic ecosystem in the northern part of Salton Sink. Post–World War II hopes for a booming recreation and retirement industry have been blighted by popular concern about water quality and health conditions in the sea and along its shores (Laflin 1995, 37–43). In a booming general real estate market, and in a region with a very large demand for outdoor recreation, the Salton Sea area has a unique distinction in California: it is the only area to decline in value. The result is a landscape of abandonment and blight brilliantly captured by a collection of photographs plus text assembled by Christopher Landis (2000). Other problems besides collapsed real estate speculation ventures continue to plague the enclosed Salton Sea ecosystem (Horvitz 1999; Bergman 2002). Increasing salinity levels have long plagued the sea, a product of using the sink as a repository for salts washed from agricultural soils. These salinity increases raise the specter of a dead ecosystem as greater salinity creates a living environment hostile to the survival of fish. Increases in salinity are readily imaginable as pressure to transfer Colorado water from agricultural users in the Imperial Valley to coastal southern California communities grows and freshwater available to offset evaporation losses decreases. The Owens Valley, which lost its water to Los Angeles and its agricultural viability as well (Sauder 1994), provides one possible future for the Imperial Valley if water withdrawals and transfers elsewhere become severe. In addition to salts, drainage water carries nitrogen, phosphates, and other materials that in moderation produce rich aquatic environments favoring fish growth. In excessive amounts, these same nutrients encourage algae blooms, oxygen shortfalls, and episodic but apparently increasingly frequent fish kills, signs of an ecosystem in stress (Landis 2000, 44; Bergman 2002, 266–71). In addition to opening up fertile land for farmers, irrigation development created habitats of great benefit to migratory wildfowl and now endemic populations of fishing birds. Significant increases in these nutrient levels could seriously undermine the fish resources that attracted the birds in the first place. All of these sacrificial impacts are examples of the destructive flip side of the creative destruction associated with the metamorphosis of this portion of the Sonoran Desert into the highly productive, but potentially unsustainable, lands of the Imperial Valley. Increasing demands for the transfer of water to other users and water right holders, from Colorado's "Big Straw" proposal to send water eastward to California's intention to shift water westward to San Diego, mean that water for the Imperial Valley's agriculture will be in increasingly short supply. More careful, stringent water management, as well as the selection of crops that are less demanding users of water, is essential if sustainability is to be maintained.

Summary

While more examples of creative destruction could be cited, the simple truth is that it is easier to find instances of destructive creation. Nonetheless, sufficient examples of creative destruction do exist in low-, medium-, and high-intensity resource-use systems, and in a sufficiently wide range of cultural settings, to engender confidence that the phenomenon is universal, albeit infrequently realized.

The low-intensity systems of hunters and gatherers, shifting agriculturalists, and traditional herders are time-honored adaptive cultural ecological systems that easily have met the test of sustainability for centuries, if not millennia. The ability of these low-intensity systems to endure is a result of two factors: (1) the slow rate of change that has been demanded of them, giving low-intensity systems opportunity to adjust to and to correct for adverse impacts generated by their livelihood activities; and (2) the long time intervals available in which to effect these adaptations. Limited contact with higher-intensity systems has made their adjustments easier, because the gradual pace of experiential knowledge acquisition and transmission that characterized these low-intensity systems was not pressured or overwhelmed by alternative forms of action and cognition. Minuscule rates of population growth also favored the adaptive application of creative destruction principles within low-intensity systems, because limited growth removed a major motivating factor encouraging rapid change in system state and resource extraction level. Ironically, for many of the same reasons that were responsible for their success, low-intensity systems find it difficult to cope with the rapid change induced by moderate- and high-intensity systems. Lacking the ability to change rapidly, the efforts of low-intensity systems to adapt creatively to pressures that impact them often trigger the development of destructive creation. The insensitivity of practitioners of high-intensity systems to the needs and advantages of low-intensity systems often compounds these problems and creates intractable instances of destructive creation, some of which were noted in preceding chapters.

Despite their decline, both in numbers of practitioners and in areal extent, low-intensity systems provide important insights into the principles of creative destruction. They do this by protecting their critical zones, rotating their use of resources in order to allow for resource base recovery, creating few sacrifice zones, capturing as many of the wastes they generate as possible, and taking care to operate at all times within the constraints of their environment. It is this sensitivity to the genius loci principle that is perhaps the greatest achievement of low-intensity systems.

Moderate- and high-intensity systems are more likely than low-intensity systems to create sacrifice zones in order to increase the productivity of their critical zones. They accomplish this enhancement of productivity by increasingly powerful applications of labor and technology. Particularly as high-intensity systems substitute technology and machines for labor, the ability of the systems created to impact, often negatively, ecosystems far removed spatially from the source of the change is increased. Successful examples of creative destruction, such as the Dutch polders or Chinese integrated pond-dike systems, sacrifice some zones for the greater productivity of the new systems; but they are able to create stable new systems because they successfully enclose their new creation conceptually and pragmatically. By so doing, such systems are able to keep most wastes within the new system, where they are quickly noted and dealt with. Thus, the draining of excessive salty water into the Salton Sea, both figuratively and literally a sacrifice sink, minimizes the appearance of salinization in the Imperial Valley. The sustainability problems that the Salton Sea itself experiences are a product of an inability to connect the sea to the agricultural environment that created it. In the most successful situations, waste products in one component of the system become productive inputs in another system segment. Mud from the Chinese carp pond becomes fertilizer for the nearby dike field; and wastes from plants grown on the dike become food for the carp in the adjacent pond, the excrement from which helps to raise the nutrient level of the pond's mud. This recycling of energy and matter turns potential negatives into positives and prevents the system from tilting into a degradation mode.

Above all, moderate- and high-energy systems that are successful and sustainable, such as the low-intensity systems that preceded them developmentally, are responsive to the genius loci principle, and build their intensification actions slowly upon the achievements of previous generations. Incremental improvement is always easier to absorb without serious disruption than is an abrupt shift to another exploitation system. Dutch impolderment in the former Zuider Zee is a good example of this process, for it is built upon centuries of drainage experience. Like the Chinese utilization of carp, which involved two millennia of experimentation before culminating in the interactive agriculture-aquaculture system that we see today, polders began as microscale drainage schemes, expanded to encompass larger areas when wind power was harnessed to drive pumps, and ultimately involved efforts to control and reclaim the floor of ocean embayments. The early-twentieth-century history of the development of the Imperial Valley, from a desert with minimal economic potential into a new zone of highly productive irrigated agriculture, further illustrates the importance of time and incremental improvements in creative destruction fostered by a powerful technology. During the initial stage of converting the area into a productive agricultural system, a series of poor management decisions resulted in the destruction of large areas of productive land by flooding. The breaching of the diversion canal by the Colorado River resulted in much of the Imperial Valley being converted into an unforeseen and unintended sacrifice zone. The accidental creation of the Salton Sea, and the resulting loss of the lands submerged under its salty waters, is an example of what can occur when moderate- and high-intensity technologies are abruptly and carelessly utilized in new areas.

The destruction of a preexisting natural ecosystem and its replacement by a sustainable human-controlled environment is essential to system success, and is the hallmark of creative destruction. In all successful moderate- and high-intensity examples of creative destruction, a change in scale occurred at each major step in the transformation process. This change in scale was paralleled by a concomitant alteration and improvement in the technology required to maintain stability. The result achieved in any successful moderate- or high-intensity system is not guaranteed in perpetuity. However, if the system design is sound and it possesses reasonable flexibility, it should be able to adjust to changes within the system and to pressures brought to bear upon the system from without. This ability to continue to evolve adaptively is a key element in the sustainability of creative destruction.

Land Degradation and Creative Destruction
Retrospective

THE ADVERSE changes in the environment that lower the overall productive potential of particular places are an inevitable consequence of human use of the Earth. As people work to derive a living from the resources of a region, the interaction that results between physical and human systems inevitably produces change. When these modifications are not to the liking of humankind, when they reduce a population's long-term prospects of securing a stable existence, when they constrain present activities, and when they cannot be reversed within a fifty-year period, we term the aggregate of these alterations "land degradation." At issue is always the temporal and spatial extent as well as the magnitude of these negative developments, for some permanent degradation of the environment is acceptable if the ultimate result is a viable, long-lived human resource use system.

Nature, Society, and Variability

Nature itself is not a particularly benevolent entity. Anyone who has experienced the destructive power of a tornado, hurricane, flood, tsunami, or earthquake, or has viewed images of the devastation wrought by such events, is aware of how unkind nature can be to those unfortunate people who get caught by the dark side of natural forces. In nature a variety of other less high-intensity phenomena operate to alter the landscape. Drought, flood, soil erosion, fire, frost action, and disease operate to place constraints upon and cause changes in the quality of local environmental resources. These events can occur with or without any human interventions. Wild animals congregating around dry-season water resources, for instance, can place a great deal of stress on local vegetation and, at least temporarily, degrade it. A flood that relocates the course of a river can destroy the basis for existence of plant and animal communities. A volcanic eruption can wipe out local ecosystems and, through the material spewed into the atmosphere, have a profound impact on the climate of distant places. Erosion processes eventually will transform hills into lowlands and sea-level rises will ultimately submerge coastlines. These events take place whether or not people want them; they are an integral part of the way in which nature operates. Thus, even without human occupancy, the Earth would be a dynamic entity. In this sense, some cases of environmental degradation can be viewed as unavoidable natural processes. Given that many of these natural hazards occur in localities that often have many favorable properties that facilitate the success of a number of activities, humans often inhabit these risky settings either by choice or lack of other viable opportunities.

Degradation integral to nature is not something that humankind commonly contemplates. Low-frequency but high-intensity events, such as a hurricane or a volcanic eruption, impinge themselves upon human consciousness by virtue of their extreme character and awesome, awful impact. Despite the degree to which nature has been incorporated into human culture, and in a sense domesticated, extreme events remain outside human control and influence. Dams and flood control structures are created to cope with floods that at a maximum can be expected to take place up to once every one hundred years. Less frequent floods that are larger in magnitude are impossible to plan for effectively, because the structures that are required to cope with them are too large and too costly to justify constructing. In reality, even if built, the structures themselves would likely cause extreme economic and environmental damage. The 1993 summer flood in the central portions of the Mississippi Valley (Iowa, Illinois, and Missouri) occurred notwithstanding a multitude of flood control measures. Furthermore, to capture a sufficient percentage of such floodwaters behind dams would create such large reservoirs that as much damage might be done to structures and ecosystems behind the dams as would occur if the floodwaters had poured unchecked downstream. Humans must accept that large floods will happen occasionally and thus should plan accordingly.

Earthquakes provide a similar example. An earthquake largely destroyed San Francisco in 1906. This did not stop the survivors or new inhabitants from rebuilding on the same hazardous site. This took place because people convinced themselves that technology could cope with any anticipated future earthquake of a similar magnitude, that the likelihood of a similar event in the near future was remote, and/or that the immediate location benefits of the central California harbor site outweighed potential future costs. Moreover, the longer the time span between extreme natural events, the more likely it is that people will manage to forget the intensity of the event and to plan their livelihood around the more benign conditions that characterize periods between catastrophes.

Even less prone to easy identification are the more modestly scaled, low-intensity, higher-frequency occurrences such as drought or soil erosion. Much less spectacular, and slower to build to the point at which adverse impacts are readily perceived, low-intensity events are easier to ignore in the short-term, but likely have greater catastrophic implications for the long-term if they are allowed to accelerate and proceed unchecked. According to the FAO, current rates of soil erosion will result in an area the size of Alaska (1,530,700 km^2 or 591,000 sq miles) losing most of its agricultural value over the next twenty years (Boston

Globe 1993). With the world's increasing population, if soil losses continue, at some point there might not be enough good land to meet the world's food needs. This has already occurred at the national level in some countries, such as Haiti. The impact of these low-intensity events can be influenced by (1) modifying the event, such as implementing successful conservation strategies; or (2) by changing the way in which society prepares for and responds to the occurrence rather than by altering the event itself. Society possesses a variety of ways to cope with and minimize the impact of low-intensity events that adversely affect humans.

The prime motivation in dealing with low-intensity events is humankind's inherent dislike of variability in the natural world. While variation may add spice to life, it can also prove disastrous to human survival if drought withers crops, floods wash away seeds, and rainfall runoff strips away soil from the landscape. When these events occur, human health and happiness diminish, economic costs increase, and higher death rates are often the consequence. Variability, often in extreme forms, may be part of nature, but most individuals and societies prefer to deal with a limited range of expected, dependable conditions that guarantee security of life, property, and food supply. In nature significant buffers usually exist, the purpose of which is to serve as shock absorbers for extreme events. These buffers provide important slack and lag time in the way nature operates. Traditional human systems always maintained substantial buffers in their use of the Earth. Long fallow cycles, low population densities, spiritual values deeply respectful of nature, and the avoidance of fragile areas kept human exploitation of nature from enhancing low-intensity, high-frequency events and allowed room for natural processes to absorb the impact of high-intensity events (Grim 2001). With these buffers in place, most traditional land-use systems avoided initiating land degradation, although often at the costs of population fluctuations. Sufficient flexibility existed in these systems such that when the resiliency of the system was threatened, feedback mechanisms curtailed the probability of irreversible negative change. Trends toward decreasing crop yields were countered by fallow. The slack in the system, in this case lands with high agricultural potential not actively being cultivated, was available for use when existing cropped lands no longer met current needs.

Contemporary resource-use systems increasingly remove these cushioning barriers between humankind and nature. Trucks now transport herds to seasonal grazing in a matter of hours where once it took days or weeks. Once in place, these herds consume the grass before it has a chance to set seed. The lag time permitted by animal-based and foot transport no longer protects the regenerative capabilities of many rangelands. We call this result overgrazing. Alternate-year fallow practices once insured time for groundwater levels to decline before the next application of irrigation water raised them again. Without construction of expensive drainage systems, unless water quantity is carefully metered, irrigation every year causes a rapid rise in groundwater, which brings harmful salts into the root zone of cultivated plants. We call this outcome salinization. Economic and demographic pressures encourage farmers to cultivate lands the steepness or poor soil attributes of which make them high-risk environments. Unless soil conservation practices are employed at the same time, rapid downslope soil movement is almost inevitable. We call this process accelerated soil erosion. In some cases, even with soil conservation practices in place, the lands being cultivated are so perilous that excessive soil loss will occur.

Alterations of this type reduce the slack in the system and result in a decline in resilience. Not only do such changes increase vulnerability to the high-intensity, low-frequency event, but they also may adversely affect the ability of the resource-use systems established by humans to cope with the high-frequency, low-intensity event or process. This is ironic, because it is reasonable to expect that human systems adapted to high-frequency events would have to cope with frequent, if minor, variability in order to be successful. Problems occur in coping with high-frequency events because, despite the impressive gains humans have made during the past two centuries in knowledge about how ecological systems function, there are many gaps in our understanding. One area in which our knowledge is limited is the identification of the precise point at which a system's resilience limit will be reached. Up to a certain limit, most systems possess considerable ability to absorb change without exhibiting a pronounced change in the state and composition of the system. Under economic and demographic pressure, people are often tempted to adjust upward the upper limit to which a system can be exploited. This boundary zone is a critical one, since it is linked to system stability. The upper limit is also not a sharp line in most cases, but rather is a relatively broad and ambiguous zone with imprecise boundaries, within which considerable variability may occur. Once exceeded by continuous and intense exploitation, a system may no longer have enough slack to be able to recover quickly from disturbance, thus losing an important dimension of its stability. Pushed to this limit, a system's capacity for resilience may be exceeded suddenly by a relatively minor fluctuation in its basic processes. When this happens, the system "flips" quickly into a different state and seeks stability around a very different set of characteristics. Thus a lake that for many years shows no visible effect from herbicide and fertilizer pollutants washed in from adjacent farmlands, or from septic seepage from lakeside residents, may suddenly produce widespread algae blooms that threaten the survival of fish populations.

There are many reasons why humans push natural systems to the point at which land degradation occurs. The most common causes are twofold. First, there is a desire to remove variability from those portions of the natural world that are important to human livelihood, and to force those segments of the environment to produce at a level of intensity far beyond what they would produce under normal conditions. When efforts to do this are successful, habitat is altered in a creative and sustainable fashion and land degradation is minimized. Less successful efforts usually result in the acts of destructive creation that are a signature of much of the human use of the Earth. Second, population growth, with the concurrent demand for more goods and services, can also bring stress to bear upon natural systems. Meeting these demands exerts increasing pressures on natural resources. Eventually, a critical threshold is reached and degradational processes are initiated. With slack removed from the utilized systems, few options exist in many cases to meet the immediate needs of the inhabitants. Again, destructive creation is the result.

Nature, Society, and Technology

While our knowledge about variability and its impact on ecological systems is less than perfect, it is growing (Meyer 1996), and a great body of technical knowledge does exist about land degradation at many scales (Scow et al. 2000). This knowledge is sufficiently extensive that we can say with confidence that most major forms of land degradation are identifiable and preventable from a technical point of view. For example, soil erosion can be arrested by applying a variety of techniques, from contour plowing to contour bunds to terraces to establishment of a permanent vegetation cover. Similarly, desertification can be halted by fixing mobile soil surfaces, by limiting the concentration of people and animals, and by installing more efficient water-use technology. Salinization can be combated by less wasteful water-application methods and by the installation of better drainage systems. Both at the scales of the individual house and the larger community it is possible to design with nature in ways that minimize human impact and align with rather than contest natural processes (McHarg and Steiner 1998). If techniques are in existence with which all of these undesirable land degradation processes can be controlled, why does land degradation continue unabated? Indeed, why does land degradation appear to be accelerating in many places?

The reasons for the general failure to control land degradation are complex. However, four major factors are heavily implicated:

1. problems of social acceptability,
2. costs associated with land degradation control,
3. short-term planning horizons, and
4. subsidizing degradation by not incorporating the true environmental costs in the specific human activities that are utilizing land resource.

Many efforts to halt land degradation and build sustainable agriculture fail because they inadequately fit into the local social context. The more sophisticated the technology employed, the more likely it is to require a level of skill that the local population does not possess. Combined with a failure to build upon the very real skills and institutions of local societies, there often is little correspondence between local capabilities and the requirements of introduced technology. The varying values held by indigenous communities and different development agencies can produce land degradation control projects that are not perceived by local societies as possessing particular merit. For example, foresters often believe in the effectiveness of tree plantations and shelterbelts as a way of increasing local fuel-wood supplies as well as to oppose the spread of desertification. The techniques for establishing these plantations are well known. However, farmers and herders seldom can conceive of how these plantations will benefit them. They view the land devoted to trees as land lost to the farming and grazing uses that sustain them. As a consequence, even if such protective belts can withstand the rigors of a semiarid environment and can be successfully established, they require constant protection. As soon as the guard's back is turned, the woodlot is invaded. A difference in objectives and values fosters competition and conflict that often accelerates rather than arrests land degradation.

Similarly, Chinese pond-dike integrated systems may be a great success in modifying delta environments into extremely productive sustainable examples of creative destruction. However, the introduction of these systems into other regions with different cultural traditions is extraordinarily difficult. The lament of the technologist, that land degradation could readily be controlled if only people would adopt new practices and tools, ignores the powerful limitations placed on such innovations by local traditions, economic constraints, and differently valued outcomes. Land degradation controls that are not rooted in local experience find limited application in many settings that superficially seem appropriate for their employment.

Furthermore, what may appear to be similar environments might well be significantly different in particular critical areas. Land-clearing practices in a midlatitude setting could be totally inappropriate in a tropical environment. When soils are compacted through the use of heavy equipment, frost action during the winter months can often restore a looser structure to midlatitude soils. In a tropical habitat, where there is no frost to counter the compaction, the use of heavy equipment can have a more serious and long-lasting impact. Moreover, soil compaction could kill the worms, ants, and other subsurface-dwelling fauna that give tropical soils many of their favorable properties. Thus the transfer of what at first glance appears to be appropriate technologies between seemingly "similar" environments in order to carry out analogous activities is a far more complex problem than might first appear to be the case (Stepanek 1999).

Costs associated with arresting land degradation often inhibit successful application of existing technology. As a general rule, land degradation costs are seldom carefully calculated when a development project is proposed. If these costs were measured, many projects would never get off the drawing board. Given the analytical boundaries that are used, it is often easy to ignore the cost of land degradation. When boundaries are set narrowly, costs that fall off the project site can be easily overlooked. These distant sacrifice zones have a high potential for catastrophe because no one pays attention to them until land degradation is well under way. Unlike the sacrifice zones that are incorporated into the structure of creative destruction, such as the hillsides degraded by the Nabataeans in order to increase the productivity of the valley bottoms, distant sacrifice zones are disaggregated from corresponding benefit zones. In these instances, the losers in the process of change are so distant (including economic and social as well as geometric distance) from the populations and habitats that gain benefits that the degraded areas do not enter into the calculation of benefits and costs. In a world where the range and frequency of global interaction steadily increases, more and more local places will be impacted adversely by distant decisions. Unless this problem is explicitly addressed, land degradation will remain a troubling and growing reality.

Frequently, there is no existing agency that looks out for the welfare of the environment. Where environmental protection agencies do exist, their work is easily compromised by the need to consider political impacts as well as economic, such as potential job losses if major polluting industries are forced to change their practices. The fear that serious efforts to control land degradation will cause companies to close or move is an important inhibiting factor in paying more attention to environmental decay. If anything, this factor is increasing as a result of globalization.

Equally significant are the major financial commitments that are required if stringent environmental protection standards are enforced. New technology is often costly and its widespread adoption is a slow process under current financial practices, since land degradation costs are rarely included in the price of environmentally deleterious activities. For example, with the lack of environmental accounting, a farmer irrigates his land without having to include the effect that this activity has on the lowering of water quality. The farmers downstream will receive waters with greater salinity due to upstream irrigation activities.

Under contemporary conditions, the costs of this lower water quality, such as lower crop yields or salinization, are borne by the downstream users and not by the culprits who cause the resource to degrade. Under current accounting practices, many environmentally unsound practices are not incorporated into the cost of the activity and few incentives exist for remedial actions short of legislation. In the meantime, land degradation continues unchecked in many places.

Land degradation is quite independent of the particular political or economic philosophy that guides national planning and development. Socialist and capitalist economies alike are culprits in degrading their land, air, and water resources. Australian or American herders are as prone to overgrazing problems as are their Tanzanian or Chinese counterparts. Land degradation associated with industrial activities is as widespread in Eastern Europe, where it occurred under state socialism, as it is in Western Europe and Latin America, where it occurs largely due to private-sector business activity. Even the major driving variables are often the same regardless of culture, economy, and political ideology. Among the most important factors that promote land degradation processes are the need for short-term survival and actions that favor one segment of the population and economy over others. Most people who deforest an area for fuel wood or excessively coppice trees to provide emergency fodder for their animals are aware of the implications of their actions. They know that if they persist they will seriously degrade their habitat. However, in most instances they see no viable alternative to the practices they are following.

These degradational actions often are linked to fluctuations in the environment, such as drought. The excessive pressure placed on the habitat by traditional herders is envisaged as temporary. Cutting tree branches for fodder is intended to be a transitory practice that will be halted once pasture conditions improve. However, changes independent of this coppicing practice often increase stress on the local ecosystem. Population growth may place more herders and animals on the range. Agricultural and/or urban expansion may confine herders and animals to a smaller area. In many cases, this contraction in rangeland robs herders of the critical land reserve areas needed in their land-use system during drought stress periods. Under these new circumstances, the previously successful, time-honored practices that resulted in sustainable land use are carried out under new constraints, which undermine the viability of previous adjustments that were environmentally sound. Under contemporary conditions, the previous adjustments, which worked well when more extensive land resources were available, often fail to maintain resource base stability and may result in severe land degradation. If people see no alternative in the present because their survival is threatened, they will do what is necessary to get through the bad period and hope that both environmental and social conditions will improve in the future. Unfortunately, as lag times and unused "slack" capacity decrease within land-use systems, and degradation occurs, the likelihood of improving conditions decreases in the absence of significant, sustained, and substantial outside financial and technical assistance.

Commercial ranchers adjust similarly when drought or a fall in prices impacts their operations. They are reluctant to destock, hoping that in the near future conditions will improve. Animals are concentrated on the best portions of the range, where grass and water are most abundant. If drought does not end or meat prices do not rise, the result is serious and prolonged land pressure and degradation of the most productive portions of the habitat. When people and place collide in a struggle for survival, it is the norm that the quality of the resource base is the first victim.

Favoritism of one economic sector or segment of the population over others often compounds degradation to the physical resource base. A segmental approach to land and resource management leads to a cascade effect, whereby the changes introduced into one area spill over into other zones. These spillover effects have a particularly serious impact on the zones affected because they are seldom adequately considered when evaluations of the primary initiating area are undertaken. Thus agricultural expansion into drier zones during periods of greater than anticipated moisture can compromise sustainable use of these dry zones over the long-term. Caught without adequate vegetation cover during a more typical period of drier conditions, increasingly rapid rates of soil loss may occur. Because these zones often are important fallback reserves during drought periods, many rural populations find that traditional famine foods are no longer available. In consequence, the ability of a region and its people to cope with hard times using their own resources is sacrificed to progress in other economic sectors and populations. Creation of unintentional sacrifice zones has serious long-term implications for the stability and resilience of the environments impacted. Generally, they are the losers in the process of change, and degradation of the land that sustains them in the short run is the consequence.

To most observers, the rate of land degradation appears to be increasing in all habitats and in all socioeconomic systems. While bright spots of creative destruction that result in sustainable development can be identified, in today's environment these positive developments are more the exception than the rule. The reason for this acceleration in the rate and scale of adverse environmental change can be found in the way in which nature, society, and technology interact. Over time, humankind has improved its technological capabilities and knowledge base to an enormous degree. This has made it possible to increase the intensity with which people use particular habitats. As the intensity of environmental exploitation has shifted from the level that characterized the low-intensity systems encountered in most of human history, there has been a corresponding increase in the rate and scale of impacts that cause land degradation. Under low- and most medium-intensity systems, considerable slack existed in human use of resources. Grazing reserves that today are converted to agricultural fields once were available lands for sustaining herds during drought. Lower human population densities obviated the need to seek new ways to exploit old areas and to extend use to new regions. Less investment in rigid engineering works and infrastructure and fewer privatized tenure practices promoted greater flexibility within and between communities and livelihood systems in their use of resources. A slower pace of change meant that there was more time in which to respond to feedback that warned resource users that land degradation problems were emerging. Also in past times, when systems were severely stressed, food shortages or famines often resulted in a locally declining human population. While not a particularly benevolent adjustment, from the land degradation perspective it did contribute to lowering human-induced stresses on the natural environment. Hence checks on population growth helped to minimize land degradation. Today food aid often sustains inhabitants in areas where their existence is threatened due to food shortages that result from meteorological variability or other negative factors. One result is that traditional feedback mechanisms that worked to curtail pressures on habitat are short-circuited and critical thresholds are reached that initiate land degradation. The social engineering changes needed to offset stresses of modernity lag far behind the rapid changes in technology.

Despite considerable contemporary sophistication in understanding the interconnection of cultural ecological systems in the abstract, we are frequently surprised by where, when, and how rapidly unexpected environmental changes will appear. Our ancestors faced fewer problems because both the pace and the intensity of change were less rapid and less powerful. Modern high-intensity technologies are the primary cause of rapid and widespread land degradation. The power and scale of these technologies makes them simultaneously appealing and difficult to cope with. By their power to modify the environment for human benefit, modern technologies promise great improvements in the human condition. This lure encourages people, planners, and politicians to move rapidly to adopt changes the long-term implications of which are only partially understood. As humankind ratchets up the intensity with which land resources are exploited, the risks are increased. A nonbenevolent nature, variable and pulsating in its basic structure, can spring surprises that compromise human resource use systems. These unexpected events in both time and space have the potential to reveal hidden flaws in human use of the Earth and to stimulate abrupt land degradation and habitat transformation.

Modern technology is not the villain in a bad morality play; rather, it is the often-careless way in which human use of modern technology takes place that causes land degradation problems. These difficulties in land degradation could be avoided at all scales of resource-use intensity and the rate of unwanted negative change reduced to manageable proportions if a set of basic generic principles were followed.

Commandments for the Minimization of Land Degradation and for Sustainable Development

Eight principles extracted from our review of the linked processes of creative destruction and destructive creation govern the wise use of resources and diminish the prospects for land degradation.

The first principle is the commitment to *plan holistically*. Many land degradation problems arise because planning for environmental management and economic development is vested in an array of different agencies and institutions. Broad overview authority to reduce competition between different user groups seldom exists, even in centrally planned economies. As a consequence, unintentional sacrifice zones are created, often in distant places far removed from the sources of the original environmental impact. Thus the initial response to local air pollution was the erection of taller smokestacks that inserted the pollutants higher into the atmosphere. Here they stood a better chance of being transported away from the production facility. The consequential acid rain, falling upon distant areas, was not part of the calculation of the plant managers who built the smokestack. These managers did not intend to create acid rain and cause land degradation. In many instances the original decision makers had neither concept nor concern about how acid rain was produced or its effect on distant lands and waters. The fact that the impacted area was far removed from the original source of the contaminants simply made it easier to ignore potential bad consequences. The decision to build the taller smokestacks solved the immediate local problem of negative pollution impacts on the local environment. Furthermore, there was also the belief that atmospheric and aquatic systems had an unlimited ability to absorb refuse without negative consequences. In combination, these factors made it easy to ignore possible consequences until a serious problem existed.

The land degradation problems made possible by the absence of holistic planning are worsened by the competition that often takes place between different sectors of the economy. The approach to economic development that characterizes much of the globe is a segmental one. Each component of the economy has its set of goals and plans, and undertakes activities that may be in each segment's best interest. This pursuit of separate sectorial gain occurs with only the most limited reference to the needs and activities of other segments of the economy or other areas. For example, industrial enterprises in most countries have a higher priority in national development efforts than do other economic activities. As a result, industrial firms are able to follow waste disposal practices that do serious harm to the environment. The leaching of hazardous chemicals into surface and groundwater can not only reduce drinking water supplies but also—before being detected—can result in serious illness to the consumers. The export of wastes into the general environment in pursuit of the chimera of economic development is responsible for much of the environmental damage sustained in industrial and industrializing countries, regardless of their political and economic systems. When authoritarian political systems limit dissent, whatever planning agencies may exist find they are powerless to control and limit land degradation. The lack of serious environmental accounting in most industrial projects only exacerbates this trend. These problems of segmental development also are found within individual sectors. For instance, animal husbandry, dry farming of cereals, and irrigation farming are all agricultural activities. In most countries, each activity is administered through a separate component of the Ministry of Agriculture. However, in countries with limited capital resources, decisions that favor irrigation—often made because this form of agriculture produces the high yields that seem to merit intensive capital investment, or because external funding agencies generally favor these projects—starve other agricultural sectors of the resources they need in order to grow. In some cases, the irrigation projects actually have direct negative effects on other agricultural sectors, such as by the flooding of their lands with reservoirs or by the diversion of water from potential downstream users.

Primarily newcomers, rather than the local inhabitants, become involved in irrigation farming schemes. When excessive concentration of activity in the irrigation sector occurs, agriculturalists who practiced dry farming are squeezed out of critical space. To survive, these farmers move into less favorable, often drier zones, which were formerly occupied by nomads. This forces herders to place more pressure on the rangeland that remains in their control, usually by staying for longer periods than was traditional in seasonal pastures. The absence of development inputs into these less favored sectors is paralleled by the emergence of land degradation as each community places more pressure on its resource base without the benefit of the capital resources needed to intensify resource use productively. Failure to meet development goals and to avoid land degradation is attributed to lapses internal to the less favored sectors rather than to an inadequate, segmental planning process that is not structured in ways that view economy and environment in a holistic fashion.

The second sustainable development principle is the need to *avoid arbitrary boundaries*, which commonly are the product of a flawed, segmental planning process. Use of ecoregions as the conceptual scale of analysis is one device proposed to attain a more integrated overview of how the human use of resources impacts habitat quality (Bailey 2002). Erection of rigid analytic and pragmatic limits to development projects and planning activities establishes absolute domains that are unrealistic. In the traditional world, different livelihood activities often overlapped in space, although they seldom overlapped in time. Thus farmers cultivated fields watered by a river's annual flood while herders were engaged in exploiting distant seasonal pastures. When the crops were harvested, and the farmer had no major productive use for the land, herders returned to graze on the postharvest stubble. This nonexclusive, overlapping use of resources worked to everyone's benefit. Fixing a rigid boundary between spaces devoted to farming and those used for animal husbandry reduced the resilience of both livelihood systems. In the long run, the collapse of overlapping pastoral and agricultural ecologies may have worked to undermine the stability of sedentary farming systems, because it promoted processes that encourage overgrazing and soil erosion in pastoral areas. Increased frequency of dust storms, silted reservoirs, more extreme river regimes, and clogged canals were often the consequence—a set of feedback loops that have serious consequences for sedentary agriculture. Usually, it is essential to maintain greater flexibility in conceiving system structure and to preserve overlapping ecologies that encourage multipurpose use of the same space in different seasons. Yet avoiding the elimination of one group's critical zone in an effort to enhance growth and development in another sector of the society and economy is often the essential missing element in environmental planning and management.

The presence of rigid boundaries makes it difficult to *calculate all costs* because off-site costs are generally overlooked. Yet possessing the ability to assess, at least qualitatively, all of the costs associated with environmental change is essential to avoiding land degradation. Confining planning and environmental assessment within strictly fixed boundaries makes it relatively easy to enumerate the benefits and costs that fall within the boundary zone. However, there are always important cascade effects that occur outside the spatial zone under consideration. In theory, establishing a boundary that defines the primary object of analysis should serve not only as a device for calculating gains and losses within the bounded region but also should make it easier to note the exchanges that occur across the boundary. In practice, the erection of a boundary quickly evolves into a conceptual barrier that relegates to inconsequential status events that are taking place beyond the boundary. It then becomes easy to ignore land-degrading impacts that affect distant areas. Certainly this was the case in the Narmada River development projects, which deliberately excluded from consideration uncomfortable off-project environmental calculations that might have called the project design into question. This is a serious violation of the third principle, which insists on the importance of including all costs generated by a given set of actions as well as the benefits. Benefits close to home are easy to note; costs that are distant are easy to overlook.

It is the failure to include these costs in environmental and economic assessments that creates distant, unintended sacrifice zones. Because these sacrifice zones are not recognized, their growth can rapidly spiral out of control and produce severe land degradation before steps can be taken to arrest or reverse the process. Uncalculated land-degradation costs also have a serious impact on societies that occupy degrading, sacrificed areas. The exported costs imposed upon them create a category of human losers impoverished by changes that benefit other areas and peoples.

It is important to note that these sacrifice zones are fundamentally different from the primary sacrifice zones that are integral to creative destruction. Zones within an area of transition to more sustainable status, which are consciously sacrificed for the benefit of part of the habitat, are deliberately degraded and the losses incurred are balanced off against the gains that result. Thus Nabataean hillsides, degraded of their soil and water resources for the benefit of valley-bottom fields, or wetland habitats destroyed to create the Chinese pond-dike system, were transformations deliberately undertaken by the people involved. Because the transformation proceeded incrementally, sufficient time was allowed to make corrections if defects emerged in the evolving system. As a consequence, the losses sustained in the zones that were sacrificed were less than the gains concentrated in the primary production zones. The lesson to be learned from this experience is that it is vitally important not to export costs blindly to other parts of the ecumene. Only the careful calculation of all costs sustained within and without the boundaries of favored zones through a holistic planning process can begin to deal satisfactorily with land degradation.

Given a holistic planning process that avoids establishing arbitrary boundaries and calculates land-degradation costs for all areas, minimization of land degradation is made more likely by practices that *retain buffers*. Buffers are important recovery mechanisms in most low-intensity traditional land-use systems and in nature. These devices permit lag time to exist as an integral part of these systems and of nature's structure. Lag time is vital because it insures that a stressed land area has sufficient time to recover before it is subjected to renewed stress. Fallow cycles are widespread and important buffers utilized in many low-intensity land-use systems. They allow sufficient time between episodes of human utilization for the recovery of soil fertility, structure, and vegetation cover. In ancient Mesopotamia, irrigation was alternated with fallow on an annual basis. This rhythm of use followed by a fallow buffer combated salinization by preventing a permanent rise in the groundwater level. The result was a sustainable system. For almost every low-intensity agricultural system, there is a minimum fallow buffer below which a system cannot fall without land degradation occurring or a major shift to a new mode of production taking place. These fallow periods were very long in the case of slash-and-burn agricultural systems in tropical forested areas, but were greatly reduced in more intensive cultivation systems. Nonetheless, in successful agricultural systems this reduction in slack time was always accompanied by sufficient inputs to enable the farming system to maintain basic fertility. Inclusion of fallow periods and fallow land within the agricultural system also left open niches that other groups, mainly mobile pastoralists, could exploit. This reinforced the principle of nonrigid, flexible boundaries that permit overlapping ecologies to survive and flourish side by side.

Increasingly in the modern world, intensification of production systems removes buffers in the interest of increased production. This removal of nature's shock absorbers makes it difficult for livelihood systems to cope with unexpected random events that are of low frequency but high intensity. Deprived of its capacity to deal with the unexpected by the removal of buffers, sudden shifts in the system state of human habitats can occur. While recognizing the difficulty of retaining buffers in their original form, the need to structure buffer substitutes into the human-modified landscape is undeniable. Without the provision of adequate buffers, it is difficult to employ successfully the remaining four principles for the minimization of land degradation.

The most important of these principles is the need to *protect critical zones*. In every livelihood system there is a critical zone without which livelihood practices cannot easily survive. Defense of these critical resources is essential if the livelihood system is to remain viable. In some instances critical resources for one component of the ecumene are sacrificed in order to concentrate benefits in another sector. An example of this process is the decision to build the High Dam at Aswan, which trapped behind the dam freshwater and sediments formerly carried in the Nile's floodwaters. One ramification of this decision was the alteration of both the salinity and the nutrient balance of the eastern Mediterranean. This change set into motion a set of serious negative consequences for the aquatic food chains upon which the Mediterranean Sea's fishermen depended. Construction of the dam for the benefit of most of the Egyptian people was incompatible with the protection of a critical zone for the offshore fishery. In the long run, the possibility exists that waste nutrients from expanded agricultural activities made possible by the Aswan Dam will replace natural inputs previously carried to the sea by the Nile's flood. Should this be the case, the corrective balance will be a serendipitous, unplanned accident whose benefits may be corrupted by the toxins carried away from contemporary agricultural activities along with the nutrients.

When changes deemed essential do take place, it is imperative that adequate substitutes for the lost critical resources are provided. For pastoralists, for instance, retention of adequate natural grazing near dry season water sources is a particularly critical resource zone. Without access to these grazing resources, herders are unable to maintain their herds. When changes occur that deny access to such resources, or when migratory herders are confined to a portion of their normal range for the entire year, both land degradation through overgrazing and, episodically during drought, catastrophic herd losses and famine are outcomes that one can anticipate. These problems are avoidable in large measure by integrating agricultural and pastoral activities more closely. Of course, to do this could place some constraints on agricultural expansion, a policy that most governments and development agencies have not been willing to consider. When a more holistic perspective pervades planning, it is possible to make available post-harvest residues and supplemental fodder to compensate for lost critical dry season pasture zones. This can only occur through an application of the principle of critical zone protection or of its corollary, critical zone resource substitution.

Equally important is the need to *capture wastes* from one region, livelihood, or process and use these wastes as important inputs into other activities. The Chinese integrated pond-dike system is based on this principle of recycling organic wastes from production systems located on the land and using them to promote fertilization in the ponds. Carefully maintained but artificially high fertility levels in the ponds sustain the aquatic food chains upon which fish culture depends. Aquatic plants and pond mud, in turn, are used as a fertilizer source for the terrestrial agricultural system. This very tight linking of system components through the use of wastes from one part of the system as a basic input to other parts is rare. However, it establishes a model by which to avoid one of the worst effects of land degradation: the export of undesirable materials into other sectors. Nabataean run-on farming employed this principle with great success by trapping behind valley check dams soil and water transported from adjacent hillsides. The waste output from one part of the environment became the foundation input for productivity in another part. As was the case with the Nabataeans, deliberate encouragement of this waste input led to the creation of sacrifice zones that were essential to long-term system success. When wastes cannot be captured in this way, there is a very high potential for land degradation. Thus soil eroded from upslope agricultural zones can produce excessive deposition along valley bottoms that kills crops planted in these low-lying areas. Excessive lost soil material transported in streams can result in sediments that silt harbors, increase turbidity in estuaries, and suffocate coral reefs. Harbor facilities are degraded, fishing productivity is undermined, and wave erosion can accelerate as a consequence of this form of excessive waste. But stopping such sedimentation completely can have equally serious consequences, since beaches that depend on the addition of eroded material begin to waste away and the aquatic systems that depend on nutrient inputs that are transported along with the sediments begin to starve. Inattention to holistic planning and a failure to capture wastes productively produces serious environmental and economic damage. In essence, the principle of waste capture is an attempt to transform a negative—the production of wastes that pollute and degrade—into a positive—the generation of inputs for agrarian and other systems—that has a high probability of achieving long-term sustainability.

In managing resources with the minimization of land degradation, it is extremely important to balance long-term needs against short-term benefits. This is reflected in the popular expression, *look before you leap*! It is more theoretically formulated in the *precautionary principle*. The struggle for survival leads many people in high-risk habitats to engage in practices that damage and degrade the environment. Expansion of agriculture into marginal habitats, for instance, may promise short-term gains in crop production, but these actions carry with them serious implications for longer-term soil erosion, fertility decline, and, all too often, a collapse of an area's agricultural livelihood. The abrupt introduction of new technology into places for which that technology seems appropriate, but where experience in using the technology is limited, poses great risk of degradation. For this reason, the extension of mechanized dry farm cereal cultivation into pastoral zones carries with it a high risk of wind-generated soil erosion as well as a loss of primary rangeland productivity. Similarly, the exuberant adoption of large-scale groundwater extraction technology, such as central-pivot irrigation, carries with it grave risks of mining groundwater resources when extraction rates exceed recharge. The flexibility and potential land resources use of future generations are greatly constrained when short-term benefits result in a deteriorating resource base.

Application of the precautionary principle makes it possible to assess the potential impacts of proposed changes more accurately. It gives a higher degree of assurance that long-term costs will be given equal consideration with the more easily calculated short-term benefits of a given development. Proceeding cautiously also increases the likelihood that feedback messages can be observed in time to take corrective action. This implies that it is essential to build into every habitat appropriate monitoring systems that report qualitatively the rate and trend of habitat modification. In this way there is reason to believe that the feedbacks that stabilize and sustain a cultural ecological system can be supported, and that those that drive a system into rapid, destructive decline can be avoided.

The eighth and final principle is the commandment to *build on the genius loci and traditional institutions* that affect environment positively. The genius loci principle assumes that each place is special and has its own unique array of opportunities and constraints. Being attentive to the spirit of place is not a mystical appeal to a form of modern geomancy; rather, it is recognition that the diversity of cultures and environments that characterize humankind's global expression require particularistic and local attention (Lippard 1997). Important lessons can be learned outside local areas and applied, either in principle or through direct transfer, to the solution of local problems and to the enhancement of local productivity. However, for success to occur, for land degradation to be avoided, the frame of reference always must consider the local habitat and culture. Without consulting intimately and in detail with the peculiarities and potentials of particular places, serious insults to local environmental stability are the norm rather than the exception.

It follows that the starting point for efforts to exploit the local resource base more intensively must be the traditional institutions that have developed in that space over extended periods (Ostrom et al. 2001). If they offer nothing else, these local cultural ecological systems have a depth of experiential knowledge about place that transcends the limited instrumental record of modern technoscience. The principles that these resource-use systems developed form a sound basis upon which to erect new ways of using the environment. Granted that changing conditions in environment and society make it impossible to operate old systems in the traditional way, at issue is whether development must begin with a clean slate that carries no messages from the past, or whether it can build on older ways of knowing and doing to blend the traditional with the modern. The approach can work as well in other parts of the world as it has in blending farms, forests, and suburban landscapes of postindustrial New England (Donahue 1999). We believe that the record of both past experience and modern development clearly shows that land degradation can best be controlled by building on the traditional, being sensitive to the spirit of place, and garnishing the best features of the indigenous with appropriate inputs of the modern. In this way one may not attain the quantum leaps in production so fervently desired by all, but the results, while slower, likely will build incrementally toward much more sustainable ways of using our environment.

Summary

Population growth, technology transfers, and development of new technologies will continue to exert pressure on the Earth's resource base if solutions to immediate problems are not conceptualized within a long-term framework. Furthermore, if environmental accounting and off-site costs do not become integral components of the evaluation process prior to institution change, unintentional land degradation will likely continue. The eight principles just summarized need to be considered prior to making decisions affecting land and water resources. Corrective strategies are always more costly in environmental, human, and economic terms than prevention of degradation. Paying lip service to environmental concerns is becoming common. The challenge is now to convert this increasing awareness into an integral component of the planning process.

There are signs that the principles presented in this book are beginning to filter down into development strategies. There is an increasing responsiveness to incorporating environmental considerations in development schemes. Given awareness of the negative impacts of land degradation in terms of production losses, economic costs, and overall decline in quality of life, this sensitivity to land-resource use will pay off if it is actively pursued in the decision-making processes. For example, since 1987 the World Bank has begun to articulate the need to incorporate the links between economics and environment into development strategies if sustainability, the antithesis of land degradation, is to be attained. However, without further development of environmental accounting in the evaluation of land-use changes, economic and political considerations will likely remain dominant in most plans. Requirements that look good on paper are only effective if they are implemented in practice! Finally, since a significant proportion of land degradation results from a multitude of individual decisions, and significant gaps remain in our ability to integrate the demands of society, technology, and environment, land degradation will likely to continue to haunt humankind in the immediate future. This is especially true as governments look for quick fixes to immediate needs. Unfortunately, short-term fixes remain the overwhelming strategy of most governments and other human institutions. The crucial concept that must be accepted is that human systems must be constructed within the constraints of the natural systems that sustain them if the environment is not to degrade.

References

Abbey, E. 1968. *Desert Solitaire: A Season in the Wilderness*. New York: Ballantine.

Acheson, J. M. 1988. *The Lobster Gangs of Maine*. Hanover, N.H.: University Press of New England.

Ackefors, H., and M. Enell. 1990. Discharge of Nutrients from Swedish Fish Farming to Adjacent Sea Areas. *Ambio* 19 (1): 28–35.

Adams, R. McC. 1965. *Land behind Baghdad: A History of Settlement on the Diyala Plains*. Chicago: University of Chicago Press.

———. 1966. *The Evolution of Urban Society: Early Mesopotamia and Prehistoric Mexico*. Chicago: Aldine.

———. 1978. Strategies of Maximization, Stability, and Resilience in Mesopotamian Society, Settlement and Agriculture. *Proceedings of the American Philosophical Society* 122 (5): 329–35.

———. 1981. *Heartland of Cities: Surveys of Ancient Settlement and Land Use on the Central Floodplain of the Euphrates*. Chicago: University of Chicago Press.

Adams, R. McC., and H. J. Nissen. 1972. *The Uruk Countryside: The Natural Setting of Urban Societies*. Chicago: University of Chicago Press.

Adams, W. M. 1989. Dam Construction and the Degradation of Floodplain Forest on the Turkwel River, Kenya. *Land Degradation & Rehabilitation* 1 (3): 189–98.

———. 1992. *Wasting the Rain: Rivers, People and Planning in Africa*. Minneapolis: University of Minnesota Press.

———. 1995. Wetlands and Floodplain Development in Dryland Africa. In *People and Environment in Africa*, edited by T. Binns, 13–21. Chichester, England: John Wiley.

Adler, T. 1994. Squelching Gypsy Moths: What's Hot and What's Not in the Arsenal against Leaf Eaters. *Science News* 145 (12): 184–85.

Agrawal, A. 1999. *Greener Pastures: Politics, Markets, and Community among a Migrant Pastoral People*. Durham: Duke University Press.

Ahnert, F. 1970. Functional Relationships between Denudation, Relief and Uplift in Large Mid-latitude Drainage Basins. *American Journal of Science* 268 (3): 243–63.

Alderman, D. 2004. Channing Cope and the Making of a Miracle Vine. *Geographical Review* 94 (2): 157–77.

Allan, A. J. 2001. *The Middle East Water Question: Hydropolitics and the Global Economy*. London: I. B. Taurus.

Anderson, J. K. 1985. *Hunting in the Ancient World*. Berkeley: University of California Press.

Antonsson-Ogle, B. 1990. Who Uses Forest Foods? *Forests, Trees and People Newsletter,* [Swedish University of Agricultural Sciences, Uppsala] No. 8, 21.

Aral Sea Homepage. 2002. German Aerospace Center (DLR) http://www.dfd.dlr.de/app/land/aralsee/

Archer, E. 2001. Climate, Political Economy and Vegetation Change in the Semi-arid Karoo, South Africa. Unpublished Ph.D. dissertation. Worcester, Mass.: Clark University, Graduate School of Geography.

Arkell, T. 1991. The Decline of Pastoral Nomadism in the Western Sahara. *Geography* 76 (2): 162–66.

Arnold, J. E. M. 1992. *Community Forestry—Ten Years in Review*. Community Forestry Note 7. Rome: FAO.

Artz, N. E., B. I. Norton, and T. J. O'Rourke. 1986. Management of Common Grazing Lands: Tamahdite, Morocco. In *Proceedings of the Conference on Common Property Management*, edited by Panel on Common Property Management, Board on Science and Technology for International Development, Office of International Affairs, National Research Council, 259–80. Washington, D.C.: National Research Council, National Academy Press.

Babbitt, B. 1991. Age-old Challenge: Water and the West. *National Geographic* 179 (6): 2–34.

Bailey, R. G. 2002. *Ecoregion-Based Design for Sustainability*. New York: Springer-Verlag.

Baines, J., and J. Malek. 1980. *Atlas of Ancient Egypt*. New York: Facts on File Publications.

Balearics Tourism Authority. 1992. Challenges and Choices for the 90's. Unpublished paper presented at English Tourist Board, Tourism and the Environment Conference, 16–17 November 1992. London.

Balls, H., B. Moss, and K. Irvine. 1989. The Loss of Submerged Plants with Eutrophication. I: Experimental Design, Water Chemistry, Aquatic Plant and Phytoplankton Biomass in Experiments Carried out in Ponds in the Norfolk Broadland. *Freshwater Biology* 22 (1): 71–87.

Banks, T. 1997. Pastoral Land Tenure Reform and Resource Management in Northern Xinjiang: A New Institutional Economics Perspective. *Nomadic Peoples* NS 1 (2): 55–76.

Bardet, J. 1987. Civiele techniek. In *De Physique Existentie dezes Lands: Jan Blanken*, edited by F. Koens, N. B. Spruyt, and M-C. Vink. 113–40. Amsterdam: Rijksmuseum and Uitgeverij AMA boeken.

Barlow, M. 1999. *Blue Gold—The Global Water Crisis and the Commodification of the World's Water Supply*. Sausalito, Calif.: International Forum on Globalization.

Barrow, C. J. 1991. *Land Degradation*. Cambridge: Cambridge University Press.

Barrow, E. G. C. 1990. Usufruct Rights to Trees: The Role of *ekwar* in Dryland Central Turkana, Kenya. *Human Ecology* 18 (2): 163–76.

Barth, F. 1961. *Nomads of South Persia: The Basseri Tribe of the Khamseh Confederacy*. Prospect Heights, Ill.: Waveland Press (reissued 1986).

Bascom, J. B. 1990. Border Pastoralism in Eastern Sudan. *Geographical Review* 80 (4): 416–30.

Baskervill, B. 2002. Kudzu is the Enemy at U.S. Army Posts. *Telegram and Gazette*, Friday, August 23, §A, 3.

Batterbury, S. P. J. 2001. "Landscapes of Diversity: A Local Political Ecology of Livelihood Diversification in Southwestern Niger. *Ecumene* 8 (4): 437–64.

Batterbury, S. P. J. et al. 2002. Responding to Desertification at the National Scale: Detection, Explanation, and Responses. In *Global Desertification: Do Humans Cause Deserts?*, edited by J. F. Reynolds and D. M. Stafford Smith, 357–85. Berlin: Dahlem University Press.

Baum, D. 1997. Butte, America. *American Heritage* 48 (2): 57–66.

Beaumont, P. 1996. Agricultural and Environmental Changes in the Upper Euphrates Catchment of Turkey and Syria and Their Political and Economic Implications. *Applied Geography* 16 (2): 137–57.

Beck, L. 1981. Government Policy and Pastoral Land Use in Southwest Iran. *Journal of Arid Environments* 4 (3): 253–67.

Beckerman, S. 1987. Swidden in Amazonia and the Amazon Rim. In *Comparative Farming Systems*, edited by B. L. Turner II and S. B. Brush, 55–94. New York: Guilford.

Bedrani, S. 1983. Going Slow with Pastoral Cooperatives. *Ceres* 16 (4): 16–21.

———. 1991. Legislation for Livestock on Public Lands in Algeria. *Nature & Resources* 27 (4): 24–30.

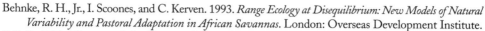

Behnke, R. H., Jr., I. Scoones, and C. Kerven. 1993. *Range Ecology at Disequilibrium: New Models of Natural Variability and Pastoral Adaptation in African Savannas.* London: Overseas Development Institute.

Bell, B. 1975. Climate and the History of Egypt: The Middle Kingdom. *American Journal of Archaeology* 7 (3): 223–69.

Bencherifa, A., and D. L. Johnson. 1991. Changing Resource Management Systems and Their Environmental Impacts in the Middle Atlas Mountains of Morocco. *Mountain Research and Development* 11 (3): 183–94.

Bennett, H. H. 1929. Facing the Erosion Problem, *Science* 69: 48.

———. 1935. Facing the Erosion Problem. *Science* 81 (2101): 321–26.

———. 1943. Adjustment of Agriculture to Its Environment. *Annals of the Association of American Geographers* 33 (4): 163–98.

Bennett, J. W. 1976. *The Ecological Transition: Cultural Anthropology and Human Adaptation.* New York: Pergamon Press.

Bergman, C. 2002. *Red Delta: Fighting for Life at the End of the Colorado River.* Golden, Colo.: Fulcrum Publishing for the Defenders of Wildlife.

Bergmann, M. 1999. *Social Impacts of Snowy Water Flow Options: An Assessment.* Discussion Paper No. 60. Canberra: The Australian National University-Integrated Catchment Assessment and Management Centre.

Berkes, F. 1986. Marine Inshore Fishery Management in Turkey. In *Proceedings of the Conference on Common Property Management,* edited by Panel on Common Property Management, Board on Science and Technology for International Development, Office of International Affairs, National Research Council, 63–68. Washington, D.C.: National Research Council, National Academy Press.

———, ed. 1989. *Common Property Resources: Ecology and Community-based Sustainable Development.* London: Belhaven Press.

Bernus, E. 1980. Desertification in the Aghazer and Azawak Region: Case Study Presented by the Government of Niger. In *Case Studies on Desertification*, edited by J. Mabbutt and C. Floret, 115–46. Paris: UNESCO.

———. 1990. Dates, Dromedaries, and Drought: Diversification in Tuareg Pastoral Systems. In *The World of Pastoralism*, edited by J. G. Galaty and D. L. Johnson, 149–76. New York: Guilford.

Berry, L. 1983. *East African Country Profile—Sudan.* Worcester, Mass.: Program for International Development; Clark University.

El-Bihbety, H., and H. Lithwick. 1998. Cost-Benefit Analysis of Water Management Mega Projects in India and China. In *The Arid Frontier: Interactive Management of Environment and Development*, edited by H. J. Bruins and H. Lithwick, 229–47. Dordrecht: Kluwer Academic.

Biswas, A. K. 1970. *History of Hydrology.* Amsterdam: North Holland.

———. 1995. Environmental Sustainability of Egyptian Agriculture: Problems and Perspectives. *Ambio* 24 (1): 16–20.

Blaikie, P., and H. Brookfield. 1987. *Land Degradation and Society.* London: Methuen.

Blakeslee, S. 2002. Restoring an Ecosystem Torn Asunder by a Dam. *New York Times,* June 11, §D, 1, 5.

Bloom, A. L. 1997. *Geomorphology: A Systematic Analysis of Late Cenozoic Landforms.* 4th ed. Upper Saddle River, N.J.: Prentice Hall.

Bock, C. E., and J. H. Bock. 2000. *The View from Bald Hill: Thirty Years in an Arizona Grassland.* Berkeley: University of California Press.

Bock, J. H., and C. E. Bock. 2002. Exotic Species in Grasslands. In *Invasive Exotic Species in the Sonoran Region*, edited by Barbara Tellman, 147–64. Tucson: The University of Arizona Press and the Arizona-Sonora Desert Museum.

Borchert, J. R. 1989. *America's Northern Heartland.* Minneapolis: University of Minnesota Press.

Borgstrom, G. 1973. *World Food Resources.* New York: Intext Educational Publishers.

Boston Globe. 1993. Failing Soil Threatens Food Supply, UN Warns. *The Boston Globe,* July 14, 9.

Bourn, D. 1978. Cattle, Rainfall and Tsetse in Africa. *Journal of Arid Environments* 1 (1): 49–61.

Bowden, M. J. 1975. Desert Wheat Belt, Plains Corn Belt: Environmental Cognition and Behavior of Settlers in the Plains Margin, 1850–99. In *Images of the Plains: the Role of Human Nature in Settlement*, edited by B. W. Blouet and M. P. Lawson, 189–201. Lincoln: University of Nebraska Press.

Bowerstock, G. W. 1983. *Roman Arabia*. Cambridge, Mass.: Harvard University Press.

Bowonder, B., and K. V. Ramana. 1986. *Environmental Degradation and Economic Development: A Case Study of a Marginally Productive Area*. Centre for Energy, Environment & Technology, Administrative Staff College of India, Bella Vista, Hyderabad.

Braeman, J. 1986. The Dust Bowl: An Introduction. *Great Plains Quarterly* 6 (2): 67–68.

Brandon, K., K. H. Redford, and S. E. Sanderson, eds. 1998. *Parks in Peril: People, Politics, and Protected Areas*. Washington, D.C. and Covelo, Calif.: The Nature Conservancy and Island Press.

Breckle, S-W., W. Wucherer, O. Agachanjanz, and B. Geldyev. 2001. The Aral Sea Crisis Region. In *Sustainable Land Use in Deserts*, edited by S-W. Breckle, M. Veste, and W. Wucherer, 27–37. Berlin: Springer-Verlag.

Breman, H., and C. T. de Wit. 1983. Rangeland Productivity and Exploitation in the Sahel. *Science* 221 (4618), 1341–47.

Brewsher Consulting P/L. 1999. *A Review of Water Efficiency Savings Available in NSW*. Epping, NSW, Australia.

Bromley, P. 1997. *Nature Conservation in Europe: Policy and Practice*. London: E & FN Spon.

Brookfield, H. 1999. Environmental Damage: Distinguishing Human from Geophysical Causes. *Environmental Hazards* 1: 3–11.

Brown, L. D., and J. E. Oliver. 1976. Vertical Crustal Movements from Leveling Data and Their Relation to Geologic Structure in the Eastern United States. *Review of Geophysics and Space Physics* 14: 13–35.

Brown, L. R. 1992. Launching the Environmental Revolution. In *State of the World 1992*, edited by L. R. Brown, 174–90. New York: W. W. Norton.

———. 1998. The Future of Growth. In *State of the World 1998*, edited by L. R. Brown, C. Flavin, and H. French, 3–20. New York: W. W. Norton.

———. 2001. *Eco-Economy: Building an Economy for the Earth*. New York: W. W. Norton.

Brown, L. R., and E. C. Wolf. 1984. *Soil Erosion: Quiet Crisis in the World Economy*. Worldwatch Paper 60, Washington, D.C.: Worldwatch Institute.

Bunyard, P. 1986. The Death of the Trees. *The Ecologist* 16 (1): 4–13.

Búrquez-Montijo, A., M. E. Miller, and A. Martínez-Yrízar. 2002. Mexican Grasslands, Thornscrub, and the Transformation of the Sonoran Desert by Invasive Exotic Buffelgrass (*Pennisetum ciliare*). In *Invasive Exotic Species in the Sonoran Region*, edited by Barbara Tellman, 126–46. Tucson: University of Arizona Press and Arizona-Sonora Desert Museum.

Butzer, K. W. 1971. *Environment and Archaeology: An Ecological Approach to Prehistory*. 2nd ed. Chicago: Aldine-Atherton.

———. 1976. *Early Hydraulic Civilization in Egypt: A Study in Cultural Ecology*. Chicago: University of Chicago Press.

Callenbach, E. 1996. *Bring Back the Buffalo! A Sustainable Future for America's Great Plains*. Berkeley: University of California Press.

Campbell, D. J. 1981. Land-use Competition at the Margins of the Rangelands: An Issue in Development Strategies for Semi-arid Areas. In *Planning African Development*, edited by G. Norcliffe and T. Pinfold, 39–61. Boulder, Colo.: Westview Press.

———. 1986. The Prospect for Desertification in Kajiado District, Kenya. *Geographical Journal* 152 (1): 44–55.

Carlson, L. W. 2001. *Cattle: An Informal Social History*. Chicago: Ivan R. Dee.

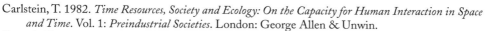

Carlstein, T. 1982. *Time Resources, Society and Ecology: On the Capacity for Human Interaction in Space and Time*. Vol. 1: *Preindustrial Societies*. London: George Allen & Unwin.

Carr, A. F. 1967. So *Excellent a Fishe: A Natural History of Sea Turtles*. Garden City, N.Y.: Natural History Press.

Carr. C. J. 1977. *Pastoralism in Crisis: The Dasanetch and their Ethiopian Lands*. Research Paper No. 180. Chicago: Department of Geography, University of Chicago.

Carter, W. E. 1969. *New Lands and Old Traditions: Kekchi Cultivators in the Guatemalan Lowlands*. Gainesville: University of Florida Press.

Cassells, D. S., and K. MacKinnon. 2002. Working for a Better Future for Forests and Forest-Dependent People. *Environment Matters at the World Bank*. Annual Review 2002, 24–25. Washington, D.C.: The World Bank Group.

Cave, S. 1991. Tourism and the Mediterranean Environment. *Our Planet* 3 (2): 4–7.

CAZRI [Central Arid Zone Research Institute]. 1976. *International Cooperation to Combat Desertification: Luni Development Block—A Case Study on Desertification*. Jodhpur, India: Central Arid Zone Research Institute.

Center for Rural Affairs. 1988. *Beneath the Wheel of Fortune: The Economic and Environmental Impacts of Center Pivot Irrigation Development on Antelope County, Nebraska*. Water Policy and Practices Project, Whitehill, Neb.

Chaudhuri, A. 2003. Three Gorges Dam: Fortune or Folly? *Murj* 8: 31–36.

Child, R. D., H. F. Heady, R. A. Peterson, R. D. Pieper, and C. D. Pouolton. 1987. *Arid and Semiarid Rangelands: Guidelines for Development*. Morrilton, Ark: Winrock International.

Chorley, R. J., S. A. Schumm, and D. E. Sugden. 1984. *Geomorphology*. New York: Methuen.

Clow, D. W., and M. A. Mast. 1999. *Trends in Precipitation and Stream-Water Chemistry in the Northeastern United States, Water Years 1984–1996*. USGS Fact Sheet 117–99. Washington, D.C.: U.S. Department of the Interior, U.S. Geological Survey.

Colacicco, D., T. Osborn, and K. Alt. 1989. Economic Damage from Soil Erosion. *Journal of Soil and Water Conservation* 44 (1): 35–39.

Collier, J. G., and L. M. Davies. 1989. *Chernobyl*. London: Central Electricity Generating Board (UK).

Collins, R. O. 2002. *The Nile*. New Haven: Yale University Press.

Colman, D. 1989. Economic Issues from the Broads Grazing Marshes Conservation Scheme. *Journal of Agricultural Economics* 40 (3): 336–44.

Committee on Selected Biological Problems in the Humid Tropics. 1982. *Ecological Aspects of Development in the Humid Tropics*. Washington, D.C.: National Academy Press.

Committee on Water. 1968. *Water and Choice in the Colorado River Basin*. Washington, D.C.: National Academy of Sciences.

Committee on Western Water Management, Water Science and Technology Board, National Research Council. 1992. *Water Transfers in the West: Efficiency, Equity, and the Environment*. Washington, D.C.: National Academy Press.

Conklin, H. C. 1954. An Ethnological Approach to Shifting Cultivation. *Transactions of the New York Academy of Science*. Series 2, 17: 133–42.

———. 1980. *Ethnographic Atlas of Ifugao: A Study of Environment, Culture, and Society in Northern Luzon*. New Haven: Yale University Press.

Conzen, M. P., and K. J. Carr, eds. 1988. *The Illinois & Michigan Canal National Heritage Corridor: A Guide to Its History and Sources*. DeKalb: Northern Illinois University Press.

Cook, K. A. 1989. The Environmental Era of U.S. Agricultural Policy. *Journal of Soil and Water Conservation* 44 (5): 362–66.

Craven, A. O. 1926. *Soil Exhaustion as a Factor in the Agricultural History of Virginia and Maryland, 1606–1860*. Urbana: University of Illinois Press.

Cronon, W. 1983. *Changes in the Land: Indians, Colonists, and the Ecology of New England*. New York: Hill and Wang.

Dahl, G. 1991. The Beja of Sudan and the Famine of 1984–1986. *Ambio* 20 (5): 189–91.

Darby, H. C. 1940. *The Draining of the Fens*. Cambridge: Cambridge University Press.

Davis, S. H. 1977. *Victims of the Miracle*. Cambridge: Cambridge University Press.

Day, G. M. 1935. The Indian as an Ecological Factor in the Northeastern Forest. *Ecology* 32 (2): 329–46.

Deegan, C. 1995. The Narmada in Myth and History. In *Toward Sustainable Development: Struggling Over India's Narmada River*, edited by William F. Fisher, 47–68. London: M. E. Sharpe.

Deevey, E. S., Jr. 1960. The Human Population. *Scientific American* 203 (3): 194–204.

Demographia. 2001. International Urbanized Area Analysis and Data Product. www.demographia.com

Denevan, W. M. 1981. Swiddens and Cattle versus Forest: The Imminent Demise of the Amazon Rain Forest Reexamined. In *Where Have All the Flowers Gone? Deforestation in the Third World*, edited by V. H. Sutlive, N. Altshuler, and M. D. Zamora, 25–44. Williamsburg, Va.: Department of Anthropology, College of William and Mary.

———. 2001. *Cultivated Landscapes of Native Amazonia and the Andes*. Oxford: Oxford University Press.

Denison, D. C. 2001. Old Theory Helps Explain New Reality. *Boston Globe*, Friday, July 20, §C, 1.

Diamond, J. 1999. *Guns, Germs, and Steel: The Fates of Human Societies*. New York: W. W. Norton.

Disch, Robert. 1970. *The Ecological Conscience: Values for Survival*. Englewood Cliffs, N.J.: Prentice-Hall.

Donahue, B. 1999. *Reclaiming the Commons: Community Farms & Forests in a New England Town*. New Haven: Yale University Press.

Douglas, Ian. 1999. Needs and Opportunities in Evaluating Land Degradation and Erosion in South-East Asia. In *Land Degradation*, edited by A. J. Conacher, 223–35. Dordrecht, Boston, and London: Kluwer Academic Publishers.

Dowd, M. 2002. Rummy Runs Rampant. *New York Times*, Oct 30, §A, 27.

Draz, O. 1974. *Report to the Government of the Syrian Arab Republic on Range Management and Fodder Development*. NDP No. Ta 3292. Rome: FAO.

———. 1977. *Role of Range Management in the Campaign against Desertification: The Syrian Experience as an Applicable Example for the Arabian Peninsula*. Regional Preparatory Meeting for the Mediterranean Area, Algarve, 28 March–1 April 1977, International Cooperation to Combat Desertification (UNCOD), UNCOD/MISO13. Mimeo.

———. 1990. The Hema System in the Arabian Peninsula. In *The Improvement of Tropical and Subtropical Rangelands*, edited by National Research Council, Office of International Affairs, Board on Science and Technology for International Development, 321–31. Washington, D.C.: National Academy Press.

Dregne, H. E. 1976. Desertification: Symptom of a Crisis. In *Desertification: Process, Problems, Perspectives*, edited by P. Paylore and R. A. Haney Jr., 11–24. Tucson: Arid/Semi-arid Natural Resources Program, University of Arizona.

———. 1977a. Generalized Map of the Status of Desertification of Arid Lands. In *Status of Desertification in the Hot Arid Regions*, UN Conference on Desertification, Nairobi.

———. 1977b. Development Strategies for Arid Land Use. In *Arid Zone Development: Potentialities and Problems*, edited by Y. Mundlak and S. F. Singer, 255–61. Cambridge, Mass.: Ballinger.

Dreze, J., M. Samson, and S. Singh, eds. 1997. *The Dam and the Nation: Displacement and Resettlement in the Narmada Valley*. Delhi: Oxford University Press.

van Duin, R. H. A., and G. de Kaste. 1984. *The Pocket Guide to the Zuyder Zee Project*. Lelystad, Netherlands: Ministry of Transport and Public Works/ Rijksdienst voor de IJsselmeerpolders.

Durning, A. B., and H. B. Brough. 1991. *Taking Stock: Farming and the Environment*. Worldwatch Paper 103. Washington, D.C.: Worldwatch Institute.

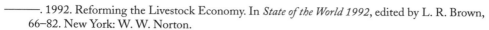

————. 1992. Reforming the Livestock Economy. In *State of the World 1992*, edited by L. R. Brown, 66–82. New York: W. W. Norton.

Eckholm, E. 1975. *The Other Energy Crisis: Fuelwood*. Worldwatch Paper 1. Washington, D.C.: Worldwatch Institute.

————. 2003. A River Diverted, the Sea Rushes In. *New York Times*, Tuesday, April 22, § D, 1.

Eckholm, E., and L. R. Brown. 1977. *Spreading Deserts—The Hand of Man*. Worldwatch Paper 13. Washington, D.C.: Worldwatch Institute.

Ehrenman, G. 2003. Digging Deeper in New York: If the Third Largest Port in the U.S. is too Shallow, then Dredge We Must. *Mechanical Engineering*, November, 51–53.

Elder, J. 1998. *Reading the Mountains of Home*. Cambridge, Mass.: Harvard University Press.

Ellis, W. S. 1990. A Soviet Sea Lies Dying. *National Geographic* 177 (2): 73–93.

Elmgren, R. 1989. Man's Impact on the Ecosystem of the Baltic Sea: Energy Flows Today and at the Turn of the Century. *Ambio* 18 (6): 326–32.

————. 2001. Understanding Human Impact on the Baltic Ecosystem: Changing Views in Recent Decades. *Ambio* 30 (4–5): 222–31.

Elton, C. S. 2000. *The Ecology of Invasions by Animals and Plants*. Foreword by D. Simberloff. Chicago: University of Chicago Press. First published in 1958.

Emel, J., and E. Brooks. 1988. Changes in Form and Function of Property Rights Institutions under Threatened Resource Scarcity. *Annals of the Association of American Geographers* 78 (2): 241–52.

Emel, J., and R. Krueger. 2003. Spoken but not Heard: The Promise of the Precautionary Principle for Natural Resource Development. *Local Environment* 8 (1): 9–25.

Emel, J., R. Roberts, and D. Sauri. 1992. Ideology, Property, and Ground-water Resources: An Exploration of Relations. *Political Geography* 11 (1): 37–54.

Englebert, E. A., and A. F. Scheuring. 1982. *Competition for California Water: Alternative Resolutions*. Berkeley: University of California Press.

————. 1984. *Water Scarcity: Impacts on Western Agriculture*. Berkeley: University of California Press.

English, J., M. Tiffen, and M. Mortimore. 1994. *Land Resource Management in Machakos District, Kenya 1930–1990*. World Bank Environment Paper No. 5. Washington, D.C.: World Bank.

Erickson, C. L. 1994. Methodological Considerations in the Study of Ancient Andean Field Systems. In *The Archaeology of Garden and Field*, edited by N. F. Miller and K. L. Gleason, 111–52. Philadelphia: University of Pennsylvania Press.

————. 2000. The Lake Titicaca Basin: A Precolumbian Built Landscape. In *Imperfect Balance: Landscape Transformations in the Precolumbian Americas*, edited by D. L. Lentz, 311–56. New York: Columbia University Press.

Esque, T. C., and C. R. Schwalbe. 2002. Alien Annual Grasses and Their Relationships to Fire and Biotic Change in Sonoran Desertscrub. In *Invasive Exotic Species in the Sonoran Region*, edited by Barbara Tellman, 165–94. Tucson: University of Arizona Press and Arizona-Sonora Desert Museum.

EST [Environmental Science & Technology]. 2001. "Black Triangle" is Greening Up. *Environmental Science & Technology* 35 (15): 315A; 317A.

Esterbrook, D. J. 1999. *Surface Processes and Landforms*. 2nd ed. Upper Saddle River, N.J.: Prentice Hall.

Estioko-Griffin, A., and P. B. Griffin. 1981. Woman the Hunter: The Agta. In *Woman the Gatherer*, edited by F. Dahlberg, 121–51. New Haven: Yale University Press.

Evans, N. A. 1972. Transmountain Water Diversion for the High Plains. In *The High Plains: Problems of Semiarid Environments*, edited by D. MacPhail, 42–59. Fort Collins: Colorado State University.

Evenari, M. 1977. Ancient Desert Agriculture and Civilizations: Do They Point the Way to the Future? In *Arid Zone Development: Potentialities and Problems*, 83–97. Cambridge, Mass.: Ballinger.

————. 1981. Twenty-five Years of Research on Runoff Desert Agriculture in the Middle East. In *Settling the Desert*, edited by L. Berkofsky, D. Faiman, and J. Gale, 5–27. Sede Boqer: Jacob Blaustein Institute for Desert Research, Ben Gurion University of the Negev.

Evenari, M., L. Shanan, and N. Tadmor. 1971. *The Negev: The Challenge of a Desert*. Cambridge, Mass.: Harvard University Press.

————. 1982. *The Negev: The Challenge of a Desert*. 2nd ed. Cambridge, Mass.: Harvard University Press.

Evenari, M., L. Shanan, N. Tadmor, and Y. Aharoni. 1961. Ancient Agriculture in the Negev. *Science* 133 (3457): 979–96.

Falloux, F., and A. Mukendi, eds. 1987. *Desertification Control and Renewable Resource Management in the Sahelian and Sudanian Zones of West Africa*. World Bank Technical Paper No. 70. Washington, D.C.: World Bank.

Fenneman, N. M. 1938. *Physiography of the Eastern United States*. New York: McGraw-Hill.

Fernea, R. A. 1970. *Sheikh and Effendi: Changing Patterns of Authority among the El Shabana of Southern Iraq*. Cambridge, Mass.: Harvard University Press.

Ferrians, O. J., Jr., R. Kachadoorian, and G. W. Greene. 1969. *Permafrost and Related Engineering Problems in Alaska*. Geological Survey Professional Paper 678. Washington, D.C.: U.S. Government Printing Office.

Fiege, M. 1999. *Irrigated Eden: The Making of an Agricultural Landscape in the American West*. Seattle: University of Washington Press.

Fine, L. 2003. Hawk Patrol Gets NYC Pigeons to Move Along. *Boston Globe*, April 21, § A, 3.

Fisher, I. 1992. Digging Out and Digging In. *New York Times*, June 18, §B, 2.

Fisher, W. B. 1993. Egypt: Physical and Social Geography. In *The Middle East and North Africa 1982–1983*, 360–408. London: Europa.

Fisher, W. F., ed. 1995. *Toward Sustainable Development: Struggling Over India's Narmada River*. Armonk, N.Y.: M. E. Sharpe.

Flawn, P. T. 1970. *Environmental Geology: Conservation, Land-use Planning, and Resource Management*. New York: Harper & Row.

Fleischer, S., S. Hamran, T. Kindt, L. Rydberg, and L. Stibe. 1987. Coastal Eutrophication in Sweden: Reducing Nitrogen in Land Runoff. *Ambio* 16 (5): 246–51.

Ford, M. A., and J. B. Grace. 1998. Effects of Vertebrate Herbivores on Soil Processes, Plant Biomass, Litter Accumulation and Soil Elevation Changes in a Coastal Marsh. *Journal of Ecology* 86 (6): 974–82.

Forests Industries. 1984. Miracle Vine Threatens Southern U.S. Forests. *Forests Industries* 111: 58.

Forrester, F. 1978. Land Subsidence. In *Geology in the Urban Environment*, edited by R. O. Utgard, G. D. McKenzie, and D. Foley, 82–84. Minneapolis: Burgess.

Fortes, M. D. 1988. Mangrove and Seagrass Beds of East Asia: Habitats under Stress. *Ambio* 17 (3): 207–13.

Foster, R. H., and S. Kaplan. 2001. *Creative Destruction: Why Companies That Are Built to Last Underperform the Market—and How to Successfully Transform Them*. New York: Currency.

Frankfort, H. 1956. *The Birth of Civilization in the Near East*. Garden City, N.Y.: Doubleday Anchor.

Frederick, D., W. L. Howenstine, and J. Sochen. 1972. *Destroy to Create: Interaction with the Natural Environment in the Building of America*. Hinsdale, Ill.: Dryden Press.

Frome, M. 1992. *Regreening the National Parks*. Tucson: University of Arizona Press.

Fuchs, V. 1986. *The Health Economy*. Cambridge, Mass.: Harvard University Press.

Fukuda, H. 1976. *Irrigation in the World: Comparative Developments*. Tokyo: University of Tokyo Press.

Gadgil, M., and P. Iyer. 1989. On the Diversification of Common-property Resource Use by Indian Society. In *Common Property Resources: Ecology and Community-based Sustainable Development*, edited by F. Berkes, 240–55. London: Belhaven.

Galaty, J.G. 1999. Grounding Pastoralists: Law, Politics, and Dispossession in East Africa. *Nomadic Peoples* NS 3 (2): 56–73.

Galaty, J. G., and D. L. Johnson, eds. 1990. *The World of Pastoralism*. New York: Guilford.

Gallais, J. 1972. Essai sur la situation actuelle des relations entre pasteurs et paysans dans le Sahel ouest-africain. In *Études de géographie tropicale offertes à Pierre Gourou*, 301–13. Paris: Mouton.

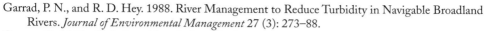

Garrad, P. N., and R. D. Hey. 1988. River Management to Reduce Turbidity in Navigable Broadland Rivers. *Journal of Environmental Management* 27 (3): 273–88.

Geertz, C. 1966. *Agricultural Involution: The Process of Ecological Change in Indonesia*. Berkeley: University of California Press.

Geist, H. J., and E. F. Lambin. 2001. *What Drives Tropical Deforestation?* UCC Report Series No. 4. Louvain-la-Neuve, Belgium: CIACO.

George, M. 1992 *The Land Use, Ecology and Conservation of Broadland*. Chichester: Packard.

Gibson, M. 1974. Violation of Fallow and Engineered Disaster in Mesopotamian Civilization. In *Irrigation's Impact on Society*. Anthropological Papers of the University of Arizona, No. 25, edited by T. E. Downing and M. Gibson, 7–19. Tucson: University of Arizona Press.

Gilles, J. L., A. Hammoudi, and M. Mahdi. 1986. Oukaimedene, Morocco: A High Mountain *Agdal*. In *Proceedings of the Conference on Common Property Resource Management*, edited by the Panel on Common Property Management, Board on Science and Technology for International Development, Office of International Affairs, National Research Council, 281–304. Washington, D.C.: National Academy Press.

Glantz, M. H. 1990. Running on Empty. *The Sciences* 30 (6): 16–20.

Glenn, E., F. Zamora-Arroyo, P. L. Nagler, M. Briggs, W. Shaw, and K. Flessa. 2001. Ecology and Conservation Biology of the Colorado River Delta, Mexico. *Journal of Arid Environments* 49 (1): 5–15.

Goldammer, J. G. 1993. Historical Biogeography of Fire: Tropical and Subtropical. In *Fire in the Environment: The Ecological, Atmospheric, and Climatic Importance of Vegetation Fires*, edited by P. J. Crutzen and J. G. Goldammer, 297–314. Chichester: John Wiley.

Goudie, A. 1981. *The Human Impact: Man's Role in Environmental Change*. Oxford: Blackwell.

Grainger, A. 1990. *The Threatening Desert: Controlling Desertification*. London: Earthscan.

Graves, W., ed. 1990. *Atlas of the World*. 6th ed. Washington, D.C.: National Geographic Society.

Greenberg, M., D. Krueckenberg,, K. Lowrie, H. Mayer, D. Simon, A. Isserman, and D. Sorenson. 1998. Socioeconomic Impacts of U.S. Nuclear Weapons Facilities: A Local-scale Analysis of Savannah River, 1950–1993. *Applied Geography* 18 (2): 101–16.

Grim, J. A. 2001. *Indigenous Traditions and Ecology: The Interbeing of Cosmology and Community*. Cambridge, Mass.: Harvard University Press for the Center for the Study of World Religions, Harvard Divinity School.

Grove, A. T. 1982. Egypt Has Too Much Water. *Geographical Magazine* 54 (8): 437–41.

Hack, J. T. 1960. Interpretation of Erosional Topography in Humid Temperate Regions. *American Journal of Science* 256A (3): 80–97.

Haglund, K. 2003. *Inventing the Charles*. Cambridge, Mass.: MIT Press.

Hallberg, R. 0. 1991. Environmental Implications of Metal Distribution in Baltic Sea Sediments. *Ambio* 20 (7): 309–16.

Hambidge, G. 1938. Soils and Men—A Summary. In *Soils and Men, Yearbook of Agriculture*, 1–44. Washington, D. C.: U.S. Department of Agriculture.

Hansson, S. 1987. Effects of Pulp and Paper Mill Effluents on Coastal Fish Communities in the Gulf of Bothnia, Baltic Sea. *Ambio* 16 (6): 344–48.

Hansson, S., and L. G. Rudstam. 1990. Eutrophication and Baltic Fish Communities. *Ambio* 19 (3): 123–25.

Hardin, G. 1968. The Tragedy of the Commons. *Science* 162 (3859): 1243–48.

Harmon, D., and A D. Putney, eds. 2003. *The Full Value of Parks: From Economics to the Intangible*. Lanham, Md.: Rowman and Littlefield.

Harris, B. L., J. N. Habiger, and Z. L. Carpenter. 1989. The Conservation Title: Concerns and Recommendations from the Great Plains. *Journal of Soil and Water Conservation* 44 (5): 371–75.

Hart, J. F. 1972. The Middle West. *Annals of the Association of American Geographers* 62 (2): 258–82.

———. 2001. Half a Century of Cropland Change. *Geographical Review* 91 (3): 525–43.

———. 2003. Specialty Cropland in California. *Geographical Review* 93 (2): 153–70.

Healy, R. G., and R. E. Sojka. 1985. Agriculture in the South: Conservation's Challenge. *Journal of Soil and Water Conservation* 40 (2): 189–94.

Heath, R. C., B. L. Fosworthy, and P. Cohen. 1966. The Changing Pattern of Ground-water Development on Long Island, New York. USGS Circular 524. Washington, D.C.: U.S. Geological Survey.

Hecht, S., A. B. Anderson, and P. May. 1988. The Subsidy from Nature: Shifting Cultivation, Successional Palm Forests, and Rural Development. *Human Organization* 47 (1): 25–35.

Hecht, S., and A. Cockburn. 1989. *The Fate of the Forest*. London: Verso.

Hecht, S. B. 1981. Deforestation in the Amazon Basin: Magnitude, Dynamics and Soil Resource Effects. In *Where Have All the Flowers Gone? Deforestation in the Third World*, edited by V. H. Sutlive, N. Altschuler, and M. D. Zamora, 61–108. Williamsburg, Va.: Department of Anthropology, College of William and Mary.

Hefny, K. 1982. Land-use and Management Problems in the Nile Delta. *Nature and Resources* 18 (2): 22–27.

Helbaek, H. 1960. Ecological Effects of Irrigation in Ancient Mesopotamia. *Iraq* 22: 186–96.

Hellier. C. 1988. The Mangrove Wastelands. *The Ecologist* 18 (2): 77–79.

Henderson, G. L. 1999. *California and the Fictions of Capitol*. New York: Oxford University Press.

Hewes, L. 1951. The Northern Wet Prairie of the United States: Nature, Sources of Information, and Extent. *Annals of the Association of American Geographers* 41 (4): 307–23.

Hewes, L., and P. E. Frandson. 1952. Occupying the Wet Prairie: The Role of Artificial Drainage in Story County, Iowa. *Annals of the Association of American Geographers* 42 (1): 24–50.

Hierneau, P. H. Y., and M. D. Turner. 2002. The Influence of Farmer and Pastoralist Management Practices on Desertification Processes in the Sahel. In *Global Desertification: Do Humans Cause Deserts?*, edited by J. F. Reynolds and D. M. Stafford Smith, 135–48. Berlin: Dahlem University Press.

Hinrichsen, D., B. Robey, and U. D. Upadhyay. 1998. *Solutions for a Water-Short World*. Population Reports, Series M, No. 14. Baltimore, Md.: Johns Hopkins School of Public Health, Population Information Program.

Hjort af Ornas, A. 1990. Pastoral and Environmental Security in East Africa. *Disasters* 14 (2): 115–22.

Hobbs, J. J. 1989. *Bedouin Life in the Egyptian Wilderness*. Austin: University of Texas Press.

———. 1995. *Mount Sinai*. Austin: University of Texas Press.

Holling, C. S. 1973. Resilience and Stability of Ecological Systems. *Annual Review of Ecology and Systematics* 4: 1–23.

Honey, M. 1999. *Ecotourism and Sustainable Development: Who Owns Paradise?* Washington, D.C.: Island Press.

Horowitz, M. M. 1981. Social Analysis of Desertification and Its Control in Kordofan and Darfur Provinces, Republic of the Sudan, Khartoum. Cited in L. Olsson, *An Integrated Study of Desertification*. Lund, Sweden: Department of Geography, Lund University, 1885.

Horsley, A. D. 1986. *Illinois: A Geography*. Boulder, Colo.: Westview Press.

Horvitz, S. 1999. *Salton Sea 101: An Introduction to the Issues of the Salton Sea—California's Greatest Resource*. North Shore, Calif.: Sea and Desert Interpretive Association.

Howe, J., E. McMahon, and L. Propst. 1997. *Balancing Nature and Commerce in Gateway Communities*. Washington, D.C.: Island Press.

Howlett, D. 2003. Around Great Lakes, Governments Try to Agree on Who Can Draw How Much. *USA Today*, June 23, §A, 3.

Huggett, R. 1990. *Catastrophism*. London: Edward Arnold.

Humphrey, C., and D. Sneath. 1999. *The End of Nomadism? Society, State and the Environment in Inner Asia*. Durham: Duke University Press.

Hundley, N., Jr. 1992. *The Great Thirst*. Berkeley: University of California Press.

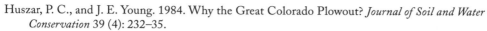

Huszar, P. C., and J. E. Young. 1984. Why the Great Colorado Plowout? *Journal of Soil and Water Conservation* 39 (4): 232–35.

al-Ibrahim, A. 1991. Excessive Use of Groundwater Resources in Saudi Arabia: Impacts and Policy Options. *Ambio* 20 (1): 34–37.

Ibrahim, F. 1978. Anthropogenic Causes of Desertification in Western Sudan. *GeoJournal* 2: 243–54.

———. 1984. *Ecological Imbalance in the Republic of the Sudan—with Reference to Desertification in Darfur.* Bayreuther Geowissenschaftliche Arbeiten Vol. 6. Bayreuth: Druckhaus Bayreuth.

Ibrahim, F., and B. Ibrahim. 2003. *Egypt: An Economic Geography.* London: I. B. Taurus.

Igler, D. 2001. *Industrial Cowboys: Miller & Lux and the Transfomation of the Far West, 1850–1920.* Berkeley: University of California Press.

Illinois Department of Agriculture. 1988. *Illinois Agricultural Statistics—Annual Summary.* Springfield, Ill.

Ingold, T. 1987. *The Appropriation of Nature: Essays on Human Ecology and Social Relations.* Iowa City: University of Iowa Press.

INPE [Instituto National de Pesquisas Espaciais / National Institute of Space Research]. 2003. Co-ordenacão – Geral de Observacão de Terra –Estimativas Anuals [in Portugese]. São Paulo, Brazil: Ministry of Science and Technology. For more information see: http://www.obt.inpe.br/prodes/prodes_1988_2003.htm Last accessed April 2003.

Jackson, D. L., and L. L. Jackson. 2002. *Farm as Natural Habitat: Reconnecting Food Systems with Ecosystems.* Washington, D.C.: Island Press.

Jacobsen, T. 1960. The Waters of Ur. *Iraq* 22: 174–85.

Jacobsen, T., and R. M. Adams. 1958. Salt and Silt in Ancient Mesopotamian Agriculture. *Science* 128 (3334): 1251–58.

Jain, L. C. 2001. *Dam vs. Drinking Water: Exploring the Narmada Judgement.* Pune, India: Parisar.

James, B. 1993. Redefining Rural France. *France Magazine,* No. 27 (Summer), 14–17.

James, S. R. 1989. Hominid Use of Fire in the Lower and Middle Pleistocene: A Review of the Evidence. *Current Anthropology* 30 (1): 1–26.

Jansson, A. M., and B. O. Jansson. 1988. Energy Analysis Approach to Ecosystem Redevelopment in the Baltic Sea and Great Lakes. *Ambio* 17 (2): 131–6.

Janzen, J. 1983. The Modern Development of Nomadic Living Space in Southeast Arabia—The Case of Dhofar (Sultanate of Oman). *Geoforum* 14 (3): 289–309.

Jennings, J. N. 1952. *The Origin of the Broads.* Royal Geographical Research Series, No. 2. London: Royal Geographical Society.

Jenny, H. 1984. My Friend the Soil. *Journal of Soil and Water Conservation* 39 (3): 158–61.

Jiang, H. 2002. Culture, Ecology, and Nature's Changing Balance: Sandification on Mu Us Sandy Land, Inner Mongolia, China. In *Global Desertification: Do Humans Cause Deserts?,* edited by J. F. Reynolds and D. M Stafford Smith, 181–96. Berlin: Dahlem University Press.

Jodha, N. S. 1985. Population Growth and the Decline of Common Property in Rajasthan, India. *Population and Development Review* 11 (2): 247–64.

———. 1988. The Effects of Climatic Variation on Agriculture in Dry Tropical Area Regions of India. In *The Impact of Climatic Variations on Agriculture,* Vol. 2, edited by M. L. Parry, T. R. Carter, and N. T. Konijn, 503–16. Dordrecht, Netherlands: Kluwer Academic.

Johansson, P. O. 1990. Valuing Environmental Damage. *Oxford Review of Economic Policy* 6 (1): 34–50.

Johnson, D. L. 1969. *The Nature of Nomadism: A Comparative Study of Pastoral Migrations in Southwestern Asia and Northern Africa.* Research Paper No. 118, Department of Geography, University of Chicago.

———. 1973. The Response of Pastoral Nomads to Drought in the Absence of Outside Intervention. UN Special Sahelian Office (ST/SSO/18), New York.

———. 1993a. Pastoral Nomadism and the Sustainable Use of Arid Lands. *Arid Lands Newsletter* 33 (Spring/Summer): 26–34.

———. 1993b. Nomadism and Desertification in Africa and the Middle East. *GeoJournal* 31 (1): 51–66.

Johnson, D. L., and T. Whitmore. 1987. Old World Population Reconstructions: The Nile and Mesopotamia. Paper presented at the 83rd Annual Meeting of the Association of American Geographers, Portland, Ore.

Johnson, H. B. 1975. Rational and Ecological Aspects of the Quarter Section: An Example from Minnesota. *Geographical Review* 47 (3): 66–87.

———. 1976. *Order Upon the Land*. Oxford: Oxford University Press.

Jordan, A., and T. O'Riordan. 1999. The Precautionary Principle in Contemporary Environmental Policy and Politics. In *Protecting Public Health and the Environment: Implementing the Precautionary Principle*, edited by C. Raffensperger and J. Tickner, 15–35. Washington, D.C.: Island Press.

Jordan, C. F., ed. 1989. *An Amazonian Rain Forest*. Man and the Biosphere Programme. Vol. 2. Paris: UNESCO.

Jordan, W. R., III, M. E. Gilpin, and J. D. Aber, eds. 1988. *Restoration Ecology: A Synthetic Approach*. Cambridge: Cambridge University Press.

Judson, S. 1981. Erosion of the Land. In *Use and Misuse of the Earth's Surface*, edited by B. J. Skinner, 130–39. Los Altos, Calif.: William Kaufman.

Kaatz, M. R. 1955. The Black Swamp: A Study in Historical Geography. *Annals of the Association of American Geographers* 45 (1): 1–35.

Kassas, M. 1972. Impact of River Control Schemes on the Shoreline of the Nile Delta. In *The Careless Technology: Ecology and International Development*, edited by M. T. Farvar and J. P. Milton, 179–88. Garden City, N.Y.: Natural History Press.

Kates, R. W., and V. Haarmann. 1992. Where the Poor Live: Are the Assumptions Correct? *Environment* 34 (4): 4–11, 25–28.

Kay, P., and D. L. Johnson. 1981. Estimation of Tigris-Euphrates Streamflow from Regional Paleoenvironmental Proxy Data. *Climatic Change* 3 (3): 251–63.

Kemp, W. B. 1971. The Flow of Energy in a Hunting Society. *Scientific American* 225 (3): 104–15.

Kennedy, D., and D. Riley. 1990. *Rome's Desert Frontier from the Air*. Austin: University of Texas Press.

al-Khashab, W. H. 1958. *The Water Budget of the Tigris and Euphrates Basin*. Research Paper No. 54, Department of Geography, University of Chicago.

Khogali, M. M. 1991. Desertification, Famine, and the 1988 Rainfall—The Case of Umm Ruwaba District in the Northern Kordofan Region. *GeoJournal* 25 (1): 81–89.

Kiersch, G. A. 1965. The Vaiont Reservoir Disaster. *Mineral Information Service* 18 (7): 129–38.

Kipuri, N. O. 1991. Age, Gender and Class in the Scramble for Maasailand. *Nature & Resources* 27 (4): 10–17.

Kirkby, J. 2001. Saving the Gash Delta, Sudan. *Land Degradation & Development* 12 (3): 225–36.

Kisangani, E. 1986. A Social Dilemma in a Less Developed Country: The Massacre of the African Elephant in Zaire. In *Proceedings of the Conference on Common Property Management*, edited by Panel on Common Property Management, Board on Science and Technology for International Development, Office of International Affairs, National Research Council, 137–60. Washington, D.C.: National Research Council, National Academy Press.

Kishk, M. A. 1986. Land Degradation in the Nile Valley. *Ambio* 15 (4): 226–30.

Klepeis, P. J. 2000. Deforesting the Once Deforested: Land Transformation in Southeastern Mexico. Ph.D. Dissertation, Clark University, Worcester, Mass.

Knapp, P. A. 1991. The Response of Semi-arid Vegetation Assemblages Following the Abandonment of Mining Towns in South-western Montana. *Journal of Arid Environments* 20 (2): 205–22.

Knight, R., W. C. Gilgert, and E. Marston, eds. 2002. *Ranching West of the 100th Meridian: Culture, Ecology, and Economics*. Washington, D. C.: Island Press.

Kohl, L. 1986. The Oosterschelde Barrier—Man against the Sea. *National Geographic* 170 (4): 526–37.

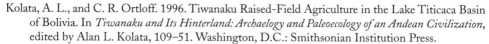

Kolata, A. L., and C. R. Ortloff. 1996. Tiwanaku Raised-Field Agriculture in the Lake Titicaca Basin of Bolivia. In *Tiwanaku and Its Hinterland: Archaelogy and Paleoecology of an Andean Civilization*, edited by Alan L. Kolata, 109–51. Washington, D.C.: Smithsonian Institution Press.

Kolata, A. L., O. Rivera, J. C. Ramirez, and E. Gemio. 1996. Rehabilitating Raised-Field Agriculture in the Southern Lake Titicaca Basin of Bolivia: Theory, Practice, and Results. In *Tiwanaku and Its Hinterland: Archaeology and Paleoecology of an Andean Civilization*, edited by A. L. Kolata, 203–30. Washington, D.C.: Smithsonian Institution Press.

Korn, M. 1996. The Dike-Pond Concept: Sustainable Agriculture and Nutrient Recycling in China. *Ambio* 25 (1): 6–14.

Kotlyakov, V. M. 1991. The Aral Sea Basin: A Critical Environmental Zone. *Environment* 33 (1): 4–6.

Kramer, S. N. 1959. *History Begins at Sumer*. Garden City, N.Y.: Doubleday Anchor.

———. 1961. Mythology of Sumer and Akkad. In *Mythologies of the Ancient World*, edited by S. N. Kramer, 93–137. Garden City, N.Y.: Doubleday Anchor.

———. 1963. *The Sumerians: Their History, Culture, and Character*. Chicago: University of Chicago Press.

Kuhlmann, D. H. H. 1988. The Sensitivity of Coral Reefs to Environmental Pollution. *Ambio* 17 (1): 13–21.

Kuletz, V. L. 1998. *The Tainted Desert: Environmental and Social Ruin in the American West*. New York: Routledge.

Kummer, D., R. Concepcion, and B. Cañizares. 2003. Image and Reality: Exploring the Puzzle of "Continuing" Environmental Degradation in the Uplands of Cebu, the Philippines. *Philippine Quarterly of Culture & Society* 31: 135–55.

Kunstadter, P. 1987. Swiddeners in Transition: Lua' Farmers in Northern Thailand. In *Comparative Farming Systems*, edited by B. L. Turner II and S. Brush, New York: Guilford.

Laflin, P. 1995. Salton Sea: California's Overlooked Treasure. *The Periscope*. Indio, Calif.: Coachella Valley Historical Society.

Lal, R. 1988. Soil. In *Soil Research Methods*, edited by R. Lal, 150–60. Ankeny, Iowa: Soil and Water Conservation Society.

Lambert, A. M. 1971. *The Making of the Dutch Landscape*. London: Seminar Press.

Lambert, J. M. 1960. *The Making of the Broads: A Reconsideration of Their Origin in the Light of New Evidence*. London: John Murray.

Lamprey, H. F. 1975. Report on the Desert Encroachment Reconnaissance in Northern Sudan. Nairobi: UNESCO/UNEP.

Landis, C. 2000. *In Search of Eldorado: The Salton Sea*. Palm Springs, Calif.: Palm Springs Desert Museum.

Lapidus, I. M. 1981. Arab Settlement and Economic Development of Iraq and Iran in the Age of the Umayyad and Early Abbasid Caliphs. In *The Islamic Middle East, 700–1900: Studies in Economic and Social History*, edited by A. L. Udovitch, 177–208. Princeton, N.J.: Darwin Press.

———. 1988. *A History of Islamic Societies*. Cambridge: Cambridge University Press.

Larsen, C. E., and G. Evans. 1978. The Holocene Geological History of the Tigris-Euphrates-Karun Delta. In *The Environmental History of the Near and Middle East Since the Last Ice Age*, edited by W. C. Brice, 227–44. London: Academic Press.

Lawson, W. E., L. M. Walsh, B. A. Stewart, and D. H. Boelter, eds. 1981. *Soil and Water Resources: Research Priorities for the Nation*. Madison, Wis.: Soil Science Society of America.

Leach, G., and R. Mearns. 1990. *Beyond the Fuelwood Crisis*. London: IIED.

Lee, R. B. 1968. What Hunters Do for a Living, or, How to Make Out on Scarce Resources. In *Man the Hunter*, edited by R. B. Lee and I. DeVore, 30–42. Chicago: Aldine-Atherton.

———. 1969. !Kung Bushman Subsistence: An Input-Output Analysis. In *Environment and Cultural Behavior: Ecological Studies in Cultural Anthropology*, edited by A. P. Vayda, 47–79. Garden City, N.Y.: Natural History Press.

Lee, R. B., and I. DeVore. 1968. *Man the Hunter*. Chicago: Aldine-Atherton.

Lefort, R. 1996. Down to the Last Drop. *UNESCO Sources*, No. 84: 7.

Leopold, L. B., W. W. Emmett, and R. M. Myrick. 1966. *Channel and Hillslope Processes in a Semiarid Area, New Mexico*. U.S. Geological Survey Professional Paper 352–G. Washington, D.C.: U.S. Geological Survey.

Le Strange, G. 1905. *The Lands of the Eastern Calilphate: Mesopotamia, Persia, and Central Asia from the Moslem Conquest to the Time of Timur*. Cambridge: Cambridge University Press.

Lewis, J. 1990. The Ogallala Aquifer: An Underground Sea. *EPA Journal* 16 (6): 42–44.

Lewis, L. A. 1981. The Movement of Soil Materials During a Rainy Season in Western Nigeria. *Geoderma* 25 (1): 13–25.

———. 1985. Assessing Soil Loss in Kiambu and Murangi Districts, Kenya. *Geografiska Annaler* 67A (3–4): 273–84.

———. 1992. Terracing and Accelerated Soil Loss on Rwandan Steeplands: A Preliminary Investigation of the Implications of Human Activities Affecting Soil Movement. *Land Degradation & Rehabilitation* 3 (4): 241–46.

Lewis, L. A., and L. Berry. 1988. *African Environments and Resources*. Boston: Unwin Hyman.

Lewis, L. A., and V. Nyamulinda. 1989. Les relations entre les cultures et les unites topographiques dans les regions agricoles de la Bordure du Lac Kivu et de l'Impara au Rwanda: quelques strategies pour une agriculture soutenue. *Bulletin Agricole Rwanda* 22 (3): 143–49.

Lewis, L. A., G. Verstraeten, and H. Zhu. 2005. RUSLE Applied in a GIS Framework: Calculating the LS Factor and Deriving Homogeneous Patches for Estimating Soil Loss. *International Journal of GIS* 19 (7): 809–29.

Lewis, M. W. 1992. *Wagering the Land: Ritual, Capital, and Enviromental Degradation in the Cordillera of Northern Luzon, 1900–1986*. Berkeley: University of California Press.

Lewis, N., and K. L. Brubaker. 1989. *Bring Back the Blackstone*. Providence, R.I.: Save the Bay.

Light, A. 2000. Ecological Restoration and the Culture of Nature: A Pragmatic Perspective. In *Restoring Nature: Perspectives from the Social Science and Humanities*, edited by P. H. Gobster and R. B. Hull, 49–70. Washington, D.C.: Island Press.

Linear, M. 1985. The Tsetse War. *The Ecologist* 15 (1/2): 27–35.

Little, P. D. 1987. Land Use Conflicts in the Agricultural/Pastoral Borderlands: The Case of Kenya. In *Lands at Risk in the Third World: Local-level Perspectives*, edited by P. D. Little and M. M. Horowitz with A. E. Nyerges, 195–212. Boulder, Colo.: Westview Press.

Lippard, L. R. 1997. *The Lure of the Local: Senses of Place in a Multicentered Society*. New York: New Press.

Lo, C. P. 1990. People and Environment in the Zhu Jiang Delta of South China. *National Geographic Research* 6 (4): 400–417.

Long, M. 1998. The Vanishing Prairie Dog. *National Geographic* 193 (4): 116–31.

Lynn, W. S. 1998. Animals, Ethics, and Geography. In *Animal Geographies: Place, Politics, and Identity in the Nature-Culture Borderlands*, edited by J. R. Wolch and J. Emel, 280–97. London: Verso.

MacGillivrary, A. 1995. Tourism and Recreation. In *Europe's Environment: The Dobříš Assessment*, edited by D. Stanners and P. Bourdeau, 489–501. Copenhagen: European Environment Agency.

Machlis, G. E., and D. R. Field. 2000. *National Parks and Rural Development: Practice and Policy in the United States*. Washington, D.C.: Island Press.

Malayang, B. S., III. 1991. Tenure Rights and Exclusion in the Philippines. *Nature & Resources* 27 (4): 18–23.

Mainguet, M., and R. Létolle. 1998. Human-made Desertification in the Aral Sea Basin: Planning and Management Failures. In *The Arid Frontier: Interactive Management of Environment and Development*, edited by H. J. Bruins and H. Lithwick, 129–42. Dordrecht: Kluwer Academic Publications.

Malingreau, J. P., and C. J. Tucker. 1987. The Contribution of AVHRR Data for Measuring and Understanding Global Processes: Large-scale Deforestation in the Amazon Basin. *Proceedings of the International Geoscience and Remote Sensing Society* 1: 443–50.

Manshard, W. 1974. *Tropical Agriculture: A Geographical Introduction and Appraisal*. London: Longman.

Marsh, G. P. 1965 [1864]. *Man and Nature: Or, Physical Geography as Modified by Human Action*. Edited and annotated by David Lowenthal. Cambridge, Mass.: Belknap Press of Harvard University Press.

Martin, P. 1973. The Discovery of America. *Science* 179 (4077): 969–94.

Martin, R. 1989. *A Story that Stands Like a Dam*. New York: Henry Holt.

Mason, J., and P. Singer. 1990. *Animal Factories*. New York: Harmony Books.

McCabe, J. T. 1990. Turkana Pastoralism: A Case Against the Tragedy of the Commons. *Human Ecology* 18 (1): 81–103.

McClanahan, T. 2002. The Near Future of Coral Reefs. *Environmental Conservation* 29 (4): 460–83.

McClelland, L. F. 1998. *Building the National Parks: Historic Landscape Design and Construction*. Baltimore: Johns Hopkins University Press.

McCully, P. 1996. *Silenced Rivers: The Ecology and Politics of Large Dams*. London: Zed Books.

McGibben, B. 1990. *The End of Nature*. New York: Anchor Doubleday.

McHarg, I. L., and F. R. Steiner. 1998. *To Heal the Earth: Selected Writings of Ian L. McHarg*. Washington, D.C.: Island Press.

McKnight, L. W., P. M. Vaaler, and R. L. Katz, eds. 2001. *Creative Destruction: Business Survival in the Global Internet Economy*. Cambridge, Mass.: MIT Press.

McKnight, T. L. 1969. Barrier Fencing for Vermin Control in Australia. *Geographical Review* 59 (3): 330–47.

McKnight, T. L., and D. Hess. 2005. *Physical Geography: A Landscape Appreciation*. 8th ed. Upper Saddle River, N.J.: Pearson Prentice Hall.

McNeill, W. H. 1976. *Plagues and Peoples*. Garden City, N.Y.: Anchor Doubleday.

Meggers, B. J. 1971. *Amazonia: Man and Culture in a Counterfeit Paradise*. Arlington Heights, Ill.: AHM Publishing.

van Meijgaard, C. H. 1987. Jan Blanken en de landsverdediging. In *De Existentie dezes Lands: Jan Blanken*, edited by F. Koens, N. B. Spruyt, and M-C. Vink, 41–58. Amsterdam: Rijksmuseum and Uitgeverij AMA boeken.

Meyer, A. H. 1936. The Kankakee "Marsh" of Northern Indiana and Illinois. *Michigan Papers in Geography* 6: 359–96.

Meyer, W. B. 1996. *Human Impact on the Earth*. Cambridge: Cambridge University Press.

Micklin, P. P. 1991. Touring the Aral: Visit to an Ecologic Disease Zone. *Soviet Geography* 32 (2): 90–105.

Miller, G. T. 1993. *Environmental Science*. 4th ed. Belmont, Calif.: Wadsworth.

Milliman, J. D., J. M. Broadus, and F. Gable.1989. Environmental and Economic Implications of Rising Sea Level and Subsiding Deltas: The Nile and Bengal Examples. *Ambio* 18 (6): 340–45.

Ministry for Economic Policy and Development of the Czech Republic. 1991. *Northern Bohemia Revitalization Project*. Prague.

Molinelli, J. A. 1984. Geomorphic Processes along the Autopista Las Americas in North Central Puerto Rico. Ph.D. dissertation, Clark University, Worcester, Mass.

Moore, J. K. 2001. *Mining and Quarrying Trends in 1999*. Washington, D.C.: U.S. Department of Interior.

Moorehead, R. 1989. Changes Taking Place in Common-property Resource Management in the Inland Niger Delta of Mali. In *Common Property Resources: Ecology and Community-based Sustainable Development*, edited by F. Berkes, 256–72. London: Belhaven Press.

Moroney, M. G. 1984. *Iraq After the Muslim Conquest*. Princeton, N.J.: Princeton University Press.

Moroz, A. 1988. Morye prosit vody. *Komosomolskaya Pravda*. January 28, cited in N. Precoda. 1991. Requiem for the Aral Sea. *Ambio* 20 (3–4): 109–14.

Morse, B., and T. Berger. 1995. Findings and Recommendations of the Independent Review. In *Toward Sustainable Development*, edited by W. F. Fisher, 371–80. Armonk, N.Y.: M. E. Sharpe.

Mortimore, M. 1998. *Roots in the African Dust: Sustaining the Sub-Saharan Drylands*. Cambridge: Cambridge University Press.

Mortimore, M., and W. Adams. 1999. *Working the Sahel: Environment and Society in Northern Nigeria*. London: Routledge.

Moss, B. 1979. An Ecosystem Out of Phase. *Geographical Magazine* 52 (1): 47–49.

Mould, R. F. 1988. *Chernobyl: The Real Story*. Oxford: Pergamon Press.

Mumford, L. 1961. *The City in History: Its History, Its Origins, Its Transformations, and Its Prospects*. New York: Harcourt, Brace and World.

Munro, D. C., and H. Touron. 1997. The Estimation of Marshland Degradation in Southern Iraq Using Multitemporal Landsat TM Images. *International Journal of Remote Sensing* 18 (7): 1597–1606.

Munslow, B., Y. Katarere, A. Ferf, and P. O'Keefe. 1987. *The Fuelwood Trap*. London: IIED.

Narmada Bachao Andolan. [1998]. *Narmada: The Struggle for Life, Against Destruction*. Baroda, Gujarat: Chittaroopa Palit/Narmada Bachao Andolan.

Narmada Planning Group, Narmada Development Department, Government of Gujurat. 1989. *Planning for Prosperity: Sardar Sarovar Development Plan*. Gandhinagar, Gujarat: Sardar Sarovar Narmada Nigam.

National Geographic Society. 1992. Geographica. *National Geographic* 181 (2): xiv.

National Park Service. 2001. *The National Parks: Shaping the System* (online book). Washington, D.C.: U.S. Department of the Interior.

National Research Council. 1996. *River Resource Management in the Grand Canyon*. Committee to Review the Glen Canyon Environmental Studies. Washington, D.C.: National Academy Press.

Neel, J. V. 1970. Lessons from a "Primitive" People. *Science* 170 (3960): 815–22.

Negev, A. 1986. *Nabatean Archaeology Today*. New York: New York University Press.

Nelson, R. 1990. *Dryland Management: The "Desertification" Problem*. World Bank Technical Paper No. 116. Washington, D.C.: The World Bank.

Neumann, R. 1998. *Imposing Wilderness: Struggles over Livelihood and Nature Preservation in Africa*. Berkeley: University of California Press.

Nicholson, S. E., J. Kim, and J. Hoopingarner. 1988. *Atlas of African Rainfall and Its Interannual Variability*. Tallahasse, Fla.: Department of Meteorology, Florida State University.

Nietschmann, B. 1972. Hunting and Fishing Focus among the Miskito Indians, Eastern Nicaragua. *Human Ecology* 1 (1): 41–67.

———. 1973. *Between Land and Water: The Subsistence Ecology of the Miskito Indians, Eastern Nicaragua*. New York: Seminar Press.

Nissen, H. J. 1988. *The Early History of the Ancient Near East 9000–2000 B.C.* Translated by E. Lutzeier and K. J. Northcott. Chicago: University of Chicago Press.

Nolan, R. L., and D. C. Croson. 1995. *Creative Destruction: A Six-Stage Process for Transforming the Organization*. Boston: Harvard University Press.

OECD [Organization for Economic Cooperation and Development]. 1987. *The Radiological Impact of the Chernobyl Accident in OECD Countries*. Paris: Nuclear Energy Agency, OECD.

OIA [Office of International Agriculture]. 1978. *Dryland Agriculture: What is It?* Corvallis, Ore.: Office of International Agriculture, Oregon State University.

O'Conner, P. 2002. Government Restores Water Flow to Legendary Snowy River. Associated Press, http//www.enn.com/news/wire-stores/2002/08/08292002/ap_48295.asp.

Ogar, N. 2001. Vegetation Dynamics on the Syrdarya Delta and Modern Land Use. In *Sustainable Land Use in Deserts*, edited by S-W. Breckle, M. Veste, and W. Wucherer, 74–83. Berlin: Springer-Verlag.

O'Keefe, P., P. Raskin, and S. Bernow. 1984. *Energy Development in Kenya: Opportunities and Constraints*. Energy, Environment, and Development in Africa Series, Vol. 1. Stockholm: Scandinavian Institute of African Studies and Beijer Institute.

O'Leary, M. 1984. Ecological Villains or Economic Victims: The Case of the Rendille of Northern Kenya. *Desertification Control Bulletin* 11: 17–21.

Olsson, K., and A. Rapp. 1991. Dryland Degradation and Conservation for Survival. *Ambio* 20 (5): 192–95.

Olsson, L. 1985. *An Integrated Study of Desertification*. Lund, Sweden: Department of Geography, University of Lund.

———. 1998. Desertification and Climate Change: Impact on Global Food Security. *International Seminar on Nuclear War and Planetary Emergencies: 22nd Session*, edited by K. Goebel, 241–52. Singapore: World Scientific.

———. 1999. Why is the Received Wisdom on Desertification so Difficult to Change?. In *International Seminar on Nuclear War and Planetary Emergencies: 23rd Session*, edited by K. Goebel, 148–59. Singapore: World Scientific.

O'Riordan, T. 1979. Signs of Disaster and a Policy for Survival. *Geographical Magazine* 52 (1): 50–59.

———. 1990. Environmental Assessment and a Future Strategy for the Broads. *Planner* 76 (7): 7–10.

Orlovsky, N., M. Glantz, and L. Orlovsky. 2001. Irrigation and Land Degradation in the Aral Sea Basin. In *Sustainable Land Use in Deserts*, edited by S-W. Breckle, M. Veste, and W. Wucherer, 115–25. Berlin: Springer-Verlag.

Ostrom, E., T. Dietz, N. Dolsak, P. C. Stern, S. Stonich, and E. U. Weber, eds. 2003. *The Drama of the Commons*. National Research Council, Division of Behavioral and Social Sciences and Education, Committee on the Human Dimensions of Global Change. Washington, D.C.: National Academy Press.

Oswalt, W. H. 1973. *Habitat and Technology: The Evolution of Hunting*. New York: Holt, Rinehart and Winston.

OTA [Office of Technology Assessment]. 1988. *Enhancing Agriculture in Africa: Role for U.S. Development Assistance*. Washington, D.C.: U.S. Government Printing Office.

Oterbridge, T. 1987. The Disappearing Chinampas of Xochimilco. *The Ecologist* 17 (2): 76–83.

OTTI [Office of Travel and Tourism Industries]. 2002. Monthly Tourism Statistics. Washington, D.C.: U.S. Dept. of Commerce. http://tinet.ita.doc.gov.

Page, M. 1999. *The Creative Destruction of Manhattan, 1900–1940*. Chicago: University of Chicago Press.

Parayil, G., and F. Tong. 1998. Pasture-led to Logging-led Deforestation in the Brazilian Amazon: The Dynamics of Socio-environmental Change. *Global Environmental Change* 8 (1): 63–79.

Parmenter, B. 1996. Endangered Wetlands and Environmental Management in North Africa. In *The North African Environment at Risk*, edited by W. D. Swearingen and A. Bencherifa, 155–74. Boulder, Colo.: Westview Press.

Parsons, J. J. 1962. *The Green Turtle and Man*. Gainesville: University of Florida Press.

Partow, H. 2001. *The Mesopotamian Marshlands: Demise of an Ecosystem*. Geneva, Switzerland: UNEP/DEWA/GRID.

Pattison, W. D. 1957. *Beginnings of the American Rectangular Land Survey System, 1784–1800*. Research Paper No. 50, Department of Geography, University of Chicago.

Pawluk, R. R., J. A. Sandor, and J. A. Tabor. 1992. The Role of Indigenous Soil Knowledge in Agricultural Development. *Journal of Soil and Water Conservation* 47 (4): 298–302.

Pearce, F. 2001. Iraqi Wetlands Face Total Destruction. *New Scientist* 170 (2291): 4–5.

Pereira, H. C. 1973. *Land Use and Water Resources*. Cambridge: Cambridge University Press.

Pereira, L. S., and J. Gowing, eds. 1998. *Water and the Environment: Innovation Issues in Irrigation and Drainage*. London: E. & F. N. Spon.

Perevolotsky, A. 1987. Territoriality and Resource Sharing among the Bedouin of Southern Sinai: A Socio-ecological Interpretation. *Journal of Arid Environments* 13 (2): 153–61.

Peters, E. L. 1968. The Tied and the Free: An Account of a Type of Patron-client Relationship among the Bedouin Pastoralists of Cyrenaica. In *Contributions to Mediterranean Sociology: Mediterranean Rural Communities and Social Change*, edited by J.-G. Peristiany, 167–88. Paris: Mouton.

Pilkey, O. H., and W. J. Neal. 1992. Save Beaches, not Buildings. *Issues in Science and Technology* 8 (3): 36–41.

Pinchot, G. 1947. *Breaking New Ground*. New York: Harcourt Brace.

Piper, S. 1989. Measuring Particulate Pollution Damage from Wind Erosion in the Western United States. *Journal of Soil and Water Conservation* 44 (1): 70–75.

Pond, W. C. 1983. Modern Pork Production. *Scientific American* 248 (5): 96–103.

Postal, S. 1992. *Lost Oasis: Facing Water Scarcity*. Worldwatch Environmental Alert Series. New York: W. W. Norton.

———. 1996. Forging a Sustainable Water Strategy. In *State of the World 1966*, edited by L. R. Brown, C. Flavin, and L.Starke, 40–59. New York: W. W. Norton.

Postel, S., and J. C. Ryan. 1991. Reforming Forestry. In *State of the World 1991*, edited by L. R. Brown, C. Flavin, S. Postel, and L.Starke, 74–92. New York: W. W. Norton.

Power, J. F., and R. F. Follett. 1987. Monoculture. *Scientific American* 256 (3): 79–86.

Powers, W. L. 1987. The Ogallala's Bounty Evaporates. *Science of Food and Nutrition* 5 (3): 2–5.

Precoda, N. 1991. Requiem for the Aral Sea. *Ambio* 20 (3–4): 109–14.

Press, F., and R. Siever. 1974. *Earth*. San Francisco: W. H. Freeman.

Pretty, J. 1998. *The Living Land: Agriculture, Food and Community Regeneration in Rural Europe*. London: Earthscan.

———. 2002. *Agri-Culture: Reconnecting People, Land and Nature*. London: Earthscan.

Pretty, J. H., J. Thompson, and J. K. Kiara. 1995. Agricultural Regeneration in Kenya: The Catchment Approach to Soil and Water Conservation. *Ambio* 24 (1): 7–15.

Prince, H. C. 1997. *Wetlands of the United States: A Historical Geography of Changing Attitudes*. Chicago: University of Chicago Press.

Proctor, J. 1983. Mineral Nutrients in Tropical Forests. *Progress in Physical Geography* 7 (3): 424–31.

Prosterman, R. L., T. Hanstad, and L. Ping. 1996. Can China Feed Itself? *Scientific American* 275 (5): 90–96.

Pyne, S. J. 1993. Keeper of the Flame: A Survey of Anthropogenic Fire. In *Fire in the Environment: The Ecological, Atmospheric, and Climatic Importance of Vegetation Fires*, edited by P. J. Crutzen and J. G. Goldammer, 245–66. Chichester: John Wiley.

———. 2001. *Fire: A Brief History*. Seattle: University of Washington Press.

Raish, Carol. 2000. Lessons for Restoration in Traditions of Stewardship: Sustainable Land Management in Northern New Mexico. In *Restoring Nature: Perspectives from the Social Science and Humanities*, edited by P. H. Gobster and R. B. Hull, 281–97. Washington, D.C.: Island Press.

Raynaut, C. 1997. *Societies and Nature in the Sahel*. London: Routledge and Stockholm Environment Institute.

Redman, C. L. 1978. *The Rise of Civilization: From Early Farmers to Urban Society in the Ancient Near East*. San Francisco: W. H. Freeman.

Reed, E. A. 1920. *Tales of a Vanishing River*. New York: John Lane.

Reining, D. 1967. Rock, Sand and Gravel Resources—Strong but Challenged. Paper presented at annual meeting of the American Institute of Mining and Metallurgical Engineers, Los Angeles, February 19–23.

Reisner, M. 1993 [1986]. *Cadillac Desert*. Revised ed. New York: Penguin.

Rekacewicz, P. 2001. *Vital Signs 2000*. Washington, D.C.: Worldwatch Institute.

Reynolds, J. F., and D. M. Stafford Smith. 2002. Do Humans Cause Deserts? In *Global Desertification: Do Humans Cause Deserts?*, edited by J. F. Reynolds and D. M. Stafford Smith, 1–21. Berlin: Dahlem University Press.

Richards, J. F. 1990. Agricultural Impacts in Tropical Wetlands: Rice Paddies for Mangroves in South and Southeast Asia. In *Wetlands: A Threatened Landscape*, edited by M. Williams, 217–33. Oxford: Blackwell.

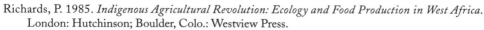

Richards, P. 1985. *Indigenous Agricultural Revolution: Ecology and Food Production in West Africa*. London: Hutchinson; Boulder, Colo.: Westview Press.

Richards, P. W. 1952. *The Tropical Rainforest*. Cambridge: Cambridge University Press.

Richards, W. Q. 1984. Reconstructing the National Conservation Program. *Journal of Soil and Water Conservation* 39 (3): 156–57.

Riebsame, W. E. 1986. The Dust Bowl: Historical Image, Psychological Anchor, and Ecological Taboo. *Great Plains Quarterly* 6 (2): 127–36.

Riehl, H., and J. Meitin. 1979. Discharge of the Nile River: A Barometer of Short Period Climatic Variation. *Science* 206 (4423): 1178–79.

Rifkin, J. 1992. *Beyond Beef: The Rise and Fall of the Cattle Culture*. New York: Penguin.

Robbins, P. F. 1998. Authority and Environment: Institutional Landscapes in Rajasthan, India. *Annals of the Association of American Geographers* 88 (3): 410–35.

———. 2000. The Practical Politics of Knowing: State Environmental Knowledge and Local Political Economy. *Economic Geography* 76 (2): 126–44.

———. 2001a. Fixed Categories in a Portable Landscape: The Causes and Consequences of Land-cover Categorization. *Environment and Planning* A 33 (1): 161–79.

———. 2001b. Tracking Invasive Land Covers in India, or Why Our Landscapes Have Never Been Modern. *Annals of the Association of American Geographers* 91 (4): 637–59.

———. 2003. Beyond Ground Truth: GIS and the Environmental Knowledge of Herders, Professional Foresters, and Other Traditional Communities. *Human Ecology* 31 (2) 233–53.

Robbins, P. F., et al. 2002. Desertification at the Community Scale: Sustaining Dynamic Human-Environment Systems. In *Global Desertification: Do Humans Cause Deserts?*, edited by J. F. Reynolds and D. M Stafford Smith, 325–55. Berlin: Dahlem University Press.

Robinson, A. H. W. 1953. The Storm Surge of 31st January–1st February, 1953. *Geography* 38: 134–41.

Robinson, F. E., and J. N. Luthin. 1976. *Adaptation to Increasing Salinity of the Colorado River*. California Water Resources Center, University of California, Davis, Contribution No. 160.

Robinson, G. D., and A. M. Speiker. 1978. *Nature to be Commanded. . . .* U.S. Geological Survey Professional Paper 950, Washington, D.C.: U.S. Government Printing Office.

Roe, E., L. Huntsinger, and K. Labnow. 1998. High Reliability Pastoralism. *Journal of Arid Environments* 39 (1): 39–55.

van Rompaey, A., J. Krasa, T. Dostal, and G. Govers. 2003. Modelling Sediment Supply to Rivers and Reservoirs in Eastern Europe during and after the Collectivisation Period. *Hydrobiologia* 1 (1): 1–8.

Rostlund, E. 1957. The Myth of a Natural Prairie Belt in Alabama: An Interpretation of Historical Records. *Annals of the Association of American Geographers* 47 (4): 392–411.

Roy, Arundhati. 1999. *The Cost of Living*. New York: Modern Library.

Rubin, R. 1991. Settlement and Agriculture on an Ancient Desert Frontier. *Geographical Review* 81 (2): 197–205.

Ruddle, K. 1974. *The Yukpa Cultivation System: A Study of Shifting Cultivation in Colombia and Venezuala*. Ibero-Americana Vol. 52. Berkeley: University of California Press.

———. 1989. Solving the Common-property Dilemma: Village Fisheries Rights in Japanese Coastal Waters. In *Common Property Resources*, edited by F. Berkes, 168–84. London: Belhaven Press.

Ruddle, K., J. I. Furtado, G. F. Zhong, and H. Z. Deng. 1983. The Mulberry Dike-Carp Pond Resource System of the Zhujiang (Pearl River) Delta, People's Republic of China: I. Environmental Context and System Overview. *Applied Geography* 3 (1): 45–62.

Ruddle, K., and W. Manshard. 1981. *Renewable Natural Resources and the Environment: Pressing Problems in the Developing World*. Dublin: Tycooly International for the United Nations University.

Ruddle, K., and G. Zhong. 1988. *Integrated Agriculture-Aquaculture in South China: The Dike-Pond System of the Zhujiang Delta*. Cambridge: Cambridge University Press.

Russell, M. 1999. *A Story that Stands Like a Dam: Glen Canyon and the Struggle for the Soul of the West*. Salt Lake City: University of Utah Press.

Rutkowski, A. 1991 Unnatural Disasters. *World Press Review*. June, 45.

Rutman, S., and L. Dickson. 2002. Management of Buffelgrass on Organ Pipe Cactus National Monument, Arizona. In *Invasive Exotic Species in the Sonoran Region*, edited by Barbara Tellman, 311–18. Tucson: University of Arizona Press and Arizona-Sonora Desert Museum.

Sächsische Landesanstalt für Forsten. 1994. *In Favour of Saxony's Forests*. Graupa, Germany.

Salati, E., M. J. Dourojeanni, F. C. Novaes, A. E. De Oliveira, R. W. Perritt, H. O. R. Schubart, and J. C. Umana. 1990. Amazonia. In *The Earth as Transformed by Human Action*, edited by B. L. Turner II, W. C. Clark, R. W. Kates, J. F. Richards, J. T. Mathews, and W. B. Meyer, 479–93. Cambridge: Cambridge University Press.

Salzman, P. C. 2000. *Black Tents of Baluchistan*. Washington, D.C.: Smithsonian Institution Press.

Sanders, D. 2000. The Implementation of Soil Conservation Programs. In *Rangeland Desertification*, edited by Olafur Arnalds and Steve Archer, 143–51. Dordrecht; Boston; London: Kluwer Academic Publishers.

Sandström, A., B. K. Eriksson, P. Karås, M. Isaeus, and H. Schreiber. 2005. Boating and Navigation Activities Influence the Recruitment of Fish in a Baltic Sea Archipelago Area. *Ambio* 34 (2): 125–30.

Sangvai, S. 2002. *The River and Life: People's Struggle in the Narmada Valley*. Mumbai: Earthcare Books.

Sauder, R. A. 1994. *The Lost Frontier: Water Diversion in the Growth and Destruction of Owens Valley Agriculture*. Tucson: University of Arizona Press.

Sauer, C. O. 1956. The Agency of Man on the Earth. In *Man's Role in Changing the Face of the Earth*, edited by W. L. Thomas, Jr., 49–69. Chicago: University of Chicago Press.

Savory, A. 1999. *Holistic Management: A New Framework for Decision Making*. 2nd ed. Washington, D.C.: Island Press.

Schama, S. 1996. *Landscape and Memory*. New York: Vintage Books.

Schell, O. 1978. *Modern Meat*. New York: Vintage Books.

Schmal, H. 1987. De ontwikkeling van de infrastructuur van het einde van de achttiende tot het midden van de 19de eeuw. In *de Physique Existentie dezes Lands: Jan Blanken*, edited by F. Koens, N. B. Spruyt, and M-C. Vink, 95–112. Amsterdam: Rijksmuseum and Uitgeverij AMA boeken.

Schumpeter, J. A. 1939. *Business Cycles: A Theoretical, Historical, and Statistical Analysis of the Capitalist Process*. New York: McGraw-Hill.

Sclove, R. B., and M. L. Scammell. 1999. Practicing the Principle. In *Protecting Public Health and the Environment: Implementing the Precautionary Principle*, edited by C. Raffensperger and J. Tickner, 252–65. Washington, D.C.: Island Press.

Scow, K. M., G. E. Fogg, D. E. Hinton, and M. L. Johnson, eds. 2000. *Integrated Assessment of Ecosystem Health*. Boca Raton: Lewis Publishers.

Seasholes, N. S. 2003. *Gaining Ground: A History of Landmaking in Boston*. Cambridge, Mass.: MIT Press.

Seckler, D. 1971. *California Water*. Berkeley: University of California Press.

Sellars, R. W. 1997. *Preserving Nature in the National Parks: A History*. New Haven: Yale University Press.

Serageldin, I. 1995. *Toward Sustainable Management of Water Resources*. Washington, D.C.: World Bank.

Shah, A. A. 1995. A Technical Overview of the Flawed Sardar Sarovar Project and a Proposal for a Sustainable Alternative. In *Toward Sustainable Development: Struggling Over India's Narmada River*, edited by William F. Fisher, 319–67. Armonk, N.Y.: M. E. Sharpe.

Shaw, J. M. 1986. Managing Change—The Broadlands Experience. *Planner* 72 (10): 15–17.

———. 1990. Environmental Government: The Example of the Broads. *Planner* 76 (7): 11–14.

Shaxson, T. F., N. W. Hudson, D. W. Sanders, E. Roose, and W. C. Moldenhauer. 1989. *Land Husbandry, a Framework for Soil and Water Conservation*. Ankeny, Iowa: Soil and Water Conservation Society.

Sheppard, T. 1912. *The Lost Towns of the Yorkshire Coast*. London: A. Brown.

Sheridan, D. 1981. *Desertification of the United States*. Washington, D.C.: Council on Environmental Quality.

Shipp, D., ed. 1993. *Loving Them to Death? Sustainable Tourism in Europe's Nature and National Parks*. Grafenau, Germany: Federation Nature and National Parks in Europe.

Shoumatoff, A. 1997. Among the Cowboys. *American Heritage* 48 (5): 78–91.

Shoup, J. 1990. Middle Eastern Sheep Pastoralism and the Hima System. In *The World of Pastoralism*, edited by J. G. Galaty and D. L. Johnson, 195–215. New York: Guilford.

Sigalov, V. 1987. Aral: vzglyad iz kosmosa (Aral: A View from Space). *Izvestiya*, June 23.

Simmons, I. G. 1989. *Changing the Face of the Earth: Culture, Environment, History*. Oxford: Blackwell.

Simon, A., and C. R. Hupp. 1986. Channel Evolution in Modified Tennessee Channels. In *Nevada Proceedings of the 4th Federal Interagency Sedimentation Conference, Las Vegas*, 5–71–5–82. Reston, Va.: Interagency Advisory Committee on Water Data, U.S. Geological Survey.

Simons, M. 1992. Pollution Blights Investment, too, in East Europe. *New York Times*, May 13, §A, 1, 12.

Skempton, A. W., and J. N. Hutchinson. 1969. Stability of Natural Slopes and Embankment Foundations. In *Proceedings, 7th International Conference on Soil Mechanics and Foundation Engineering*. State of the Art Volume, Mexico City.

Skinner, B. J., and S. C. Porter. 1987. *Physical Geology*. New York: John Wiley.

Sleeper, B. 1990. Out on a Ledge. *Animals* 123 (4): 18–25.

Smil, V. 1997. Global Population and the Nitrogen Cycle. *Scientific American* 277 (1): 74–81.

Smiles, M., and A. H. L. Huiskes. 1981. Holland's Eastern Scheldt Estuary Barrier Scheme: Some Ecological Considerations. *Ambio* 10 (4): 158–65.

Smith, N. J. J. 1982. *Rainforest Corridors: The Transamazonian Colonization Scheme*. Berkeley: University of California Press.

Smits, H. 1970. Land Reclamation in the former Zuyder Zee in the Netherlands. *Geoforum* 4: 37–44.

———. 1988. *Land Reclamation in the Former Zuyder Zee in the Netherlands*. Revised by J. de Jong. Lelystad, Netherlands: Ministerie van Verkeer en Waterstaat.

Sollod, A. E. 1990. Rainfall, Biomass and the Pastoral Economy of Niger: Assessing the Impact of Drought. *Journal of Arid Environments* 18 (1): 97–107.

Spencer, J. E., and G. A. Hale. 1961. The Origins, Nature, and Distribution of Agricultural Terracing. *Pacific Viewpoint* 2 (1): 1–10.

Stanley, D. J., and A. G. Warne. 1993. Nile Delta: Recent Geological Evolution and Human Impact. *Science* 260 (5108): 628–34.

Stansfield, J., B. Moss, and K. Irvine. 1989. The Loss of Submerged Plants with Eutrophication III. Potential Role of Organochlorine Pesticides: A Palaeoecological Study. *Freshwater Biology* 22 (1): 109–32.

Starrs, P. F. 1998. *Let the Cowboy Ride: Cattle Ranching in the American West*. Baltimore: Johns Hopkins University Press.

Stepanek, J. H. 1999. *Wringing Success from Failure in Late-Developing Countries: Lessons from the Field*. Westport, Conn.: Praeger.

Stern, M. 1985. Census from Heaven? Ph.D. dissertation, Department of Physical Geography, University of Lund, Sweden.

Stewart, Doug. 2000. Kudzu: Love It—or Run. *Smithsonian* 31 (7): 64–70.

Stocks, B. J., and W. S. W. Trollope. 1993. Fire Management: Principles and Options in the Forested and Savanna Regions of the World. In *Fire in the Environment: The Ecological, Atmospheric, and Climatic Importance of Vegetation Fires*, edited by P. J. Crutzen and J. G. Goldammer, 315–26. Chichester: John Wiley.

Stockton, C. W., and D. M. Meko. 1983. Drought Recurrence in the Great Plains as Reconstructed from Long-term Tree Ring Records. *Journal of Climate and Applied Meteorology* 22 (1): 17–29.

Stolgitis, J. 1991. *Briefings: Rhode Island Division of Fish and Wildlife*. Providence: Rhode Island Division of Fish and Wildlife.

Stuth, J. W., and B. G. Lyons, eds. 1993. *Decision Support Systems for the Management of Grazing Lands: Emerging Issues*. Paris: UNESCO and Parthenon Publishing Group.

Sutton, R. K. 1977. Circles on the Plains: Center Pivot Irrigation. *Landscape* 22 (1): 3–10.

Swank, W. G., and G. A. Petrides. 1954. Establishment and Food Habits of Nutria in Texas. *Ecology* 35 (2): 172–76.

SWCS [Soil and Water Conservation Society]. 1984. Out of the Dust Bowl. *Journal of Soil and Water Conservation* 39 (1): 6–17.

Tank, R. W., ed. 1983. *Environmental Geology*. New York: Oxford University Press.

Taylor, J. 2002. *Petra and the Lost Kingdom of the Nabataeans*. Cambridge, Mass.: Harvard University Press.

Tellman, B., ed. 2002. *Invasive Exotic Species in the Sonoran Region*. Tucson: University of Arizona Press and Arizona-Sonora Desert Museum.

TG [Telegram & Gazette]. 2002. Bounty Offered for Cajun Country Rodents. *Telegram & Gazette*, Wednesday, November 20, §A, 6.

Thesiger, W. 1964. *The Marsh Arabs*. London: Longman.

Thiam, A. 1998. GIS and Remote Sensing Methods for Assessing and Monitoring Land Degradation in the Sahel Region, The Case of Southern Mauritania. Unpublished Ph. D. dissertation. Worcester, Mass.: Clark University.

Thompson, J. 2002. *Wetlands Drainage, River Modification, and Sectoral Conflict in the Lower Illinois Valley, 1890–1930*. Carbondale: Southern Illinois University Press.

Thurman, H. V. 1985. *Introductory Oceanography*. Columbus, Ohio: Charles E. Merrill.

Tiffen, M., M. Mortimore, and F. Gichuki. 1994. *More People, Less Erosion: Environmental Recovery in Kenya*. Chichester: John Wiley.

Todd, K. 2001. *Tinkering with Eden: A Natural History of Exotics in America*. New York: W. W. Norton.

Torrens, I. M. 1984. What Goes up Must Come Down: The Acid Rain Problem. *The OECD Observer*, No. 127 (July): 9–19.

Tricart, J. 1965. *Le modelé des régions chaudes, forêts et savanes*. Paris: Société d'édition d'enseignement supérieur.

Trimble, S. W. 1974. *Man-induced Soil Erosion on the Southern Piedmont*. Ankeny, Iowa: Soil Conservation Society of America.

Trussell, D. 1989. The Arts and Planetary Survival. *The Ecologist* 19 (5): 170–76.

Tuan, Y.-F. 1970. Our Treatment of the Environment in Ideal and Actuality. *American Scientist* 58 (3): 244–49.

Turner, B. L., II, J. Geohegan, and D. Foster, eds. 2003. *Integrated Land-change Science and Tropical Deforestation in the Southern Yucatan: Final Frontiers*. Oxford: Clarendon Press of Oxford University Press.

UNCSD [United Nations Commission on Sustainable Development]. 1997. Comprehensive Assessment of the Freshwater Resources of the World. *Report of the Secretary General*. New York: United Nations.

UNEP [United Nations Environment Program]. 1991. *Environmental Data Report*. 3rd ed. Cambridge, Mass.: Blackwell.

UNEP/GEMS [United Nations Environment Program/Global Environment Monitoring System]. 1991. Earthwatch Global Environment Monitoring System WHO/UNEP. In *Report on Water Quality: Progress in the Implementation of the Mar del Plata Action Plan and a Strategy for the 1990s*, Geneva/Nairobi: UNEP/GEMS.

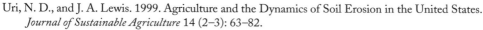

Uri, N. D., and J. A. Lewis. 1999. Agriculture and the Dynamics of Soil Erosion in the United States. *Journal of Sustainable Agriculture* 14 (2–3): 63–82.

U.S. Census Bureau. 1987. *1987 Census of Agriculture*, Vol. 1, *Geographic Area Series, Part 5, California*. Washington, D.C.: U.S. Department of Commerce, Bureau of the Census.

USDA [United States Department of Agriculture]. 1983. *Conversion of Southern Cropland to Southern Pine Plantings: Conversion for Conservation Feasibility Study*. Washington, D.C.: USDA.

USEPA [U.S. Environmental Protection Agency]. 1999. Work Assignment Contract No. 68–D5–0038, May 1999 http://www.epa.gov/region3/mtntop/pdf/workasst.pdf.

USGAO [U.S. General Accounting Office]. 1984. *Department of Energy Acting to Control Hazardous Wastes at Its Savannah River Nuclear Facilities* (GAO/RCED–85–23), Washington, D.C.: The U.S. General Accounting Office.

USGS [U.S. Geological Survey]. 2000. *Geologic and Geophysical Characterization Studies of Yucca Mountain Nevada, Potential High-level Radioactive Repository*. Denver, Colo.: USGS Information Services.

Valentin, C. 1985. Effects of Grazing and Trampling on Soil Deterioration Around Recently Drilled Water Holes in the Sahelian Zone. In *Soil Erosion and Conservation*, edited by S. A. el-Sawify, W. C. Moldenhauer, and A. Lo, 51–65. Ankey, Iowa: Soil Conservation Society of America.

Varady, R. G., K. B. Hankins, A. Kaus, E. Young, and R. Meredith. 2001. . . . to the Sea of Cortés: Nature, Water, Culture, and Livelihood in the Lower Colorado River Basin and Delta—an Overview of Issues, Policies, and Approaches to Environmental Restoration. *Journal of Arid Environments* 49 (1): 195–209.

Vargo, B. D., and S. Gallagher. 2002. Dirty Little Secrets. *Delaware Today* 41 (5): 64.

van de Ven, G. P. 1987. Blanken en de Waterstaat. In *de Physique Existentie dezes Lands: Jan Blanken*, edited by F. Koens, N. B. Spruyt, and M-C. Vink, 59–94. Amsterdam: Rijksmuseum and Uitgeverij AMA boeken.

Verstraeten, G., and J. Poesen. 2001. Factors Controlling Sediment Yield from Small Intensively Cultivated Catchments in a Temperate Humid Environment. *Geomorphology* 40 (1–2): 123–44.

Vogel, H. 1987. Terrace Farming in Yemen. *Journal of Soil and Water Conservation* 42 (1): 18–21.

Volker, A. 1982. Polders: An Ancient Approach to Land Reclamation. *Nature and Resources* 18 (4): 2–13.

Wagstaff, J. M. 1985. *The Evolution of Middle Eastern Landscapes: An Outline to A.D. 1840*. Totowa, N.J.: Barnes & Noble.

Walker, H. J., J. M. Coleman, H. H. Roberts, and R. S. Tye. 1987. Wetland Loss in Louisiana. *Geografiska Annaler* 69A (1): 189–200.

Walls, J. 1980. *Land, Man, and Sand*. New York: Macmillan.

Walters, S. E. 1970. *Water for Larsa: An Old Babylonian Archive Dealing with Irrigation*. New Haven: Yale University Press.

Ward, F. 1990. Florida's Coral Reefs are Imperiled. *National Geographic* 178 (1): 115–32.

Warren, A. 2002. Land Degradation is Contextual. *Land Degradation and Development* 13 (6): 449–59.

Warren, A., and J. K. Maizels. 1977. Ecological Change and Desertification. In *Desertification: Its Causes and Consequences*, compiled and edited by the UNCOD Secretariate, 169–260. New York: Pergamon Press.

Waterbury, J. 1979. *Hydropolitics of the Nile Valley*. New York: Syracuse University Press.

Watson, A. M. 1983. *Agricultural Innovation in the Early Islamic World: the Diffusion of Crops and Farming Techniques 700–1100*. Cambridge: Cambridge University Press.

Watson, R. M. 1989. The Green Menace Creeps North. *Garden* 13: 8–9.

Watson, T. 2002. Mountaintop Mining Halted. *USA Today* May 9, §A, 2.

Watters, R. F. 1971. *Shifting Cultivation in Latin America*. FAO Forestry Paper No. 17. Rome: FAO.

White, G. F. 1960. *Science and the Future of Arid Lands*. Paris: UNESCO.

————. 1988. The Environmental Effects of the High Dam at Aswan. *Environment* 30 (7): 5–11; 34–40.

White, J. E. M. 1991. *The Journeying Boy: Scenes from a Welsh Childhood*. New York: Atlantic Monthly Press.

White, L., Jr. 1967. The Historical Roots of Our Ecological Crisis. *Science* 155 (3767): 1203–207.

Whitmore, T. M., B. L. Turner II, D. L. Johnson, R. W. Kates, and T. R. Gottschang. 1990. Long-term Population Change. In *The Earth as Transformed by Human Action*, edited by B. L. Turner II, W. C. Clark, R. W. Kates, J. F. Richards, J. T. Mathews, and W. B. Meyer, 25–39. Cambridge: Cambridge University Press.

Whittington, D., and G. Guariso. 1983. *Water Management Models in Practice: A Case Study of the Aswan High Dam*. Amsterdam: Elsevier.

WHO [World Health Organization]. 1992. *Our Planet, Our Health*, Geneva.

Wiessner, P. 1982. Risk, Reciprocity and Social Influences on !Kung San Economics. In *Politics and History in Band Societies*, edited by E. Leacock and R. Lee, 61–84. Cambridge: Cambridge University Press/Paris: Editions de la Maison des Sciences de l'Homme.

Wikjman, A., and L. Timberlake. 1985. Is the African Drought an Act of God or of Man? *The Ecologist* 15 (1/2): 9–18.

Wilkie, R. W., and J. Tager. 1990. *Historical Atlas of Massachusetts*. Amherst: University of Massachusetts Press.

Willcocks, W. 1889. *Egyptian Irrigation*. London: Spon.

Williams, O. B. 1978. Desertification in the Pastoral Rangelands of the Gascoyne Basin, Western Australia. *Search* 9 (7): 257–61.

Winchester, S. 2003. *Krakatoa: The Day the World Exploded: August 27, 1883*. New York: HarperCollins.

Wolverton, B. C., and R. C. McDonald. 1979. The Water Hyacinth: From Prolific Pest to Potential Provider. *Ambio* 8 (1): 2–9.

Wooley, C. L. 1965. *The Sumerians*. New York: W. W. Norton.

World Commission on Dams. 2000. *Dams and Development—A New Framework for Decision-Making*. London: Earthscan Publications.

World Resources Institute. 1990. *World Resources 1990–91*. Oxford: Oxford University Press.

————. 1992. *World Resources 1992–1993*. New York: Oxford University Press.

Worster, D. 1979. *Dust Bowl*. New York: Oxford University Press.

————. 1985. *Rivers of Empire: Water, Aridity, and the Growth of the American West*. New York: Pantheon.

Wu, B., and J. C. Long. 2002. Landscape Change and Desertification Development in the Mu Us Sandland, Northern China. *Journal of Arid Environments* 50 (3): 429–44.

Wyatt, A. W. 1988. Estimated Net Depletion Shown. *The Cross Section* 34 (1): 2.

Wyckoff, W. 1995. Postindustrial Butte. *Geographical Review* 85 (4): 478–96.

Xu, S. S. W., and C. Y. Jim. 2003. Using Upland Forest in Shimentai Nature Reserve, China. *Geographical Review* 93 (3): 308–27.

Yafa, S. 2005. *Big Cotton: How a Humble Fiber Created Fortunes, Wrecked Civilizations and Put America on the Map*. New York: Viking.

Yair, A. 2001. Water-Harvesting Efficiency in Arid and Semiarid Areas. In *Sustainable Land Use in Deserts*, edited by S-W. Breckle, M. Vest, and W. Wucherer, 289–302. Berlin: Springer-Verlag.

Yee, A. W. C. 1999. New Developments in Integrated Dike-Pond Agriculture-Aquaculture in the Zhujiang Delta, China: Ecological Implications. *Ambio* 28 (6): 529–33.

Young, J. E. 1992. Mining the Earth. In *State of the World 1992*, edited by L. R. Brown et al. New York: W. W. Norton.

Younkin, L. M. 1974. *Prediction of the Increase in Suspended Sediment Transport Due to Highway Construction*. Lewisburg, Pa.: Bucknell University.

Index

About the Authors

DOUGLAS L. JOHNSON is professor of geography at Clark University. His research concentrates on land degradation issues in arid lands, particularly desertification and change in traditional pastoral communities in North Africa and the Middle East, as well as the historical geography of successful, sustainable agroecosystems. He has held visiting appointments at the University of California, Berkeley and al-Akhawayn University, Morocco. He is the author of *The Nature of Nomadism*, coeditor (with John Galaty) of *The World of Pastoralism* and (with Alec Murphy) of *Cultural Encounters with the Environment*, and currently serves as coeditor (with Viola Haarmann) of *The Geographical Review* and (with Viola Haarmann and Merrill Johnson) of the ninth edition of *World Regional Geography: A Development Approach*.

LAURENCE A. LEWIS is professor of geography at Clark University. His research has concentrated on minimizing soil erosion on agricultural lands, including both the human and fluvial causes of land degradation. Fieldwork for this research took place mostly in the tropics (Puerto Rico, Haiti, Jamaica, Rwanda, Nigeria, and Kenya). In addition he has worked on soil erosion problems within government ministries in Kenya and Rwanda. His recent work deals with modeling soil erosion and deposition utilizing GIS. He has been a visiting professor at the University of Ibadan (Nigeria) and the University of Leuven (Belgium) and coauthored *African Resources and Environments* (with Len Berry) as well as numerous government publications dealing with land degradation problems in Africa.